P9-BHZ-187

BARRON'S
BUSINESS
REVIEW
SERIES

Business Statistics

Third Edition

Douglas Downing, Ph.D.

School of Business and Economics
Seattle Pacific University

Jeffrey Clark, Ph.D.

Mathematics Department
Elon College

BARRON'S EDUCATIONAL SERIES, INC.

Copyright © 1997 by Barron's Educational Series, Inc.

Previous editions copyright © 1992, 1985 by
Barron's Educational Series, Inc.

All rights reserved.
No part of this book may be reproduced in any form,
by photostat, microfilm, xerography, or any other
means, or incorporated into any information retrieval
system, electronic or mechanical, without the written
permission of the copyright owner.

All inquiries should be addressed to:
Barron's Educational Series, Inc.
250 Wireless Boulevard
Hauppauge, NY 11788

Library of Congress Catalog Card No. 96-30454

International Standard Book No. 0-8120-9658-4

Library of Congress Cataloging-in-Publication Data

Downing, Douglas.
    Business statistics / Douglas Downing, Jeffrey Clark.
— 3rd ed.
        p.   cm. — (Barron's business review series)
    Includes index.
    ISBN 0-8120-9658-4
    1. Commercial statistics. 2. Statistics. I. Clark, Jeff.
II. Title. III. Series.
HF1017.D65   1997
519.5—dc20   96-30454
                                                    CIP

PRINTED IN THE UNITED STATES OF AMERICA
11 10 9 8 7 6

CONTENTS

PREFACE

In this book you will learn how to apply statistical methods for analyzing data. Chapter 1 lists some of the ways in which statistical methods serve as valuable tools for business decision makers. We will cover many statistical techniques with examples of calculations and interpretations.

It takes practice to learn statistics. Each chapter starts with a list of key terms, with definitions, that you will learn in that chapter. The book contains a total of 137 discussion questions and 275 exercises at the ends of the chapters. The discussion questions cause you to think about the concepts you have just learned, and they expand your understanding of the applications of those concepts. The exercises give you practice in performing statistical calculations. The answers for the discussion questions and the exercises are also included at the ends of the chapters; complete solutions for the computational problems are included where appropriate. In some other cases all formulas necessary to work the problems are given. Where there are several problems of the same type, complete or sample calculations are shown for the first problem.

Several Greek letters and other symbols are used frequently in statistics. You may check the List of Symbols section to become familiar with these. You can learn the meanings of the new terms introduced in this book and set in **boldface type** by checking the Glossary in Appendix 1.

You will quickly realize that you will need to use a computer to perform your calculations. If you will be regularly working with statistics, you should obtain and learn to work a computer statistics program. Appendix 2 discusses some of the essential features of calculations, computer statistics programs, and the spreadsheet Microsoft Excel.

Appendix 3 contains frequently used statistical tables for the normal, chi-square, t, and F distributions.

While doing statistical calculations, be sure you are aware of the effect of rounding. For example, even a simple fraction such as 1/3 cannot be expressed exactly as a decimal fraction with a finite number of digits. Therefore, we will frequently use decimal approximations to the quantities that we calculate in this book. So, for example, if you see an expression such as "1/3 = .333," be warned that this is a short form of the complete statement which would be ".333 is a decimal approximation of the fraction 1/3, rounded to three decimal places." In this case, the first nonincluded digit is less than .5, so we simply drop it. However, if the first nonincluded digit is greater than or equal to .5, then we need to increase the previous digit by 1. For example, 2/3 rounded to three decimal places is .667.

The third edition of the book has been updated and improved over the first two editions in several ways. Also, portions of this book have been published in our other book, *Statistics the Easy Way*, which contains a more complete treatment of the mathematical development of statistics.

Douglas Downing
Jeffrey Clark

LIST OF SYMBOLS AND ABBREVIATIONS

POPULATIONS AND SAMPLES

Σ (capital sigma) summation

for example: ΣX_i or $\displaystyle\sum_{i=1}^{n} X_i$

N number of items in a population

n number of items in a sample

i, j used as subscripts

for example: X_i or X_{ij}

μ (mu) population mean

$\bar{x}$ sample average (in general, a bar over a quantity signifies average)

σ (lower case sigma) population standard deviation

σ^2 population variance

$s_1^{\,2}$ sample variance, version 1

$$s_1^2 = \frac{\Sigma_{i=1}^{n} (X_i - \bar{x})^2}{n}$$

$s_2^{\,2}$ sample variance, version 2

$$s_2^2 = \frac{\Sigma_{i=1}^{n} (X_i - \bar{x})^2}{n - 1}$$

f_i (frequency); number of items in category i

PROBABILITY

$\Pr(A)$ probability that event A will occur

$\Pr(A \mid B)$ conditional probability that event A will occur, given that event B has occurred

$N(A)$ number of outcomes in event A

s	total number of possible outcomes
A^c	A complement; consists of all outcomes that are not in event A
!	factorial
	for example: $6! = 6 \times 5 \times 4 \times 3 \times 2 \times 1 = 720$
$\binom{n}{j}$	number of combinations of n things taken j at a time;

$$\binom{n}{j} = \frac{n!}{j!(n-j)!}$$

$_nC_j$	same as $\binom{n}{j}$
$_nP_j$	number of permutations of n things taken j at a time;

$$_nP_j = \frac{n!}{(n-j)!}$$

RANDOM VARIABLES

capital letters, such as X, Y, and Z, are used to represent random variables

$f(a)$	probability function for a random variable; for discrete random variables $f(a) = \Pr(X = a)$
$F(a)$	cumulative distribution function for a random variable; $F(a) = \Pr(X \le a)$
$E(X)$	expectation (expected value) of X
μ	mean (same as expectation)
$\mathrm{Var}(X)$	variance of X
σ^2	variance
$\mathrm{Cov}(X, Y)$	covariance of X and Y
λ	(lambda) used as parameter for Poisson distribution
n	number of trials for binomial distribution
p	probability of success for binomial distribution
Z	represents standard normal random variable (mean 0, variance 1)
$\Phi(z)$	cumulative distribution function for standard normal random variable; $\Phi(z) = \Pr(Z < z)$
χ^2	chi-square distribution
t	t distribution
F	F distribution

STATISTICAL ESTIMATION

a hat over a symbol indicates that this quantity is being used as an estimator for an unknown population parameter

for example: $\hat{m}$

ANALYSIS OF VARIANCE AND REGRESSION

b	vertical intercept of simple regression line
$b_1, \ldots, b_{m-1}$	estimated multiple regression coefficients
b_m	estimated multiple regression constant term
$B_1, \ldots, B_{m-1}$	true multiple regression coefficients
B_m	true multiple regression constant term
COLSS	column sum of squares
e	error term in regression model
ERSS	error sum of squares
m	slope of simple regression line
MSE	mean square error
r	correlation coefficient
r^2	coefficient of determination for simple linear regression
R^2	coefficient of multiple determination
RGRSS	regression sum of squares
ROWSS	row sum of squares
$s(b_i)$	standard error of coefficient b_i
SE_{line}	squared error about the line
SE_{av}	squared error about the average
TSS	total sum of squares
TRSS	treatment sum of squares
x	independent variable(s) in regression
y	dependent variable in regression

BUSINESS DATA

BLS	Bureau of Labor Statistics
CPI	consumer price index
GDP	gross domestic product
NNP	net national product

MATHEMATICAL SYMBOLS

{ }	braces; indicate set membership
$\cup$	union
$\cap$	intersection
$\emptyset$	empty set
π	3.14159. . .
e	2.71828. . .
log	logarithm; if $a^n = x$ then $\log_a x = n$
| |	absolute value
$\sqrt{}$	square root
$>$	greater than
$\geq$	greater than or equal
$<$	less than
$\leq$	less than or equal

HYPOTHESIS TESTING

H_o	null hypothesis
H_a	alternative hypothesis
Type 1 error	rejecting the null hypothesis when it is really true
Type 2 error	accepting the null hypothesis when it is really false
α	probability of type I error
β	probability of type II error

1
WHY STATISTICS?

KEY TERMS

population the set of all items you are interested in

sample a group of items chosen from the population and used to estimate the properties of the population

USING STATISTICAL TECHNIQUES IN BUSINESS

Why, you may be wondering, should a business manager study statistics? In this book you will learn many powerful statistical techniques. You will learn how to extract meaningful information from piles of raw data, how to make inferences about the nature of a population based on observations of a sample taken from that population, how to predict the rates of occurrences of random events, and how to understand and interpret statistical calculations performed by others.

There are times when statistics seem to take on an almost magical quality. How can pollsters make reasonably accurate predictions of the presidential vote when they ask only a sample of a few thousand voters out of the total voting population of 110 million? How can television rating services estimate the total audience for a show based on a small sample of selected households? In this book we will learn why these methods work.

Many people think of statistics as a boring and difficult subject. Admittedly some parts of statistics are quite difficult, but there are other parts that are concise and elegant. Part of the reason why statistics has a reputation for difficulty dates from the time prior to computers when statisticians were forced to perform laborious calculations by hand. Now it is possible to have a computer do most of the tedious work, leaving you much freer to understand the meaning of what is going on. Early in your statistics career you should learn how to program a computer or how to use a computer statistics package or a calculator with built-in statistics operations.

TWO MEANINGS OF THE WORD *STATISTICS*

The word **statistics** has two different (but related) meanings. In the most common usage, *statistics* means "a collection of numerical data." For example, you may look at the statistics that summarize the performance of a football team's season, the statistics that list the number of births and deaths in a city, or the statistics that describe the characteristics of a new building. There are also important national business statistics that you should become aware of, such as the gross domestic product (GDP) and the consumer price index (CPI).

The word *statistics* also refers to the branch of mathematics that deals with the analysis of statistical data. In Chapter 2 we will study **descriptive statistics**, which is the process of obtaining meaningful information from sets of numbers that are often too large to deal with directly. You are undoubtedly already familiar with some aspects of descriptive statistics, such as how to calculate the average of a set of numbers. In most of the rest of the book we will be studying **statistical inference**, which is the process of using observations from a sample to estimate the properties of a larger population.

YOU SHOULD REMEMBER

1. Powerful statistical techniques, with calculators and appropriate computer programs assisting with the calculations, are valuable tools in business.

2. The word *statistics* has two meanings: (1) a body of numerical data, and (2) a branch of mathematics, encompassing descriptive statistics and statistical inference, that deals with the analysis of statistical data.

POPULATIONS AND SAMPLES

The term **population** refers to all of the people or things in the group you are interested in. A **sample** is a group of items chosen from the population. Examples include the following:

- population: the 31 flavors of ice cream at a 31-flavor ice cream store
  sample: the five flavors that you have tested in order to determine whether this store sells good ice cream

- population: all voters in the United States
  sample: the 3,000 people who are interviewed as part of a Gallup poll

- population: all people in the United States with television sets
  sample: the people who are surveyed by the Nielson television rating service

- population: all people in the United States
  sample: the 100,000 people interviewed in the Census Bureau's monthly Current Population Survey

You may wonder why you should study samples instead of populations. If you happen to choose a sample that is very unrepresentative of the corresponding population, you will make very inaccurate predictions when you try to estimate the characteristics of the population based on that sample. You could avoid that danger by studying the entire population instead. However, there are several reasons why samples are studied instead of populations. The most important reason is the excessive cost and/or difficulty of studying the entire population. Any business decision maker is faced with the problem of limited resources. It would not be a wise use of resources to undertake the expense of studying the entire population if predictions based on a sample are almost as accurate. (The key question you are probably asking is, How do we know that a sample is likely to represent the population accurately? That will be one of the most important questions that we will answer in this book.) The U.S. government is subject to the same types of limitations. It would be very costly for the government to calculate the monthly unemployment index by actually counting the total number of unemployed people. Instead, the unemployment figure is calculated from the Current Population Survey, which is a monthly sample. There are times, though, when the entire population is investigated. Every 4 years the presidential preferences of the complete voting population are checked in an election, and every 10 years a census of the total population is conducted.

There are other reasons why samples are studied instead of populations. In some cases the process of investigating an item destroys it. For example, suppose you are trying to measure the amount of stress that a particular type of post can withstand before it breaks. Obviously you cannot investigate the entire population of posts because then you would have no unbroken posts left. There are also times when you cannot obtain access to the entire population. Sometimes it can be more accurate to deal with data from a sample rather than from a population. For example, if you are conducting a survey that requires the use of trained interviewers, you have a much better chance of obtaining accurate results if you need only to train enough interviewers to conduct a few thousand interviews instead of a few million.

YOU SHOULD REMEMBER

1. Because it is expensive, difficult, and sometimes impracticable to survey an entire population, a sample is usually selected for study.

2. To avoid inaccurate predictions, it is essential that the sample be representative of the population from which it is chosen.

APPLYING STATISTICS IN BUSINESS

Here are some examples of business applications of statistics:

- A firm preparing to introduce a new product needs to estimate the preferences of the consumers in the relevant market. It can often do this by conducting a marketing survey based on interviews with some randomly selected households. The results of the survey can then be used to estimate the preferences of the entire population.

- Statistical techniques are needed to disentangle the separate effects of several different factors. For example, the demand for ice cream in a community can be expected to depend on the price of ice cream, the level of average income, the number of children in the community, and the average temperature. If you have observations of all the different factors involved, you can use regression analysis to determine which factors have the most important effects.

- An auditor has the job of checking the books of a company to make sure that they accurately reflect the financial condition of the company. The auditor will need to check through piles of original documents such as sales slips, purchase orders, and requisitions. It would require massive amounts of work to check every single original document; instead, the auditor can check a randomly selected sample of documents and make inferences about the entire population of documents based on that sample.

- Before a new drug is marketed, it is necessary to perform extensive experiments to make sure the drug is safe and effective. The best way to test a drug is to take two groups that are as much alike as possible, give the drug to one of the groups but not to the other, and then see whether the results for the two groups are different. The group that is given the drug is called the experimental group, and the other group is called the control group. Statistical analysis is necessary to determine whether any observed differences really were caused by the drug or could have been caused by other factors.

- If you are receiving a large shipment of goods from a supplier, you will want to make sure that the goods meet the quality standards agreed upon. It would be very expensive to perform a quality control check on every single item, but once again statistical techniques come to the rescue by allowing you to make inferences about the quality of the entire lot by checking a randomly selected sample of items chosen from the lot.

RELATIONSHIP BETWEEN PROBABILITY AND STATISTICS

In order to understand statistical inference it is necessary to understand several concepts in **probability**. Probability and statistics are very closely related because they ask opposite types of questions. In probability, we know how a process works and we want to predict what the outcomes of that process will be. In statistics, we don't know how a process works but we can observe the outcomes of the process. We want to use the information about the outcomes to learn about the nature of the process.

We will develop the ideas of probability by concentrating on familiar random processes: flipping a coin, tossing dice, and drawing cards from well-shuffled decks. Learning how the principles of probability apply to situations where you know how the process works will help you to understand how statistical inference can be used to learn about the nature of an unknown process. For example, suppose you have 40 blue marbles and 60 red marbles in a jar and you draw 10 randomly selected marbles out of the jar. By using probability theory you can calculate the probability that exactly 6 of the 10 selected marbles will be red. Now suppose that before an election you ask 1,000 randomly selected people the name of their favorite presidential candidate. You cannot calculate the probability that 60 percent of the people in the sample will agree with your preference because you don't know the presidential preferences of all the people in the population. However, by using statistical inference you can estimate the presidential preferences of the population based on the presidential preferences of the people in your sample.

These examples are only a small sample drawn from the total population of ways that statistical tools can be useful in business.

YOU SHOULD REMEMBER

1. Examples of business applications of statistical techniques include the use of samples to estimate the preferences or infer the quality of an entire population, and the use of regression analysis to distinguish the separate effects of several factors.

2. Knowledge of probability concepts helps us to predict the outcome of a process.

KNOW THE CONCEPTS

DO YOU KNOW THE BASICS?

Test your understanding of Chapter 1 by answering the following questions:

1. Why do statisticians investigate samples instead of populations?
2. How can you tell whether a particular sample will accurately represent the population?

For questions 3–5, imagine that you are conducting an experiment for which you will divide your subjects into two groups: an experimental group and a control group.

3. Why should you try to have the groups as much alike as possible?
4. Should the people know which group they are in?
5. What is the best system for assigning people to the two groups?

TERMS FOR STUDY

descriptive statistics	sample
population	statistical inference
probability	statistics

ANSWERS

KNOW THE CONCEPTS

1. The main reason for using samples is that it would be too expensive to survey entire populations.

2. There is no way to know *for sure* whether the sample accurately represents the population unless you conduct a census of the entire population. However, the methods of statistical inference can be used to estimate the likelihood that the sample is representative of the population.

3. Suppose the two groups are very different to start with. Then, if you observe different experimental results for the two groups, you cannot be sure that these differences are due to the treatment you are investigating.

4. No. People may react differently if they know which group they are in.

5. Ideally the people should be divided into the two groups completely at random.

2
DESCRIPTIVE STATISTICS

KEY TERMS

frequency histogram a bar diagram that illustrates how many observations fall within each category

mean the value that is equal to the sum of a list of numbers divided by the number of numbers (same as **average**)

median if a list of numbers has an odd number of items, then the median is the middle number (when the list is arranged in order); if the list has an even number of items, then the median is the average of the two numbers closest to the middle

mode the value that occurs most frequently in a list of numbers

standard deviation the square root of the variance

variance a measure of dispersion for a list of numbers; it is symbolized by σ^2 (sigma squared)

MEASURES OF CENTRAL TENDENCY: MEAN, MEDIAN, AND MODE

If you've just been given a large pile of numbers, you have little or no hope of understanding them unless you can figure out some way to summarize them. For example, suppose you are managing a pizza parlor and you are keeping track of your daily sales figures for several different types of pizza. Suppose you have observed the following values for daily sales of pepperoni pizzas during a 9-day period:

$$40, 56, 38, 38, 63, 59, 52, 49, 46$$

• *MEAN*

A list of numbers like this is called **raw data**. One useful thing to do would be to calculate the **average** of all of the numbers. To calculate the average we just add up all the numbers and divide by the number of numbers. The average

9

is also called the **mean**. The average in the pepperoni example can be calculated like this:

$$\frac{40 + 56 + 38 + 38 + 63 + 59 + 52 + 49 + 46}{9} = \frac{441}{9}$$

$$= 49.0$$

Therefore, on average, you have sold 49 pepperoni pizzas per day.

We can derive a general formula for the average. Suppose we have n numbers. We'll call them $x_1, x_2, x_3, \ldots, x_n$. The little numbers next to and slightly below the x's are called **subscripts**. We'll use the symbol $\bar{x}$ (x with a bar written over it) to stand for average. Then we can write this formula:

$$\bar{x} = \frac{x_1 + x_2 + x_3 + \cdots + x_n}{n}$$

The average is an example of a **statistic**. A statistic is a single number that is used to summarize the properties of a larger group of numbers. We will talk about some other important statistics later.

We can save some writing by using **summation notation**. In summation notation we use the Greek capital letter sigma: Σ. The expression Σx means "add up all the values of x." The average can be written as $\bar{x} = (\Sigma x)/n$. Sometimes the summation symbol is written like this:

$$\sum_{i=1}^{n} x_i$$

to indicate that we start with $i = 1$ and keep going until $i = n$.

• MEDIAN

Another useful statistic is the **median**. The median is the halfway point of the data, that is, half of the numbers are above it and half are below. To compute the median, we must put the list of numbers in order. Here, in descending order, is the list for our pepperoni pizza example:

1. 63		6. 46	
2. 59		7. 40	
3. 56		8. 38	
4. 52		9. 38	
5. 49			

For this list the median value is 49. Four values are greater than 49, and four values are below 49. (If you have a long list, it helps to have a computer put the numbers in order. Note that you also could put the list in ascending order.)

When there is an odd number of items in the list, it is easy to find the single number that is exactly in the middle. What if we have a list with an even number of items? Suppose we have the following eight observations for daily mushroom pizza sales:

1. 53	5. 47
2. 52	6. 46
3. 49	7. 44
4. 48	8. 41

We have already placed the values in order, but this time we cannot find a single number that is exactly in the middle. Instead, the median is equal to the number halfway between the two numbers closest to the middle. In this case the two numbers closest to the middle are 47 and 48 (the fourth and fifth values), so the median is halfway between 47 and 48, or 47.5. You may also think of the median as being equal to the average of the two middle numbers: $47.5 = (47 + 48)/2$.

Like the average, the median is a measure of what is called the **central tendency** of the distribution. In other words, the mean or median often gives you a good idea about the size of the number you are likely to get if you select one value from the list at random. There are times when the median gives a better measure of the central tendency of a list than does the mean. For example, suppose that these figures represent sales of bacon/pineapple pizza for a 9-day period:

$$36, 35, 37, 29, 39, 36, 340, 35, 36$$

Note that one day a large busload of bacon/pineapple pizza lovers arrived at your pizza place, so bacon/pineapple pizza sales were much larger than usual that day. If we calculate the average for this list we get:

$$\frac{623}{9} = 69.22$$

However, we can see that none of the actual values is very close to 69.22. We can put the list in order:

1. 340	6. 36
2. 39	7. 35
3. 37	8. 35
4. 36	9. 29
5. 36	

and then determine that the median is 36. We can see that in this case the value of the median gives us a much better idea of what our sales are actually likely to be on a given day. In general, when a list contains one extreme value (either far above or far below the other numbers in the list), the average will not be very representative of most of the numbers in the list. In that case the median is a better measure of the central tendency. However, the average is often easier to calculate, so it is more often used. When a distribution of numbers is reasonably symmetric without an extremely high or low value, then the values of the median and the mean will usually be close together.

• *MODE*

Another interesting statistic is the **mode**. The mode is the value that occurs most frequently. If there is more than one value that occurs the greatest number of times, all such values are called *modes*. For the pepperoni sales the value of the mode is 38, because the number 38 occurs twice in the list while none of the other numbers occurs more than once. For the bacon/pineapple sales the mode is 36, which occurs three times. There is no mode for the mushroom pizza sales since no values occur more than once. Many distributions that arise in practice are reasonably symmetric with most of the values concentrated near the middle, in which case the mean, median, and mode are all close together. However, a distribution can have more than one mode. A distribution with two modes is called a **biomodal** distribution.

YOU SHOULD REMEMBER

1. Summation notation: the Greek capital letter sigma, Σ, means "add up the values."

example: $\displaystyle\sum_{i=1}^{n} x_i = x_1 + x_2 + \cdots + x_n$

2. Measures of central tendency:

mean (average) $= \bar{x} = \dfrac{x_1 + x_2 + \cdots + x_n}{n} = \displaystyle\sum_{i=1}^{n} \dfrac{x_i}{n}$

median: when a list is arranged in order, the value such that as many numbers in the list are above it as below it

mode: the value in a list that occurs most frequently

MEASURES OF DISPERSION: VARIANCE AND STANDARD DEVIATION

It would be very helpful if we had a way of measuring the unpredictability of the daily sales figures. We will use the term **dispersion** or **spread** to refer to the degree to which a group of numbers are scattered away from their average. If you sold a total of exactly 200 pizzas every single day, there would be no spread to the figures. On the other hand, if you sold 400 pizzas half the time and zero pizzas the other half the time, the average would still be the same but the degree of dispersion would be much greater. One way to measure the dispersion is simply to take the difference between the highest value and the lowest value. This quantity is called the **range**. For pepperoni sales the range is $63 - 38 = 25$, and for bacon/pineapple sales the range is $340 - 29 = 311$. The range gives you an idea of the spread between the highest value and the lowest value, but it is not really a very good measure of the spread of the entire distribution. The range does not convey any information about any item in the list except the highest and lowest values. Here are two lists that have the same range:

List a	List b
500	500
250	490
250	480
250	20
250	10
0	0

but it is clear that the overall spread of list *b* is much greater than the spread of list *a*.

One problem with the range is that it will be greatly affected by the presence of very large or very small values. One way to solve that problem is to ignore the bottom and top quarters of the data and calculate the range of the remaining values. This quantity is called the **interquartile range** (or IQR). To calculate the IQR:

1. Put the list in order.

2. Find the value such that 3/4 of the other values are at or below it. (This value is called the *third quartile* or *75th percentile*.)

3. Find the value such that 1/4 of the other values are below it. (This value is called the *first quartile* or *25th percentile*.)

4. Take the difference between these two values.

YOU SHOULD REMEMBER

1. The distribution of the numbers in a list can be characterized by calculating **quartiles** and **percentiles**:

   quartiles: The first quartile is the number in the list such that one-fourth of the items in the list are below it; the third quartile is the number in the list such that three-fourths of the items in the list are below it.

   percentiles: The pth percentile in a list is the number such that p percent of the numbers are below it.

2. range: the largest value minus the smallest value

   interquartile range: the value of the third quartile (75th percentile) minus the value of the first quartile (25th percentile)

We still need a measure of dispersion that takes into account all of the numbers in the list. Consider the pepperoni sales figures, which have a mean of 49. It will be helpful to find out how far away from 49 each number is. To find the distance we subtract 49 from each number and then take the absolute value (symbolized by two vertical lines: $|\quad|$).

Pepperoni Sales	Distance from Mean
40	$9 = \|40 - 49\|$
56	$7 = \|56 - 49\|$
38	$11 = \|38 - 49\|$
38	$11 = \|38 - 49\|$
63	$14 = \|63 - 49\|$
59	$10 = \|59 - 49\|$
52	$3 = \|52 - 49\|$
49	$0 = \|49 - 49\|$
46	$3 = \|46 - 49\|$

We could simply calculate the average distance:

$$\frac{9 + 7 + 11 + 11 + 14 + 10 + 3 + 0 + 3}{9} = \frac{68}{9}$$

$$= 7.556$$

This quantity is called the **mean absolute deviation**. In general, it can be found from the following formula:

$$\frac{|x_1 - \bar{x}| + |x_2 - \bar{x}| + \cdots + |x_n - \bar{x}|}{n} = \frac{\Sigma_{i=1}^{n} |x_i - \bar{x}|}{n}$$

• *VARIANCE*

The mean absolute deviation is a good measure of dispersion because it tells us the average distance from each number in the list to the mean. However, for many purposes it turns out to be most convenient to *square* each deviation and then take the average of all the squared deviations. This quantity is called the **variance**. To calculate the variance for the pepperoni sales we first add the squares of each deviation:

$$(40 - 49)^2 + (56 - 49)^2 + (38 - 49)^2 + (38 - 49)^2 + (63 - 49)^2$$
$$+ (59 - 49)^2 + (52 - 49)^2 + (49 - 49)^2 + (46 - 49)^2$$

$$= 9^2 + 7^2 + 11^2 + 11^2 + 14^2 + 10^2 + 3^2 + 0^2 + 3^2$$

$$= 81 + 49 + 121 + 121 + 196 + 100 + 9 + 0 + 9$$

$$= 686$$

Then we divide this quantity by 9, since there are nine numbers in the list:

$$(\text{variance}) = \frac{686}{9} = 76.222$$

The variance is often symbolized by σ^2 (sigma squared). The symbol σ is the lowercase Greek letter sigma. (Do not confuse the lowercase sigma, σ, with the uppercase sigma, Σ, used to represent summation. Since they do not look at all alike, this should not be difficult.) We will also use the abbreviation Var(x) to represent the variance of x.

The general formula for the variance is:

$$\text{Var}(x) = \sigma^2 = \frac{(x_1 - \bar{x})^2 + (x_2 - \bar{x})^2 + \cdots + (x_n - \bar{x})^2}{n}$$

$$= \frac{\sum_{i=1}^{n} (x_i - \bar{x})^2}{n}$$

For practical purposes there is a simpler formula for calculating the variance:

$$\text{Var}(x) = \overline{x^2} - \bar{x}^2$$

We need to explain carefully what these symbols mean. As always, $\bar{x}$ represents the mean; $\bar{x}^2$ represents the square of the value of the mean. For the pepperoni example, $\bar{x} = 49$ and $\bar{x}^2 = 49^2 = 2,401$. Also, $\overline{x^2}$ represents the average of the squares of each value of x. In our case:

$$\overline{x^2} = \frac{40^2 + 56^2 + 38^2 + 38^2 + 63^2 + 59^2 + 52^2 + 49^2 + 46^2}{9}$$

$$= \frac{1,600 + 3,136 + 1,444 + 1,444 + 3,969 + 3,481 + 2,704 + 2,401 + 2,116}{9}$$

$$= \frac{22,295}{9} = 2,477.222$$

Therefore the variance is:

$$\text{Var}(x) = \overline{x^2} - \bar{x}^2 = 2,477.222 - 2,401 = 76.222$$

which is the same result we found earlier.

• STANDARD DEVIATION

The variance is a good measure of dispersion, but it suffers from one disadvantage: it is difficult to interpret the numerical value of the variance. Does a variance of 76.222 mean that there is a lot of spread or a little spread? Part of the problem arises because of units: the variance is measured in a unit that is the square of the unit in which x is measured. In our case, we have a variance of 76.222 pizzas squared, whatever that means. If is often more convenient to calculate the square root of the variance, which is called the **standard deviation**:

$$(\text{standard deviation of } x) = \sigma = \sqrt{\text{Var}(x)}$$

$$= \sqrt{\frac{(x_1 - \bar{x})^2 + \cdots + (x_n - \bar{x})^2}{n}}$$

$$= \sqrt{\frac{\sum_{i=1}^{n} (x_i - \bar{x})^2}{n}}$$

As you might have guessed, the standard deviation is symbolized by just plain sigma σ instead of σ^2. In our case the standard deviation is $\sqrt{76.222} = 8.731$ pizzas. The standard deviation of x is measured in the same units that x is measured in.

If we want to know whether the dispersion is very large relative to the average, we can calculate a statistic known as the **coefficient of variation**:

$$(\text{coefficient of variation}) = \frac{(\text{standard deviation})}{(\text{mean})}$$

For pepperoni pizza, the coefficient of variation is $8.731/49 = 0.178 = 17.8$ percent.

We will have more to say later about how to interpret the value of the standard deviation. Here are some interesting properties worth mentioning now.

First, for any list it will always be true that at least 75 percent of the numbers in the list will be within two standard deviations of the mean. With a mean of 49 and a standard deviation of 8.731 we can calculate that two standard deviations below the mean is $49 - 2 \times 8.731 = 49 - 17.462 = 31.538$ and two standard deviations above the mean is $49 + 17.462 = 66.462$. We know then that at least 75 percent of the daily pepperoni sales figures must be between 31.538 and 66.462. (It actually turns out that 100 percent of the figures are within this range.)

Second, in general, the proportion of numbers in the list within k standard deviations of the mean must be at least $1 - 1/k^2$. This result is known as **Chebyshev's theorem** (also spelled **Tchebysheff's theorem**).

Chebyshev's theorem applies to every single possible list of numbers. If you happen to know more about the distribution of the numbers, you often can make more precise statements. In particular, when a list of numbers follows a common pattern called the **normal distribution**, then 68 percent of the numbers will be within one standard deviation of the mean and 95 percent of the numbers will be within two standard deviations of the mean.

In the variance formulas we have used so far we have known all the values in the entire population in which we are interested. Therefore, σ^2 and σ are called the **population variance** and **population standard deviation**. Now suppose that we would like to know about the daily sales of pepperoni pizza for a whole year, but we have observed pepperoni sales only on 9 randomly selected days. In this case we don't know the true value of the mean or standard deviation for pizza sales for the entire year. However, we will guess that the average for the sample ($\bar{x}$) is close to the average value for the entire population, and we will guess that the variance of the sample (**sample variance**) is close to the value of the variance of the entire population. The square root of the sample variance is called the **sample standard deviation**.

There are two different formulas for calculating the variance of a sample:

version 1:

$$\text{(sample variance)} = s_1^2 = \frac{(x_1 - \bar{x})^2 + \cdots + (x_n - \bar{x})^2}{n} = \frac{\sum_{i=1}^n (x_i - \bar{x})^2}{n}$$

$$\text{(sample standard deviation)} = s_1 = \sqrt{\frac{\sum_{i=1}^n (x_i - \bar{x})^2}{n}}$$

version 2:

$$\text{(sample variance)} = s_2^2 = \frac{(x_1 - \bar{x})^2 + \cdots + (x_n - \bar{x})^2}{n-1} = \frac{\sum_{i=1}^n (x_i - \bar{x})^2}{n-1}$$

$$\text{(sample standard deviation)} = s_2 = \sqrt{\frac{\sum_{i=1}^n (x_i - \bar{x})^2}{n-1}}$$

Note that version 2 is the same as version 1 except that we divide by $n - 1$ rather than n. We will not at this time attempt to explain why version 2 often works better when we are trying to guess the variance or standard deviation of a population based on a sample. We will cover these topics in Chapter 10 when we discuss statistical inference. When the value of n is large, it doesn't make too much difference whether we divide by n or $n - 1$, so in that case the value of s_1^2 will be close to the value of s_2^2.

In the pepperoni case we have $s_1^2 = 76.222$ and $s_1 = 8.731$. We can calculate s_2^2 and s_2:

$$s_2^2 = \frac{ns_1^2}{n-1} = 85.750$$

$$s_2 = s_1 \sqrt{\frac{n}{n-1}} = 9.261$$

YOU SHOULD REMEMBER

1. mean absolute deviation $= \dfrac{\sum_{i=1}^{n} |x_i - \bar{x}|}{n}$

2. If you have a list of N numbers $(x_1, x_2, \ldots, x_N)$ that represent an entire population:

$$\mathrm{Var}(x) = \sigma^2 = \dfrac{\sum_{i=1}^{N} (x_i - \bar{x})^2}{N}$$

$$= \overline{x^2} - \bar{x}^2$$

$$= \dfrac{\sum_{i=1}^{N} x_i^2 - N\bar{x}^2}{N}$$

$$\sigma = \sqrt{\mathrm{Var}(x)} = \sqrt{\overline{x^2} - \bar{x}^2}$$

3. If you have a list of n numbers $(x_1, x_2, \ldots, x_n)$ that represent a sample:

sample variance, version 1:

$$s_1{}^2 = \dfrac{\sum_{i=1}^{n} (x_i - \bar{x})^2}{n}$$

$$= \overline{x^2} - \bar{x}^2$$

$$= \dfrac{n-1}{n} s_2{}^2$$

sample standard deviation, version 1:

$$s_1 = \sqrt{\dfrac{\sum_{i=1}^{n} (x_i - \bar{x})^2}{n}}$$

sample variance, version 2:

$$s_2{}^2 = \dfrac{\sum_{i=1}^{n} (x_i - \bar{x})^2}{n-1}$$

$$= \dfrac{n}{n-1} s_1{}^2$$

$$= \dfrac{\sum_{i=1}^{n} x_i^2 - n\bar{x}^2}{n-1}$$

sample standard deviation, version 2:

$$s_2 = \sqrt{\frac{\sum_{i=1}^{n} (x_i - \bar{x})^2}{n - 1}}$$

$$= s_1 \sqrt{\frac{n}{n - 1}}$$

FREQUENCY HISTOGRAMS

Matters become more complicated if we are given a longer list of numbers. For example, suppose we are managing a marching band and need to order uniforms. We need to know the heights (in inches) of the members:

67, 65, 68, 67, 67, 64, 69, 66, 66, 66,
68, 71, 67, 67, 70, 65, 65, 66, 70, 64,
67, 68, 66, 68, 64, 65, 67, 66, 69, 68,
65, 69, 68, 67, 68, 67, 67, 67, 66, 66

Keeping track of the entire list is cumbersome. It would help to summarize the list by counting the number of people at each height and making a table like this:

Height	Frequency
64	3
65	5
66	8
67	11
68	7
69	3
70	2
71	1

The number of people at each height is called the frequency of occurrence of that height, so this table is called a **frequency table**. Notice that we get a much clearer picture of the nature of the distribution of heights from the frequency table than we do from the original list. We can obtain a still clearer view by drawing a picture of the frequency table (called a **frequency histo-**

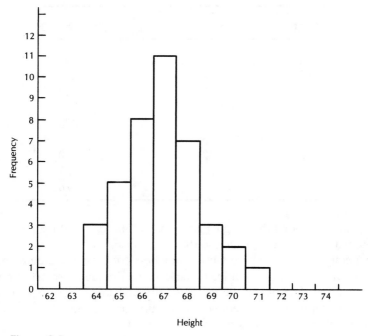

Figure 2-1

gram, see Figure 2-1). The height of each bar is the number of people who have the indicated height.

We can use the information from the frequency table to calculate the average. We can calculate the total of all the heights like this:

64 + 64 + 64 +
65 + 65 + 65 + 65 + 65 +
66 + 66 + 66 + 66 + 66 + 66 + 66 + 66 +
67 + 67 + 67 + 67 + 67 + 67 + 67 + 67 + 67 + 67 + 67 +
68 + 68 + 68 + 68 + 68 + 68 + 68 +
69 + 69 + 69 +
70 + 70 +
71
   = 2,676

As you look at all those numbers, you can quickly see an easier way: all we need to do is multiply each height value times its frequency of occurrence, and then add up all of these products. Here is how it works:

Height	Frequency	Height × Frequency
64	3	192
65	5	325
66	8	528
67	11	737
68	7	476
69	3	207
70	2	140
71	1	71
total:		2,676

Since the total of all the heights is 2,676, and there are 40 people, we can calculate that the average height is 66.9.

In general, suppose we have a list of numbers containing m different values (call them $x_1, x_2, \ldots, x_m$). We'll let f_1 stand for the frequency of occurrence of the value x_1. For example, in the list above, x_1 is 64 and f_1 is 3. In general, we will use f_i to represent the frequency of occurrence for the value x_i. Let n be the total number of observations, which is equal to

$$n = f_1 + f_2 + f_3 + \cdots + f_m$$

Then the average can be calculated from the following formula:

$$\text{(average)} = \frac{f_1 x_1 + f_2 x_2 + f_3 x_3 + \cdots + f_m x_m}{n}$$

$$= \frac{\sum_{i=1}^{m} f_i x_i}{n}$$

We can also calculate the median from the frequency table data. Since there are 40 people altogether, the median height is halfway between the heights of the 20th person and the 21st person. From the frequency table we can see that 27 people have heights less than or equal to 67, and 24 people have heights greater than or equal to 67. Therefore, both the 20th person and the 21st person must have heights of 67, which is the median.

We know that 50 percent of the people in the band have heights of 67 or less. Suppose we also wish to know what percent of the heights are 65 or less. From the frequency table we can calculate that 8 people are in that category, so $8/40 = 20$ percent of the band members have heights of 65 or less. This calculation is an example of the more general concept of *percentiles*. We can calculate many different percentiles for a set of numbers. For example, the 25th percentile is the number such that 25 percent of the items in the list are at or below it; the 60th percentile is the number such that 60 percent of the items in the list are below it; and so on.

Percentiles are commonly used when you want to compare yourself to the rest of the population. For example, growth charts for children are often shown as percentiles. If your child's height at a certain age is at the 60th percentile, then you know that the child is taller than 60 percent of the other children of that age. Results for standardized test scores are often given as percentiles. If your score is at the 65th percentile, then your score is higher than the scores of 65 percent of the people taking the test.

We can calculate the variance of the band members' heights using the frequency information from one of these formulas:

$$\text{Var}(x) = \frac{\sum_{i=1}^{m} f_i(x_i - \bar{x})^2}{n}$$

$$= \frac{\sum_{i=1}^{m} f_i x_i^2}{n} - \left(\frac{\sum_{i=1}^{m} f_i x_i}{n}\right)^2$$

In our case we have a variance of 2.69 and a standard deviation of 1.64.

GROUPED DATA

Here's our next problem: we need to analyze the adjusted gross income figures for all 1993 U.S. tax returns. Obviously, we do not want to see a list of all 114,601,815 different values for gross income. Instead, we would like to have the information in frequency table form. However, even then we don't want a list of the frequencies for every possible value of income. A portion of that type of frequency table would look like this:

Income	Frequency
$10,000.00	32
10,000.01	29
10,000.02	43
10,000.03	17
10,000.04	25
•	•
•	•
•	•

We don't really care how many households have incomes of $10,000.02 as opposed to $10,000.03. Instead, we would like the data presented to us as a frequency table that groups the data according to broad categories. We can call data in that form **grouped data**. Here is the income frequency table:

Adjusted Gross income (in thousands of dollars)	Number of Returns
0–1	3,550,455
1–3	6,472,017
3–5	5,748,112
5–7	5,825,575
7–9	5,963,864
9–11	5,701,001
11–13	5,496,355
13–15	5,210,087
15–17	4,859,885
17–19	4,306,292
19–22	6,252,720
22–25	5,426,552
25–30	7,783,772
30–40	12,358,342
40–50	9,072,138
50–75	12,248,446
75–100	4,224,878
100–200	3,107,998
200–500	786,038
500–1,000	140,803
>1,000	66,485
	Total 114,601,815

What if someone has an income of exactly $11,000.00? We will put that person in the 9–11 thousand class, so technically the label for each class should be written in a form similar to this: "9 thousand and over but less than 11 thousand."

We would like to calculate the average value for adjusted gross income. We used this formula to calculate the average height of the band members:

$$\bar{x} = \frac{\sum_{i=1}^{m} f_i x_i}{n}$$

However, there are some complications now that we have the data grouped by classes rather than individual values. For example, the table tells us that 5,701,001 people have incomes between $9,000 and $11,000. However, we don't know anything about the exact distribution within this range. All 5,701,001 people might have incomes of $9,000.57, or they might all have incomes of $10,999.71. Both of these possibilities, though, seem unlikely. We guess that the incomes of the 5,701,001 people are probably spread uniformly over the interval from $9,000 to $11,000. Therefore, we can assume that, if we considered all 5,701,001 people in this class, we would find that their average income is $10,000 (halfway between the two boundaries of the interval). The two end points of an interval are called the **class limits**, and the point halfway between them is called the **class mark**.

If we let f_i represent the number of observations in class i and x_i represent the class mark for class i, then we can calculate the average $\bar{x}$ from this formula:

$$\bar{x} = \frac{f_1 x_1 + f_2 x_2 + \cdots + f_m x_m}{n}$$

$$= \frac{\sum_{i=1}^{m} f_i x_i}{n}$$

We can start using this formula to calculate the average income:

$$3{,}550{,}455 \times \quad 500 +$$
$$6{,}472{,}017 \times 2{,}000 + \cdots$$

but we will run into a problem when we come to the last class:

$$>1{,}000{,}000 \quad 66{,}485$$

There is no upper limit for this class. All 66,485 people in the class might have incomes of $1,000,001, or all might have incomes of $30,000,000.

Our frequency distribution does not even provide us with enough information to guess what the average income for this class might be. This type of class is called an **open-ended class**. Open-ended classes may occur at either the top or the bottom of grouped data, and they cause problems for calculating some statistics, including the average. In this case we are bailed out by the IRS because it will tell us that the average gross income for all people in the million-dollar-and-over class is 2,566,600. (The IRS cannot give us too much detailed information on the distribution of people in the high-income category without violating the confidentiality of the tax returns.) If we cannot obtain the average value for an open-ended class, then we cannot calculate the average for the data. The best we can do in that case is to estimate the average. In our present problem we can now proceed to calculate the average income:

$$\bar{x} = \frac{(3{,}550{,}455 \times 500) + (6{,}472{,}017 \times 2{,}000) + (5{,}748{,}112 \times 4{,}000) + \cdots + (66{,}485 \times 2{,}566{,}600)}{114{,}601{,}815}$$

$$= 3{,}934{,}860{,}330.5/114{,}601{,}815 = 34.34 \text{ thousand}$$

(This result overstates the true value because the actual average for each class tends to be less than the midpoint of the class, particularly for the higher classes.)

We can also calculate the median adjusted gross income. Since there are 114,601,815 total returns, we must find one return such that 57,300,907 have lower incomes and 57,300,907 have higher incomes. We could start with a trial-and-error process. By adding 3,550,455 + 6,472,017 + 5,748,112, we find that 15,770,584 people (13.8 percent of the total) have incomes less than $5,000. Clearly the median income must be much greater than $5,000. By continuing this calculation (easy to do on a spreadsheet), we eventually find that 53,133,643 (46.4 percent of the total) have incomes below $19,000 and that 59,386,363 households (51.8 percent of the total) have incomes below $22,000.

This information does not tell us exactly where the median income is, but it does provide a very important clue. We know that the median income must be above $19,000 and below $22,000. We can narrow our search for the median income to the 6,252,720 households with incomes in the $19,000 to $22,000 class. The frequency table alone does not provide us with enough information to calculate the exact value of the median, so once again we must make an assumption. We will assume that the 6,252,720 households in this class are distributed evenly in the class. That is, one-third of the households in that class have incomes between $19,000 and $20,000, another third have incomes between $20,000 and $21,000, and so on. Since 57,300,907 households in the total population have incomes less than the median income, we can calculate as follows:

$$
\begin{array}{rl}
57{,}300{,}907 & \text{households have incomes less than the median} \\
-\,53{,}133{,}643 & \text{households have incomes less than \$19,000} \\
\hline
4{,}167{,}264 & \text{households above \$19,000 but less than the median}
\end{array}
$$

Therefore, 4,167,264/6,252,720 = 66.6 percent of the people in the $19,000 to $22,000 class have incomes below the median. Since we assumed that income in that class is evenly distributed, the median income must be:

$$\$19{,}000 + .666 \times 3{,}000 = \$20{,}998$$

Here is a general formula for calculating the median once the median class has been identified (assuming that the values within the median class are uniformly distributed):

$$\text{Median} = x_L + \frac{(N/2 - n_L)w}{n_m}$$

where x_L is the lower class limit of the median class
($19,000 in our example)

N is the total number of items in the population
(example: 114,601,815)

n_L is the total number of items *below* the median class
(example: 53,133,643)

n_m is the number of items *in* the median class
(example: 6,252,720)

w is the width of the median class
(example: $3,000)

(Note that when the number of items in the list is large we don't really need to worry about whether the total number is an even number or an odd number. Also note that the presence of open-ended classes poses no problem at all for calculating the median. We don't need to know the actual values of the extreme levels of income to find the median.)

• *THE HISTOGRAM*

It will also be very helpful to draw a frequency diagram for the income data. We might draw a diagram such as the one in Figure 2-2. However, this diagram is misleading. Although there is a big jump in the frequency diagram for incomes above $30,000, this does not mean that there are very many more people with incomes above that level than below it. The diagram looks the way it does because the classes above $30,000 are all much wider than the classes below $30,000. There are 12,358,342 people in the class directly above $30,000 and only 7,783,772 people in the class directly below $30,000. However, this great disparity occurs because the upper class is $10,000 wide ($30,000 to $40,000), whereas the lower class is only $5,000 wide ($25,000 to $30,000). Therefore, we must be very careful when we are drawing a frequency histogram for a set of grouped data where the classes have unequal widths.

Here is how to draw the histogram. Let's first try to draw a frequency diagram for the income data using classes that all have the same width of $1,000. This immediately raises the problem that we don't know how many people have incomes from $20,000 to $21,000, or from $21,000 to $22,000, and so on. However, we can once again assume that all of the incomes in the $20,000 to $25,000 class are distributed evenly, so that 1/5 of the people in the class have incomes from $20,000 to $21,000, 1/5 have incomes from $21,000 to $22,000, and so on. A portion of the new frequency diagram is shown in Figure 2-3.

You can see that this distribution is not symmetric. Instead, there are more very high incomes than very low incomes. In this case it is said that the distribution has a long right **tail**, or that it is **skewed** to the right. In general, when a distribution is skewed to the right, the value of the mean will be pulled up by the extreme high values, so the value of the mean will be above the value of the median.

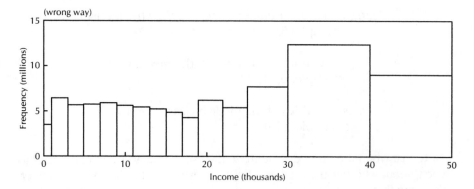

Figure 2-2

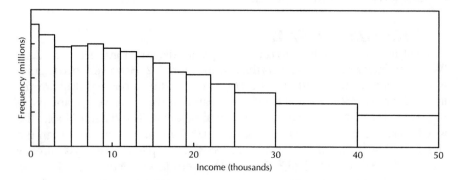

Figure 2-3. Adjusted Gross Income, 1993.

Here is the general procedure to draw a histogram:
1. The width of each rectangle in the diagram should be proportional to the width of the class it represents. (For example, in our income problem the rectangles representing the classes with width $5,000 must be five times wider than the rectangles representing the classes with width $1,000.)
2. The height of each rectangle should be proportional to the number of items in its class divided by the width of the class. Here is another way to state this condition: The *area* of each rectangle should be proportional to the number of items in the corresponding category. For example, there are 7,783,772 people in the $25,000 to $30,000 category, so the height of the rectangle should be proportional to 7,783,772/5,000 = 1,556.75. Also, there are 12,358,342 people in the $30,000 to $40,000 category, so the height of that rectangle should be proportional to 12,358,342/10,000 = 1235.83. In other words, the $25,000 to $30,000 rectangle should be 1556.75/1235.83 = 1.26 times as high as the $30,000 to $40,000 rectangle. The actual values of the heights, of course, depend on the size of your paper.

• OTHER GRAPHS

There are four more types of graphs that we will mention here. Suppose we were given the following grouped data for the daily withdrawals at a bank:

Withdrawal	Frequency
500–600	12
600–700	36
700–800	63
800–900	81
900–1,000	77
1,000–1,100	42
1,100–1,200	24

We can plot this information on a histogram (see Figure 2-4)

THE FREQUENCY POLYGON

The same information can be shown on a diagram called a **frequency polygon**. Instead of using a bar to represent each class, we place a dot above the class midpoint. Then the dots are connected with line segments (see Figure 2-5).

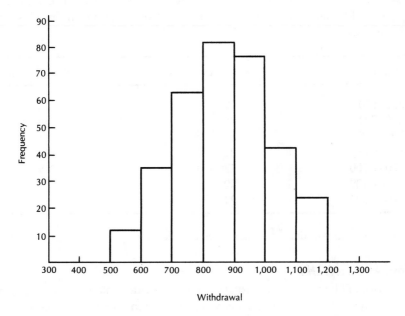

Figure 2-4

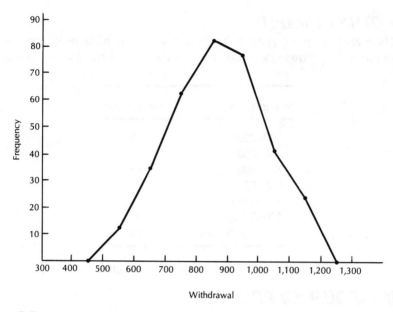

Figure 2-5

THE OGIVE

It also can be useful to calculate the cumulative frequencies. The cumulative frequency for each class is the total number of observations in that class or in a lower class.

Withdrawal	Frequency	Cumulative Frequency
500–600	12	12
600–700	36	48
700–800	63	111
800–900	81	192
900–1,000	77	269
1,000–1,100	42	311
1,100–1,200	24	335

A graph of the cumulative frequencies is called an **ogive** (see Figure 2-6).

THE PIE CHART

If we would like to see how a total quantity is divided into different categories, the best type of diagram to draw is a **pie chart**. Each category is represented by a wedge in the pie. The angular size of each wedge is proportional to the fraction of the total that belongs to that category. Here is a table showing U.S. government spending by category in 1994. Figure 2-7 illustrates the pie chart.

Category	Amount (billion $)	Percent
Social Security	320	21.9
Defense	282	19.3
Income security	214	14.6
Interest	203	13.9
Medicare	145	9.9
Health	107	7.3
Other	190	13.0
Total	1461	100.0

Source: *Economic Report of the President, 1996.*

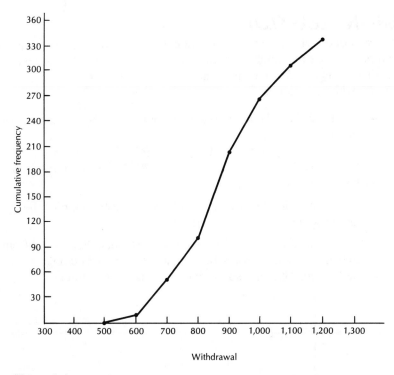

Figure 2-6

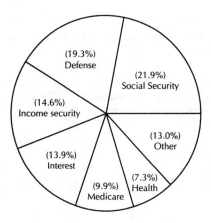

Figure 2-7. U.S. Budget 1994

STEM-AND-LEAF PLOT

A clever way of drawing a picture to describe the distribution of some numbers in a sample is to use the numbers themselves. The following process, invented by John Tukey, is called a stem-and-leaf plot. As always, it's easier to describe if we work with an example.

GIVEN   We have the following 40 scores (out of 100) on a standardized mathematics test:

78 59 86 94 43 56 78 84 57 89 96 68 67 65 75 73 67 87 84 45 56 94 87 56 85 76 86 79 78 77 59 76 68 49 86 87 83 94 85 96

DESIRED   We want a simple way of describing the distribution of the data, giving some sense of the range and median.

SOLUTION   Notice that the data all have two digits. The first digit ranges from 4 to 9; if we were to group the data, a natural way of doing so would be into 40s, 50s, and so on. We'll do that in a picture:

```
4 |
5 |
6 |
7 |
8 |
9 |
```

Instead of counting up the number of data in the 40s, 50s, and so on, we'll put the second digit for each data point in the appropriate row. For example, using the first five data points, our picture would look like:

```
4 | 3
5 | 9
6 |
7 | 8
8 | 6
9 | 4
```

If we finish off the first row of 20 data points, our picture would look like:

```
4 | 35
5 | 967
6 | 8757
7 | 8853
8 | 64974
9 | 46
```

We can quickly finish off the second row of data, and our complete picture looks like:

```
4 | 359
5 | 967669
6 | 87578
7 | 885369876
8 | 649747566735
9 | 46446
```

What we've done is akin to making a frequency count using tick marks, but using the last digits of the data as the tick marks. We can easily count up the number of data points in each row of our picture:

```
(3)   4 | 359
(6)   5 | 967669
(5)   6 | 87578
(9)   7 | 885369876
(12)  8 | 649747566735
(5)   9 | 46446
```

Now we have a sense for the distribution (note the peak in the 80s) and the range (lowest is 43, highest is 96), and we can quickly find the median. Since there are 40 data points, the median lies between the 20th and 21st ordered data points. We haven't completely ordered the data points, but since there are $3+6+5=14$ data points in the first three rows, we know that the median is between the 6th and 7th largest number in the 70s row, i.e., between 78 and 78; i.e., the median for the data is 78.

A **stem-and-leaf plot** is a frequency chart constructed for two- or three-digit sample points using the last digit as a counter for the frequency class corresponding to the beginning digits. Typically stem-and-leaf plots are drawn with the stem (the beginning digits) listed vertically and the last digits (forming the leaves) listed horizontally. Often, the length of each leaf is recorded to the left of the stem.

The strength of the stem-and-leaf plot lies in its ability to give you the information quickly and easily, without having to perform calculations: a sample of 100 data points can be analyzed in only a couple of minutes.

If more than two or three digits are present in the values of the data points, you would probably want to round them off before proceeding.

It may happen that the stem-and-leaf plot as described above groups the data into too small a number of leaves:

GIVEN   The following 30 measurements in inches are the heights of the students in a statistics class:

68 62 64 65 67 59 68 63 63 72 71 62 65 65 58
65 66 67 63 68 63 65 65 67 69 65 71 70 68 67

DESIRED   We want a way of describing the distribution, indicating the range and median.

SOLUTION   This set of data has been rounded to two digits. The first digit for each data point is a 5, 6, or 7. A three-leafed stem-and-leaf plot for this data set looks like:

```
(2)   5 | 98
(24)  6 | 824578332555673835579587
(4)   7 | 2110
```

which tells us about the peak in the 60s but otherwise doesn't help us to get a sense for the distribution. Instead of grouping all of the 60s together, we can split the leaves up into early 50s (50–54), late 50s (55–59), early 60s (60–64), late 60s (65–69), early 70s (70–74), and late 70s (75–79):

```
(2)   5 | 98
(7)   6 | 2433233
(17)  6 | 85785556785579587
(4)   7 | 2110
```

(Note that because the students in the class are neither very short nor very tall, the first and last rows are not affected.) We can see that the peak in heights lies in the upper 60s, the range is from 58 to 72, and the median (which lies between the 15th and 16th ordered data points) lies between the 6th and 7th largest data points in the third row, i.e., between 65 and 65; i.e., the median is 65.

* * * *

Creating graphs by hand is a lot of work. Fortunately, computer programs are now available with a wide variety of graphics capabilities; these will be discussed later.

Note to Chapter 2: In some books the term "average" is used as a generic term for measures of central tendency; in this sense it includes the mean, median, and mode. However, in everyday usage the term "average" is always used as a synonym for "mean."

The mean discussed in this chapter is also sometimes called the "arithmetic mean" to distinguish it from a quantity called the "geometric mean." The geometric mean of two numbers is the square root of their product.

YOU SHOULD REMEMBER

1. For very large lists it would be too cumbersome to keep track of the entire list. The data can be summarized by defining categories and then counting the number of data points in each category (called the *frequency*). If we have m groups, and we let f_i be the frequency for group i, then we can approximate the mean with this formula:

$$(\text{mean}) = \frac{\sum_{i=1}^{m} f_i x_i}{\sum_{i=1}^{m} f_i} = \frac{\sum_{i=1}^{m} f_i x_i}{n}$$

   where $n = \sum_{i=1}^{m} f_i$ is the total number of data items, and x_i is the midpoint for class i.

2. It is also very informative to represent data in graphical form. For grouped data the distribution can be displayed on a frequency diagram, using bars of different heights to represent the frequencies. Other types of diagrams are *frequency polygons* (formed by placing a dot at the midpoint of each category at a height proportional to the number of items in that category), *ogives* (graphs of the cumulative frequency distribution), *pie charts* (where each category is represented by a wedge-shaped sector of a circle whose size depends on the proportion of items in that category), and *stem-and-leaf plots* (where the last digit of a number is plotted on a frequency chart for the beginning digits).

KNOW THE CONCEPTS
DO YOU KNOW THE BASICS?

Test your understanding of Chapter 2 by answering the following questions:

1. When does the median work better than the mean as a measure of the typical value in a group?

2. Suppose you conduct a survey in which people indicate how much they like quiche (using a scale of 1 to 10). If half the people love quiche and the other half detest quiche, what will the frequency distribution look like? What will be the best measure of central tendency for this distribution?

3. Why do you need to know the variance of your daily sales figures if you are managing a firm?

4. Why are open-ended classes sometimes used?

5. Do open-ended classes pose problems for calculating the mean? For calculating the median?

6. If there were a sudden arrival of a group of very tall people in your population, what would happen to the frequency diagram for height? To the mean? To the median? To the variance?

7. When is a pie chart better than a bar diagram?

8. Suppose you are a clothing manufacturer. You do not design garments for extremely tall or extremely short people. If you would like your range of products to appeal to half of the population, what measure of population heights do you most need to know?

TERMS FOR STUDY

average	ogive
bimodal	open-ended class
central tendency	percentile
Chebyshev's theorem	pie chart
class limits	population standard deviation
class mark	population variance
coefficient of variation	quartile
dispersion	range
frequency diagram	raw data
frequency polygon	sample standard deviation
frequency table	sample variance
grouped data	skewed
histogram	standard deviation
interquartile range	statistic
mean	subscript
mean absolute deviation	summation notation
median	tail
mode	variance
normal distribution	

PRACTICAL APPLICATION

COMPUTATIONAL PROBLEMS

For each list of numbers calculate the mean, median, range, and both versions of the sample standard deviation.

1.	68	53	55	63	80
	51	62	79	65	50
	60	74			
2.	227	231	204	210	203
	211	235	232	222	222
	239	203	236	231	234
	232				
3.	22	22	21	28	25
	23	24	21	22	25
4.	168	172	169	171	168
	160	163	167	166	164
	170	172	165	175	170
	170	171			
5.	108	100	102	113	98
	101	98	106	100	114
	102	102	114	106	106
	97	103	103		

For each exercise below you are given a frequency table. Calculate the mean, median, and population standard deviation.

6.

	Frequency		Frequency
0– 1,000	12	6,000– 7,000	48
1,000– 2,000	15	7,000– 8,000	40
2,000– 3,000	19	8,000– 9,000	30
3,000– 4,000	22	9,000–10,000	16
4,000– 5,000	30	10,000–11,000	4
5,000– 6,000	56	11,000–12,000	2

7.

	Frequency		Frequency
0– 10	122	80– 90	180
10– 20	180	90–100	175
20– 30	256	100–110	143
30– 40	350	110–120	120
40– 50	311	120–130	106
50– 60	278	130–140	99
60– 70	250	140–150	97
70– 80	211	150–160	75

8.

	Frequency		Frequency
0– 100	69	600– 700	250
100– 200	133	700– 800	219
200– 300	174	800– 900	170
300– 400	211	900–1,000	136
400– 500	255	1,000–1,100	68
500– 600	288		

9.

	Frequency		Frequency
0– 500	34	5,500– 6,000	78
500– 1,000	62	6,000– 6,500	83
1,000– 1,500	75	6,500– 7,000	96
1,500– 2,000	98	7,000– 7,500	119
2,000– 2,500	160	7,500– 8,000	140
2,500– 3,000	125	8,000– 8,500	156
3,000– 3,500	119	8,500– 9,000	130
3,500– 4,000	96	9,000– 9,500	111
4,000– 4,500	84	9,500–10,000	95
4,500– 5,000	76	10,000–10,500	84
5,000– 5,500	71		

ANSWERS

KNOW THE CONCEPTS

1. The median is better when there are a few very large or very small values in the population.

2. This type of distribution is called a *bimodal distribution*; the frequency diagram resembles a two-humped camel. Neither the mean nor the median provides a good indicator of the typical value. In other words, there is no "best measure."

3. If the sales figures have a relatively small variance, it will be easier to make plans for your business. If the variance is large, you must develop a way to cope with this uncertainty. For example, you may be able to let your inventory level fluctuate. You must determine the consequences of the two types of problems that can occur with erratic sales: (a) some days there will be unsold merchandise remaining; (b) other days you may run out of merchandise. The decisions you make will depend on the relative seriousness of these two types of problems.

4. It is not always possible to obtain precise data for the extremely high or low values in a distribution.

5. Open-ended classes make it difficult to calculate the mean, but they pose no problem for calculating the median.

6. The bars at the upper end of the frequency diagram for height will become larger; the mean will be increased significantly; the median will also increase, but probably not as much as the mean; and the variance will become larger.

7. A pie chart provides a clearer picture than a bar diagram of the fraction of items in each category.

8. It would be most helpful to know the interquartile range, since half of the values are contained between the third quartile and the first quartile.

PRACTICAL APPLICATION

Here are the formulas and definitions you need for doing Exercises 1–9:

$$\text{mean} = \frac{x_1 + x_2 + \cdots + x_n}{n} = \frac{\sum_{i=1}^{n} x_i}{n}$$

median: the value such that half the numbers in a list are at or above it and half are at or below it.

range: the largest value minus the smallest value

sample variance, version 1:

$$s_1^2 = \frac{\sum_{i=1}^{n} (x_i - \bar{x})^2}{n} = \overline{x^2} - \bar{x}^2 = \frac{n-1}{n} s_2^2$$

sample standard deviation, version 1:

$$s_1 = \sqrt{\frac{\sum_{i=1}^{n} (x_i - \bar{x})^2}{n}}$$

sample standard deviation, version 2:

$$s_2 = s_1 \sqrt{\frac{n}{n-1}}$$

Complete calculations are shown below for Exercise 1 and Exercise 6. The same methods are used for Exercises 2–5 and 7–9 respectively.

1. Total = 68 + 53 + 55 + 63 + 80 + 51 + 62 + 79
    + 65 + 50 + 60 + 74 = 760
    Mean = 760/12 = 63.333   Median = (62 + 63)/2 = 62.5
    Range = 80 − 50 = 30
    $\overline{x^2}$ = (68² + 53² + 55² + 63² + 80² + 51² + 62² + 79²
        + 65² + 50² + 60² + 74²)/12 = 4,109.5
    s_1^2 = 4,109.5 − 63.333² = 98.4 s_1 = 9.919 s_2 = 10.360

2. Total: 3,572

   Mean = 223.25   Median = $(227 + 231)/2 = 229$

   Range = $239 - 203 = 36$

   $x^2 = 49,995$

   $s_1^2 = 49,995 - 223.25^2 = 154.44$ $s_1 = 12.427$ $s_2 = 12.835$

3. Mean = 23.300   Median = 22.5   Range = 7

   $s_1^2 = 4.410$ $s_1 = 2.100$ $s_2 = 2.214$

4. Mean = 168.294   Median = 169   Range = 15

   $s_1^2 = 13.502$ $s_1 = 3.674$ $s_2 = 3.788$

5. Mean = 104.056   Median = 102.5   Range = 17

   $s_1^2 = 26.941$ $s_1 = 5.191$ $s_2 = 5.341$

6. Total in population = $12 + 15 + 19 + 22 + 30 + 56$
   $$+ 48 + 40 + 30 + 16 + 4 + 2$$
   $$= 294$$

   Total of all items = $500 \times 12 + 1,500 \times 15 + 2,500 \times 19$
   $$+ 3,500 \times 22 + 4,500 \times 30$$
   $$+ 5,500 \times 56 + 6,500 \times 48$$
   $$+ 7,500 \times 40 + 8,500 \times 30$$
   $$+ 9,500 \times 16 + 10,500 \times 4$$
   $$+ 11,500 \times 2$$
   $$= 1,680,000$$

   Mean = $1,680,000/294 = 5,714.2857$

   Median = $5,000 + (294/2 - 98) \times 1,000/56 = 5,875$

   $x^2 = (500^2 \times 12 + 1,500^2 \times 15 + 2,500^2 \times 19 + 3,500^2 \times 22$
   $$+ 4,500^2 \times 30 + 5,500^2 \times 56 + 6,500^2 \times 48$$
   $$+ 7,500^2 \times 40 + 8,500^2 \times 30 + 9,500^2 \times 16$$
   $$+ 10,500^2 \times 4 + 11,500^2 \times 2)/294$$
   $$= 38,508,503$$

   Variance = $38,508,503 - 5,714.2857^2 = 5,855,442$

   Standard deviation = 2,419.8

7. Mean = 66.5   Median = 59.3

   Standard deviation = 39.6

8. Mean = 550   Median = 550.2

   Standard deviation = 257

9. Mean = 5,575   Median = 5,794.9

   Standard deviation = 2,914.2

3
INTRODUCTION TO PROBABILITY AND HYPOTHESIS TESTING

KEY TERMS

factorial for a particular whole number, the product of all the whole numbers from 1 up to that number

hypothesis testing a statistical procedure that involves collecting evidence and then making a decision as to whether a particular hypothesis should be accepted or rejected

probability the study of chance phenomena

We will investigate the act of coin tossing as an introduction to probability. Admittedly there are few occasions where a business decision depends on the outcome of a coin toss, but study of the probabilities associated with coin tosses provides valuable practice for some more difficult concepts we will encounter later on.

COIN TOSSING

If you flip a coin, the result will be either a head or a tail. However, it is practically impossible to predict correctly which outcome will occur. You could, in theory, calculate the path of the coin by using the laws of mechanics, if you knew the initial force that was applied to it and the other forces acting on it, and thus determine which side would be up when the coin comes to a stop. This would be very hard, though, so for practical purposes a coin toss is a good example of a random event: there is no accurate way to predict what the result will be.

If you toss a coin more than once, though, you can make some interesting predictions about what will happen. This is a general feature of random events, and it is one of the foundations of probability: you can't say what will happen in one occurrence, but if you watch the same thing happen many times you will be able to predict some patterns. For example, if you throw a dart at a phone book and try to guess the height of the person whose name is hit by the dart, you will probably be wrong. If you select 100 people at random and try to guess their average height, you will be much more likely to get it right.

If you flip a coin twice, there are four possibilities for the result: head-head, head-tail, tail-head, and tail-tail. (After this, H will be used for head and T for tail.) There are two possibilities for the first flip, and for each possible result of the first flip there are two possibilities for the second flip.

Now, if you need to know whether you will see no head, one head or two heads in the two flips, you can make a prediction. Assuming that the coin is a fair one, not biased toward head or tail, each of the four possible outcomes (HH, HT, TH, TT) is equally likely. You will see two heads if outcome 1 (HH) occurs, and you will see no head if outcome 4 (TT) occurs. Therefore, you have just as much chance of seeing no head as you do of seeing two heads. However, you will see exactly one head if either outcome 2 (HT) or outcome 3 (TH) occurs. Hence your best bet is to guess that one head will appear on the two flips. Of course, you still have a 50 percent chance of being wrong, but you have a much better chance of being right than if you had guessed no head or two heads.

You probably could have figured this much out for yourself without knowing anything about probability theory. Now use the same ideas to figure out how many heads are likely to appear if you toss a coin any number of times. Intuitively you will guess that if you toss a coin 92 times, the number of heads that show up will probably be about 46. At the opening of the play "Rosencrantz and Guildenstern Are Dead," Guildenstern tosses 92 heads in a row, thereby losing a lot of money to Rosencrantz. Intuitively, you know that this is extremely unlikely to occur.

• *COMPUTING PROBABILITIES*

In general, if you tossed a coin n times, n being a large number, you would expect that the number of heads would be close to $n/2$ and that you would be unlikely to get n heads in a row. First, let us consider the case of flipping the coin three times. There are eight possible outcomes:

HHH, HHT, HTH, HTT, THH, THT, TTH, TTT

That is the first step in computing a probability: find out how many possible outcomes there are. Then find out how many of those outcomes lead to the event that you are interested in. The **probability** of that event occurring is equal to the number of ways of getting that event divided by the total number of possible outcomes. For example:

Event You're Interested In	Outcomes That Lead to That Event	Number of Outcomes	Probability of Event
No head	TTT	1	1/8
One head	HTT, THT, TTH	3	3/8
Two heads	HHT, HTH, THH	3	3/8
Three heads	HHH	1	1/8

If you are forced to guess how many heads will appear on three flips, you should guess either one or two. You clearly shouldn't guess none or three.

We can follow the same method if we need to make a prediction about the results of four flips: list all the outcomes, and then count how many outcomes have no head, how many outcomes have one head, and so on. This method would get a bit tedious, though, and it would be very difficult if we tried it for a large number of flips. What is needed is a way to count the outcomes without having to list them all. That concept is one of the important ideas in probability.

First, we need to count all the possible outcomes from four flips. There are two possibilities for the first toss, then for each of those possibilities there are two for the second toss, then for each of these combinations there are two possibilities for the third toss, and so on. Altogether, there will be $2 \times 2 \times 2 \times 2 = 2^4 = 16$ possible outcomes. You should be able to convince yourself that, if you flip a coin n times, there will be 2^n possible outcomes. Now you can see why the listing method doesn't work. If you flip a coin 10 times, there will be $2^{10} = 1,024$ possible outcomes.

Next, we need to count how many outcomes have zero head, how many have one head, and so on, up to how many have four heads. The case of zero head is easy, since there is only one outcome (TTTT) that has zero head. To calculate the number of possibilities with one head, we need to figure out how many different ways we can write one H and three T's. Since there are four possible places for the H, there are four possibilities: HTTT, THTT, TTHT, TTTH. Now we need to figure out how many ways we can write two H's and two T's. There are four possible places to put the first H (call it H1):

$$H1 \_\_ \_\_ \_\_ , \_\_ H1 \_\_ \_\_ , \_\_ \_\_ H1 \_\_ , \_\_ \_\_ \_\_ H1$$

For each of these possibilities there are three places to put the second H (call it H2):

H1 H2 __ __ ,   H1__ H2 __ ,   H1 __ __ H2,
H2 H1 __ __ ,   __ H1 H2 __ ,   __ H1 __ H2,
H2 __ H1 __ ,   __ H2 H1 __ ,   __ __ H1 H2,
H2 __ __ H1,   __ H2 __ H1,   __ __ H2 H1

(Note that we have to worry only about where to put the H's, since once we've done that it will be obvious where to put the T's.) There are 12 of these possible ways to arrange the H's. However, it doesn't make any difference which is H1 and which is H2; __ H1 H2 __ is the same as __ H2 H1 __ . We must divide by 2 to avoid double-counting these duplications. That means there are six possible ways to get two heads:

HHTT, HTHT, HTTH, THHT, THTH, TTHH

Therefore, the results for four flips are as follows:

Number of Heads (h)	Number of Outcomes with h Heads	Probability That h Heads Will Occur
0	1	1/16 = .0625
1	4	4/16 = .2500
2	6	6/16 = .3750
3	4	4/16 = .2500
4	1	1/16 = .0625

(You should note that these probabilities are symmetric—that is, the probability that you will get h heads is exactly the same as the probability that you will get h tails.)

Now we need to figure out a general formula for the number of outcomes that have h heads in n tosses. In principle it is the same as what we did before: we need to figure out how many ways we can write h capital H's on n blanks. There are n possibilities for the first H that we write, then $n - 1$ places left for the second, $n - 2$ places left for the third, and so on. Altogether, there will be

$$n \times (n - 1) \times (n - 2) \times (n - 3) \times \cdots \times (n - h + 1)$$

ways of writing down all the H's. Then we must divide by

$$h \times (h - 1) \times (h - 2) \times (h - 3) \times \cdots \times 3 \times 2 \times 1$$

to eliminate the duplications caused by the different orderings. Therefore, if we flip a coin n times, the number of possible outcomes that have exactly h heads will be

$$\frac{n \times (n - 1) \times (n - 2) \times (n - 3) \times \cdots \times (n - h + 1)}{h \times (h - 1) \times (h - 2) \times \cdots \times 3 \times 2 \times 1}$$

• *USING FACTORIALS*

The quantity $h \times (h - 1) \times (h - 2) \times (h - 3) \times \cdots \times 3 \times 2 \times 1$ is interesting. It turns out that there are many times in probability theory when we need to calculate the product of all the numbers from 1 up to another particular number. This quantity is called the **factorial** of the number, and it is symbolized by an exclamation mark (!). For example:

$$3! = 3 \times 2 \times 1 = 6, \qquad 4! = 4 \times 3 \times 2 \times 1 = 24,$$

$$10! = 10 \times 9 \times 8 \times 7 \times 6 \times 5 \times 4 \times 3 \times 2 \times 1 = 3{,}628{,}800,$$

$$69! = 1.71 \times 10^{98}$$

As you can see, factorials become very big very fast. It should be obvious that $1! = 1$. There are also some formulas where we may need to find $n!$ for $n = 0$. These formulas work only if $0!$ has the value of 1, so we will make that definition.

Now we can use the factorial notation in the denominator to simplify the formula for the number of combinations with h heads:

$$\frac{n \times (n - 1) \times (n - 2) \times (n - 3) \times \cdots \times (n - h + 1)}{h!}$$

We can also write the numerator in a shorter fashion using the factorial function. We multiply and divide the numerator by $(n - h)!$:

$$
\begin{aligned}
& n \times (n - 1) \times (n - 2) \times \cdots \times (n - h + 1) \\
&= \frac{n \times (n - 1) \times (n - 2) \times \cdots \times (n - h + 1) \times (n - h)!}{(n - h)!} \\
&= \frac{n \times (n - 1) \times (n - 2) \times \cdots \times 3 \times 2 \times 1}{(n - h)!} \\
&= \frac{n!}{(n - h)!}
\end{aligned}
$$

Therefore, we can write our formula in a very compact form:

$$\frac{n!}{h!(n - h)!}$$

(We will see this formula again in Chapter 4.)

What all this means is that we are now willing to predict that, if we flip a coin n times, the probability that we will get h heads is

$$\left[\frac{n!}{h!(n-h)!} \right] 2^{-n}$$

Let's make sure that this formula works for $n = 5$ (in this case 2^{-n} is 1/32):

Table 3-1

h	Number of Combinations with h Heads		Probability of h Heads
0	$\frac{5!}{0!5!} = 1$	TTTTT	1/32 = .031
1	$\frac{5!}{1!4!} = 5$	HTTTT, THTTT, TTHTT, TTTHT, TTTTH	5/32 = .156
2	$\frac{5!}{2!3!} = 10$	HHTTT, HTHTT, HTTHT, HTTTH, THHTT, THTHT, THTTH, TTHHT, TTHTH, TTTHH	10/32 = .313
3	$\frac{5!}{3!2!} = 10$	TTHHH, THTHH, THHTH, THHHT, HTTHH, HTHTH, HTHHT, HHTTH, HHTHT, HHHTT	10/32 = .313
4	$\frac{5!}{4!1!} = 5$	HHHHT, HHHTH, HHTHH, HTHHH, THHHH	5/32 = .156
5	$\frac{5!}{0!5!} = 1$	HHHHH	1/32 = .031

We can also use the formula to see that the chance of getting 92 heads in a row is 2.019×10^{-28}. Keep that in mind if you decide to try to break Rosencrantz and Guildenstern's record.

YOU SHOULD REMEMBER

1. The study of coin tossing provides a good introduction to probability because we can easily understand the process that is occurring.

2. To find the probability that h heads will appear if we toss a coin n times, we first need to calculate the number of possible outcomes (2^n). Then we need to calculate the number of possible ways of tossing h heads:

$$\frac{n!}{h!(n-h)!}$$

3. The exclamation point represents a function called the *factorial function*, which is the product of all the whole numbers from 1 up to the number:

$$n! = n \times (n-1) \times (n-2) \times (n-3) \times \cdots \times (3) \times (2) \times (1)$$

Therefore, the probability of tossing h heads if we toss a fair coin n times is

$$\frac{\dfrac{n!}{h!(n-h)!}}{2^n}$$

HYPOTHESIS TESTING

During all of these calculations we have assumed that the coin was a fair coin—that is, during any particular flip there was a 50 percent chance of flipping a head and a 50 percent chance of flipping a tail. Now we will consider an even more perplexing problem: How can we tell whether the coin is really fair? You will especially need to know the answer to this question if you are considering playing a coin-flipping game with an unkempt-looking stranger in a strange town. To put it formally, if we let p stand for the probability that the coin will come up heads, how do we know that $p = 1/2$?

In our coin example we should first, of course, make an obvious check. If the coin has two heads, then $p = 1$; if it has two tails, then $p = 0$. Once we've done that, though, it is very difficult to tell just by looking at the coin whether it is fair. Intuitively, we can't think of any reason why it might be more likely to come up "heads" rather than "tails" (or vice versa), but it might be unbalanced in such a way as to make one outcome more likely than the other. If we flip the coin once, we won't have a clue as to whether it is fair. However, if we flip the coin many times, we will start getting some information that we can use to estimate how fair it is.

• *THE NULL HYPOTHESIS AND THE ALTERNATIVE HYPOTHESIS*

Problems of this sort are called **hypothesis testing** problems. First we decide on the hypothesis we want to test. In this case our hypothesis is that $p = 1/2$. The hypothesis that is to be tested is often called the **null hypothesis**. The only other possibility is that the null hypothesis is wrong. The hypothesis that says, "The null hypothesis is wrong" is called the **alternative hypothesis**. In our case the alternative hypothesis is that the coin is not fair ($p \neq 1/2$). We know that either the null hypothesis or the alternative hypothesis must be true, since they are the only two possibilities. The question is: Do we accept the null hypothesis and say that the coin is fair, or do we reject the null hypothesis and say that the coin is unfair?

It is clear intuitively that we should flip the coin many times; let n be the number of flips. Then, if the number of heads that appears is close to $n/2$, we should accept the hypothesis that the coin is fair. If the number of heads is very far from $n/2$, we should reject the hypothesis that the coin is fair. The big question is: How far away from $n/2$ can it be before we say the coin is unfair?

Therefore, our test procedure will work like this. We will pick a number c. If the number of heads (h) is between ($n/2 - c$) and ($n/2 + c$), we will accept the null hypothesis and say that the coin is fair; otherwise we will say that the coin is not fair. We can call the region from ($n/2 - c$) to ($n/2 + c$) the **zone of acceptance**. If h is not in the zone of acceptance, we will be very critical of the hypothesis and will reject it. Therefore, the zone for which the hypothesis will be rejected is called the **rejection region** or **critical region**. (See Figure 3-1.)

The main problem now is: How far away from $n/2$ can we let the number of heads get before we say that the coin is unfair—that is, how big should we make c?

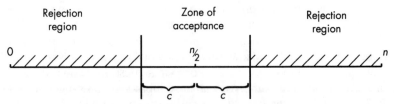

Figure 3-1. Number of heads in n flips.

• *AVOIDING TYPE 1 AND TYPE 2 ERRORS*

We would like to make the right judgment about our null hypothesis. There are two ways we can be right: we can accept the hypothesis when it is true, or we can reject it when it is false. However, that means that there are also two ways we can be wrong: we might reject the hypothesis when it is really true, or we might accept the hypothesis when it is really false. The first kind of mistake can be called a **type 1 error**, and the second kind can be called a **type 2 error**. Table 3-2 shows the possibilities.

Table 3-2

	Accept Hypothesis	Reject Hypothesis
Hypothesis is true	right	type 1 error (we want to avoid this)
Hypothesis is false	type 2 error	right

If we choose a large value for c, we will have a wide zone of acceptance and we will be more likely to accept the hypothesis than we would with a small value of c. That means that there is less chance of committing a type 1 error—that is, we're not likely to reject the hypothesis if it is really true. However, if we make the zone of acceptance large, we are increasing the risk that we will accept the hypothesis even if it is really false, thereby committing a type 2 error.

The other strategy is to choose a narrow zone of acceptance. If we do that, it is unlikely that we will commit a type 2 error (we're not likely to accept the hypothesis if it is really false) but we stand a much greater chance of committing a type 1 error (rejecting the hypothesis when it is really true). Obviously there is an inherent trade-off involved in hypothesis testing. We usually can't devise a single test procedure that will minimize the chances of committing both types of error.

Often we will be more worried about the possibility of incorrectly rejecting the hypothesis, so we will be more careful about avoiding type 1 errors. If we're going to have the courage to tell the unkempt stranger that we think his coin is unfair, then we want to make almost certain that we're right. (Otherwise he might get nasty.) In scientific work, if we decide to accept the hypothesis, we're likely to keep on searching for more evidence to see whether we can make a convincing case for it. If, on the other hand, we decide to reject the hypothesis, that means we're really convinced that the hypothesis is false, so we can stop.

What is often done in statistics is to set an upper limit to the probability of committing a type 1 error. Usually this limit is set at either 10 percent or 5 percent. At first it can be confusing to remember the difference between type 1 and type 2 errors. Just remember that our number one priority is to avoid errors of type 1, which we do by making certain that we don't reject the hypothesis unless we're pretty sure it's wrong. If we decide on a 10 percent test, then the probability that h will equal a particular value k is

$$\frac{n!}{k!(n-k)!} \times 2^{-n}$$

Once we've decided this, we need to figure out how wide to make the zone of acceptance. Let's suppose that the coin is fair. Once we make that assumption, we know exactly what the probabilities are, using the principles we developed earlier in the chapter. If n is the number of flips and h is the number of heads, that means we want to make sure there is only a 10 percent chance that our test procedure will say the coin is unfair when it is really fair.

Suppose that $n = 20$. Then we can make a table of the probabilities that $h = 0, h = 1, h = 2$, and so on (See Table 3-3). These probabilities are shown in Figure 3-2.

We want to choose our zone of acceptance so that there is roughly a 90 percent chance that h will land in the zone and only a 10 percent chance that h will land outside the zone. If we add up the probabilities for $h = 7, h = 8, h = 9, h = 10, h = 11, h = 12$, and $h = 13$, we find that if the coin is fair there is a .8847 probability that h will have one of these seven values. Therefore, we'll design our test like this: We will flip the coin 20 times and count the number of heads (h). If h is between 7 and 13, we will accept the hypothesis and say that the coin is fair. If h is less than or equal to 6, or if h is greater than or equal to 14, we will say that the coin is unfair.

In this way we can ensure that the probability of erroneously rejecting the hypothesis (type 1 error) is only about 12 percent. For example, suppose that there are 5 heads out of the 20 flips. Then we can say with a fair degree of confidence that the coin is not fair. We cannot say this with absolute certainty, because there is a 1.48 percent chance that only 5 heads will turn up in 20 flips

Table 3-3. Results of Tossing a Coin 20 Times

Number of Heads	Probability				
0	.000001				
1	.000019				
2	.000181				
3	.001087				
4	.004621				
5	.014786				
6	.036964				
7	.073929				
8	.120134				
9	.160179				
10	.176197	8 to 12:			
11	.160179	.7368	7 to 13:		
12	.120134		.8847	6 to 14:	
13	.073929			.9586	5 to 15:
14	.036964				.9882
15	.014786				
16	.004621				
17	.001087				
18	.000181				
19	.000019				
20	.000001				

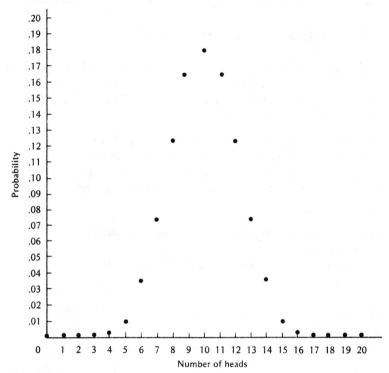

Figure 3-2

of a fair coin. Therefore there still is a possibility that we might commit a type 1 error by saying that the hypothesis is false when it is really true. However, we have made sure that the probability of this happening is less than 12 percent.

Of course, you can be still more cautious if you want to. Suppose you are very worried about erroneously rejecting the fair-coin hypothesis, so you want to make sure that the chance of that happening is about 4 percent. In that case you can change your test procedure so that you will accept the fair-coin hypothesis if h is between 6 and 14. With this procedure, you are even more sure that you won't say that the coin is unfair when it is really fair. However, by widening the zone of acceptance you are increasing the chances of committing a type 2 error—that is, saying that the coin is fair when it is really unfair. There is no way to calculate the probability of a type 2 error, since you don't know the probabilities for the different numbers of heads if the coin is unfair. Therefore, even after you've decided to accept the hypothesis, you're not sure that the coin is really fair. For example, suppose the probability of heads is .51. Then you are very likely to accept the fair-coin hypothesis even though the coin is not fair. The only way to improve on this situation is to increase the number of flips.

We'll talk more about hypothesis testing in Chapter 13.

YOU SHOULD REMEMBER

1. Many common statistical problems are *hypothesis testing* problems. The hypothesis that is being tested is called the *null hypothesis*, and the hypothesis that says, "The null hypothesis is wrong," is called the *alternative hypothesis*.

2. In general, it is not possible to prove that the null hypothesis is true or false. After some observations have been collected, however, it is possible to use statistical analysis to determine whether the null hypothesis should be accepted or rejected.

3. There are two possible types of errors:

   type 1 error: saying that the null hypothesis is *false* when it is really *true*
   type 2 error: saying that the null hypothesis is *true* when it is really *false*

   It is a common procedure to design the test so that the chance of committing a type 1 error is less than a specified amount (often 5 percent).

KNOW THE CONCEPTS

DO YOU KNOW THE BASICS?

Test your understanding of Chapter 3 by answering the following questions:

1. Why do statisticians study coin tossing?

2. If you want to determine for sure whether a coin is fair, why don't you flip the coin a few thousand times?

3. If you flip a coin eight times, which of these patterns is more likely: HTHTHTHT or HHHHHHHH?

4. Why is the probability of getting h heads the same as the probability of getting $n - h$ heads?

5. Suppose you have 100 coins that actually are fair, but you don't know whether they are fair. You will flip each coin 20 times as described in the chapter. Will your investigation lead you to conclude that all 100 coins are fair?

6. Should you make the zone of acceptance very wide in order to minimize the chance of a type 1 error?

7. In a practical hypothesis testing situation, how should you decide on the maximum probability of a type 1 error that you are willing to tolerate?

TERMS FOR STUDY

alternative hypothesis   probability
critical region   type 1 error
factorial   type 2 error
hypothesis testing   zone of acceptance
null hypothesis

PRACTICAL APPLICATION

COMPUTATIONAL PROBLEMS

For Exercises 1–3 use the formula given in the chapter to make a table showing the probabilities for the number of heads that will appear if you toss a fair coin:

1. 6 times

2. 7 times

3. 8 times

    The lists in Exercises 4–6 present the probabilities for the number of heads appearing for 50 tosses, 75 tosses, and 100 tosses of a fair coin. In each case you want to test the hypothesis that the coin you toss is fair. Calculate how wide the zone of acceptance should be so that the chance of committing a type 1 error (that is, rejecting the fair-coin hypothesis if the coin really is fair) is less than 5 percent.

4. 50 tosses.

h	$\Pr(X = h)$	h	$\Pr(X = h)$
15	.002	26	.108
16	.004	27	.096
17	.009	28	.079
18	.016	29	.060
19	.027	30	.042
20	.042	31	.027
21	.060	32	.016
22	.079	33	.009
23	.096	34	.004
24	.108	35	.002
25	.112		

5. 75 tosses:

h	Pr(X = h)	h	Pr(X = h)
25	.001	38	.091
26	.003	39	.087
27	.005	40	.078
28	.008	41	.066
29	.013	42	.054
30	.021	43	.041
31	.030	44	.030
32	.041	45	.021
33	.054	46	.013
34	.066	47	.008
35	.078	48	.005
36	.087	49	.003
37	.091	50	.001

6. 100 tosses:

h	Pr(X = h)	h	Pr(X = h)
35	.001	44	.039
36	.002	45	.048
37	.003	46	.058
38	.004	47	.067
39	.007	48	.074
40	.011	49	.078
41	.016	50	.080
42	.022	51	.078
43	.030	52	.074

h	Pr(X = h)	h	Pr(X = h)
53	.067	60	.011
54	.058	61	.007
55	.048	62	.004
56	.039	63	.003
57	.030	64	.002
58	.022	65	.001
59	.016		

ANSWERS

KNOW THE CONCEPTS

1. It is helpful to study the principles of probability in a simple situation where we understand the nature of the process; we can then apply these principles to practical situations where the process or population is unknown.

2. You don't flip a coin a few thousand times for the same reason you don't check an entire population: it would be too expensive. (Time is valuable, and it would take a lot of time to flip a coin a few thousand times.)

3. The two patterns are equally likely.

4. The probability of getting h heads is exactly the same as the probability of getting h tails.

5. Even if all the coins are fair, you will probably determine that 12 coins are unfair because our test procedure has a 12 percent chance of saying the coin is unfair when it is really fair. If you test a great many hypotheses in your career as a statistician, you will inevitably make this type of error a certain fraction of the time.

6. If you widen the zone of acceptance, you are increasing the chance of a type 2 error.

7. The decision depends on the consequences of each type of error in your particular situation.

PRACTICAL APPLICATION

Here are the formula and directions you need for doing Exercises 1–6:

To find the probability of h heads appearing if you toss a coin n times:

$$\frac{\dfrac{n!}{h!(n-h)!}}{2^n}$$

To test the hypothesis that a coin is fair, set up a zone of acceptance: $(n/2 - c)$ to $(n/2 + c)$, where c is chosen to make the chance of a type 1 error less than a specified amount.

Sample calculations are shown below for Exercise 1. The same method is used for Exercises 2 and 3.

1. 6 tosses:

$$h = 0: \dfrac{\dfrac{6!}{0!6!}}{64} = \dfrac{1}{64} = .016$$

$$h = 1: \dfrac{\dfrac{6!}{1!5!}}{64} = \dfrac{6}{64} = .094$$

$$h = 2: \dfrac{\dfrac{6!}{2!4!}}{64} = \dfrac{\dfrac{6 \times 5 \times 4 \times 3 \times 2 \times 1}{2 \times 1 \times 4 \times 3 \times 2 \times 1}}{64} = \dfrac{15}{64} = .234$$

$$h = 3: \dfrac{\dfrac{6!}{3!3!}}{64} = \dfrac{\dfrac{6 \times 5 \times 4 \times 3 \times 2 \times 1}{3 \times 2 \times 1 \times 3 \times 2 \times 1}}{64} = \dfrac{20}{64} = .313$$

$$h = 4: .234 \qquad h = 5: .094 \qquad h = 6: .016$$

2. 7 tosses:

h	$\Pr(X = h)$	h	$\Pr(X = h)$
0	.008	4	.273
1	.055	5	.164
2	.164	6	.055
3	.273	7	.008

3. 8 tosses:

h	$\Pr(X = h)$	h	$\Pr(X = h)$
0	.004	5	.219
1	.031	6	.109
2	.109	7	.031
3	.219	8	.004
4	.273		

4. Zone of acceptance: 18 to 32. With this zone of acceptance the chance of committing a type 1 error is .033. (Add all of the probabilities from 18 to 32, and then subtract the result from 1.) If the zone of acceptance is narrowed to the interval from 19 to 31, then the chance of a type 1 error increases to .065.

5. Zone of acceptance: 29 to 46.   Chance of type 1 error: .037

6. Zone of acceptance: 40 to 60.   Chance of type 1 error: .035

4
CALCULATING PROBABILITIES

KEY TERMS

combinations the number of different ways of selecting j objects from a group of n objects when the order in which the objects are chosen does not matter

permutations the number of different ways of selecting j objects from a group of n objects when each distinct way of ordering the chosen objects counts separately

probability of an event the number of outcomes that corresponds to that event divided by the total number of possible outcomes, provided all outcomes are equally likely

sampling with replacement a method for choosing a sample where an item that has been selected is put back in the population and therefore has a chance of being selected again

sampling without replacement a method for choosing a sample where an item that has been selected is not put back in the population and therefore cannot be selected again

INTERPRETATIONS OF PROBABILITY

What exactly is probability? That is a tricky (and sometimes controversial) question.

Consider the statement, "If we flip a coin, the probability is 1/2 that the result will be a head." It is a difficult philosophical question to determine exactly what this statement means. According to the **relative frequency view of probability**, the statement means that the number of heads will be close to 1/2 of the total tosses if you toss the coin a large number of times.

There are some events for which the relative-frequency interpretation is difficult. The weather report often says, "There is a 20 percent chance of rain today." However, we can't have today repeat itself 100 times to see whether it rains 20 of those times.

The **subjective view of probability** states that probability is an estimate of what an individual *thinks* is the likelihood that an event will happen. In that case two individuals might estimate the probability differently. The subjective view makes it possible to talk meaningfully about the probabilities of a wider class of events, but the probabilities become more intangible because we can't objectively specify what they are.

GAMBLER'S INTERPRETATION OF PROBABILITY

Another perspective of probability is the **gambler's view** (remember that the early mathematics of probability was done by gamblers): the probability of an event occurring can be derived by the odds that would have to be offered to you before you would bet that the event does occur. For example, if you are willing to bet at even odds (win or lose the same amount) that it is going to rain tomorrow, you must believe, perhaps unconsciously, that there is at least a 50 percent chance of it raining tomorrow. If you are willing to bet $10 that you can win a game of tennis, but only if you can win at least $30 for risking your $10, then you must believe that there is a 1 out of 4 chance of your winning and a 3 out of 4 chance of losing.

The gambler's view of probability is intuitive, but most people's intuition is inconsistent.

In the mathematical, or axiomatic, approach to probability, the term *probability* is left as an undefined term. A small set of assumptions (axioms) are made about the behavior of probability. These assumptions follow our intuitive idea of what probability means. Then, those assumptions are used to prove theorems.

PROBABILITY SPACES

Now we will develop the formal method of determining the probabilities associated with a random experiment. First, we'll make a list of all the possible results of our experiment. We're going to use the technical term set for this list. *Set* is just a formal name for a collection of objects. Here is an example:

{1, 2, 3, 4, 5, 6}

Sets can be defined by either of two methods. We can use the listing method as we just did. That just involves writing down all of the members of the set. The set members are enclosed in braces: { }. Then it's clear what is in the set and what is not.

A set can also be defined by stating a rule that makes it clear what is in the set. For example, the set above could have been defined by this rule: the set of all possible results of tossing one die.

With small sets, the listing method and the rule method will work equally well. However, with large sets the rule method works much better. For example, it would be very difficult to list all of the members of these sets:

- the set of all whole numbers from 1 to a million

- the set of all possible results of flipping a coin 12 times

We are especially interested in sets that contain all of the possible results of an experiment. The set of all possible results is called the **probability space** (or sometimes the **sample space**). Mathematicians like to use the term *space* for this type of set. Here are some examples of probability spaces:

- experiment: flip coin one time
  probability space: {H, T}

- experiment: flip coin three times
  probability space: {HHH, HHT, HTH, HTT, THH, THT, TTH, TTT}

- experiment: roll one die
  probability space: {1, 2, 3, 4, 5, 6}

- experiment: roll two dice
  probability space: {(1, 1)(1,2)(1,3)(1,4)(1,5)(1,6) (2,1)(2,2)(2,3)
     (2,4)(2,5)(2,6) (3,1)(3,2)(3,3)(3,4)(3,5)(3,6)
     (4,1)(4,2)(4,3)(4,4)(4,5)(4,6) (5,1)(5,2)(5,3)
     (5,4)(5,5)(5,6) (6,1)(6,2)(6,3)(6,4)(6,5)(6,6)}

- experiment: draw one card from a deck of 52 cards
  probability space:

     {ace hearts, ace diamonds, ace spades, ace clubs,
        two hearts, two diamonds, two spades, two clubs, etc.}

(There will be 52 elements if we list them all.)

To avoid the bother of writing "probability space" each time, we'll sometimes use the letter S (short for "space") to stand for a probability space.

The important thing is that every single possible result of the experiment must be included in the probability space. We'll call each possible result an **outcome**. We'll use a small s to represent the total number of outcomes in the probability space S. In the examples above, the first probability space has two possible outcomes and the others have 8, 6, 36, and 52 outcomes, respectively. For now, we'll assume that each outcome is equally likely. (This is called the

classical approach to probability.) Then, since there are *s* outcomes, the probability of any one outcome is $1/s$.

Now we'll use a probability space to solve a practical problem. Suppose that we're playing Monopoly, and that our marker is on North Carolina Avenue. We already own Park Place, so we will have it made if we roll a 7 and land on Boardwalk. As we saw earlier, there are 36 possible outcomes for the experiment of rolling two dice. We want to get a 7, so it doesn't make any difference to us whether we roll (1,6) or (2,5) or (3,4) or (4,3) or (5,2) or (6,1). We can put all of these outcomes together in a set. We'll often use capital letters to stand for sets, so we may as well call this one set *A*:

$$A = \{(1,6), (2,5), (3,4), (4,3), (5,2), (6,1)\}$$

We can also define set *A* by the rule method: Set *A* is the set of all possible outcomes from rolling two dice such that the sum of the numbers on the two dice is 7.

Set *A* contains 6 outcomes. If we roll the dice, we have an equal chance of getting any one of the 36 possible outcomes. Since 6 of these outcomes give us what we want (a total of 7 on the dice), the probability of getting 7 is $6/36 = 1/6$.

(1 ,1)	(1 ,2)	(1 ,3)	(1 ,4)	(1 ,5)	(1 ,6)
(2 ,1)	(2 ,2)	(2 ,3)	(2 ,4)	(2 ,5)	(2 ,6)
(3 ,1)	(3 ,2)	(3 ,3)	(3 ,4)	(3 ,5)	(3 ,6)
(4 ,1)	(4 ,2)	(4 ,3)	(4 ,4)	(4 ,5)	(4 ,6)
(5 ,1)	(5 ,2)	(5 ,3)	(5 ,4)	(5 ,5)	(5 ,6)
(6 ,1)	(6 ,2)	(6 ,3)	(6 ,4)	(6 ,5)	(6 ,6)

Figure 4-1

Set *A* is an example of what we will call an *event*. An event is a set that consists of a group of outcomes. In our case, (1,6), (2,5), (3,4), (4,3), (5,2), (6,1) are all outcomes, and the set {(1,6), (2,5), (3,4), (4,3), (5,2), (6,1)} is an event.

An event can also be defined in this way: An event is a **subset** of the probability space. A subset is a set that contains some (or possibly all) members of another set. If set *A* is a subset of set *S*, then every outcome in set *A* is contained in set *S*. For example:

• set: all people living in New York State
  subset: all people living in New York City

YOU SHOULD REMEMBER

1. Probability may be interpreted according to the relative frequency view or the subjective view.

2. A set can be defined by listing all its members or by stating a rule that makes clear what the set contains.

3. The set of all possible results of an experiment is called the *prob-ability space.*

4. Each possible result of an experiment is called an *outcome.*

5. In the *classical approach to probability* each outcome is regarded as equally likely.

PROBABILITY OF AN EVENT

• *PROBABILITY THAT AN EVENT WILL HAPPEN*

Now we're ready to give a formal definition for the **probability of an event** A. First, count the number of outcomes in A. Call that number $N(A)$. (Read this as "N of A," which is short for "number of outcomes in A.") Remember that all outcomes are equally likely and s is the total number of outcomes in the probability space S. Then we can define the probability:

Probability that event A will occur is $\dfrac{N(A)}{s}$

In other words, count the number of outcomes that give you event A, and then divide by the total number of possible outcomes.

It would be helpful to have a shorter way to write "Probability that event A occurs." We will use Pr to stand for probability, and write it like this:

Pr(A) means "probability that event A will occur"

Our result becomes:

$$Pr(A) = \frac{N(A)}{s}$$

Examples: Determining the Probability of an Event

PROBLEM   What is the probability of getting one head if we toss a coin three times?

SOLUTION   In this case, there are $2^3 = 8$ possible outcomes, so $s = 8$. There are three outcomes that give one head (HTT, THT, TTH), so if A is the event of getting one head, then $N(A) = 3$. Therefore, Pr(A) = 3/8.

PROBLEM   What is the probability of getting a total of 5 if we roll two dice?

SOLUTION   There are 36 possible outcomes, so $s = 36$. Let B be the event of getting 5 on the dice, so B contains four outcomes:

$$B = \{(1,4), (2,3), (3,2), (4,1)\}$$

Then $N(B) = 4$, so Pr(B) = 4/36 = 1/9. (Your chances of getting 5 are worse than your chances of getting 7.)

PROBLEM   What is the probability of drawing an ace if we draw one card from a deck of cards?

SOLUTION   There are 52 possible outcomes, so $s = 52$. Let C be the event of drawing an ace, so C contains four outcomes:

$$\{\text{A hearts, A diamonds, A clubs, A spades}\}$$

Then $N(C) = 4$, so $\Pr(C) = 4/52 = 1/13$.

Two special events are interesting. First, let us consider the probability that event S occurs. Remember that S contains all of the possible outcomes of the experiment. Using the formula gives this:

$$\Pr(S) = \frac{N(S)}{s} = \frac{s}{s} = 1$$

This result should be obvious. All it says is, "The probability is 100 percent that the result will be one of the possible results." (Just ask yourself: What is the probability that the result will *not* be one of the possible results?)

Another possible event is the set that contains *no* outcome. You should be able to convince yourself that this set has zero probability of occurring. This set is often called the empty set, because it doesn't have anything in it:

$$\Pr(\text{empty set}) = 0$$

The empty set is often symbolized by a zero with a slash, /, through it, like this: ∅. Therefore, we can say:

$$\Pr(\emptyset) = 0$$

• *PROBABILITY THAT AN EVENT WILL NOT HAPPEN*

Now we can develop another important result that is fortunately very obvious. Many times we will want to know the probability that an event will *not* happen. For example, returning to our Monopoly game, suppose we are on North Carolina Avenue and our opponent has a hotel on Boardwalk. In that case we're mainly interested in the probability that we will *not* get a 7. It should be clear that, if there is a 1/6 chance of getting a 7, then there is a 5/6 chance of not getting a 7. We can demonstrate that. Let A be the event of not getting a 7. Then A contains 30 outcomes:

$$\begin{array}{llllll}
\{(1,1) & (1,2) & (1,3) & (1,4) & (1,5) & \\
(2,1) & (2,2) & (2,3) & (2,4) & & (2,6) \\
(3,1) & (3,2) & (3,3) & & (3,5) & (3,6) \\
(4,1) & (4,2) & & (4,4) & (4,5) & (4,6) \\
(5,1) & & (5,3) & (5,4) & (5,5) & (5,6) \\
& (6,2) & (6,3) & (6,4) & (6,5) & (6,6)\}
\end{array}$$

Therefore $N(A) = 30$, so $\Pr(A) = 30/36 = 5/6$. In general, if p is the probability that a particular event A will occur, then $1 - p$ is the probability that the event will not occur.

We will give a special name to the set that contains all of the outcomes that are not in set A. It is called the **complement** of A, written as A^c. (The little c stands for "complement." Read the symbol as "A complement.")

Then:

$$\Pr(A^c) = 1 - \Pr(A)$$

This result is important for calculations, since sometimes it is easier to calculate the probability that an event will not occur than to calculate the probability that it will occur.

Here are some examples of complements:

- If the experiment consists of flipping a coin, and A is the event of getting a head, then A^c is the event of getting a tail.

- If the experiment consists of flipping four coins, and B is the event of getting four heads, then B^c is the event of getting at least one tail.

- If the total set is the set of all 52 cards in a deck of cards, and R is the event of drawing a red card, then R^c is the event of drawing a black card.

- If the total set is the set of all major league baseball teams, and A is the set {Yankees, Mets}, then A^c is the set of all major league teams that are not from New York.

PROBABILITY OF A UNION

Now suppose that nobody owns either Boardwalk or Park Place, and we would like to know the probability that we will land on either of them. If we're currently on North Carolina Avenue, then we need to know the probability that we will get either a 5 or a 7. It often happens in probability that we need to find out the probability that either one of two events will occur. Let's say that A is the event of getting a 7. Then A contains the outcomes

$$\{(1,6), (2,5), (3,4), (4,3), (5,2)\ (6,1)\}$$

If B is the event of getting a 5, then B contains the outcomes

$$\{(1,4), (2,3), (3,2), (4,1)\}$$

Let's say that C is the event of getting either a 5 *or* a 7. Then C contains these outcomes:

$$\{(1,6), (2,5), (3,4), (4,3), (5,2), (6,1), (1,4), (2,3), (3,2), (4,1)\}$$

C contains 10 outcomes, so $\Pr(C) = 10/36$.

(1,1)	(1,2)	(1,3)	(1,4)	(1,5)	(1,6)
(2,1)	(2,2)	(2,3)	(2,4)	(2,5)	(2,6)
(3,1)	(3,2)	(3,3)	(3,4)	(3,5)	(3,6)
(4,1)	(4,2)	(4,3)	(4,4)	(4,5)	(4,6)
(5,1)	(5,2)	(5,3)	(5,4)	(5,5)	(5,6)
(6,1)	(6,2)	(6,3)	(6,4)	(6,5)	(6,6)

Figure 4-2

There is a special name for the set that contains all of the elements in either or both of two other sets. It is called the **union** of the other two sets. In this case, C is the union of set A and set B. You can think of it in this way: set A and set B get together and join forces to form a union, and the result is set C. In mathematics the word *union* is symbolized by a little symbol that looks like a letter *u*: ∪. Therefore, we can write "C is the union of A and B" as

$$C = A \text{ union } B$$

or

$$C = A \cup B$$

Here are some examples of unions:

- If A is the set of even numbers and B is the set of odd numbers, then $A \cup B$ is the set of all whole numbers.

- If V is the set of vowels and C is the set of consonants, then $V \cup C$ is the set of all letters.

- If A is the set {AH, KH, QH, JH, 10H}, and B is the set {QH, JH, 10H, 9H, 8H}, then $A \cup B$ is {AH, KH, QH, JH, 10H, 9H, 8H}.

There is another interesting fact that we can learn from the dice example. We said that (probability of getting a 7) = $\Pr(A)$ = 6/36, (probability of getting a 5) = $\Pr(B)$ = 4/36, and (probability of getting a 5 or a 7) = $\Pr(A \text{ or } B)$ = $\Pr(A \cup B)$ = 10/36. It looks as though we could just add the probabilities for the two events to get the probability that either one of them will occur, since $10/36 = 6/36 + 4/36$. This amazingly simple rule will often work:

$$\Pr(A \text{ or } B) = \Pr(A \cup B) = \Pr(A) + \Pr(B)$$

However, this result holds only when there is no possibility that event *A* and event *B* can occur at the same time. Here is an example of incorrect reasoning: "Toss a coin twice. Since the probability of getting a head on the first toss is 1/2 and the probability of getting a head on the second toss is 1/2, the probability of getting a head on either the first toss or the second toss is 1/2 + 1/2 = 1." Obviously, that is not the case. In the dice example we could use the simple formula because there is no way to get both a 5 and a 7 on a single toss of a pair of dice. Two events that can never occur together are called **disjoint events** or **mutually exclusive events**.

PROBABILITY OF AN INTERSECTION

Now we will consider an example in which two events can happen together. Suppose you are selecting one card from a deck and you want to know the probability that you will get a red face card—that is, a red jack, queen, or king. Let's say that *F* is the event of getting a face card. Then *F* contains these outcomes:

$$\{JH, JD, JC, JS,$$
$$QH, QD, QC, QS,$$
$$KH, KD, KC, KS\}$$

(Here J = jack, Q = queen, K = king, H = hearts, D = diamonds, C = clubs, S = spades.)

Since $N(F) = 12$, the probability of getting a face card is 12/52. Let *R* be the event of getting a red card; then *R* contains these outcomes:

$$\{AH, 2H, 3H, 4H, 5H, 6H, 7H, 8H, 9H, 10H, JH, QH, KH,$$
$$AD, 2D, 3D, 4D, 5D, 6D, 7D, 8D, 9D, 10D, JD, QD, KD\}$$

R contains 26 outcomes, so $\Pr(R) = 26/52 = 1/2$.

Let *C* be the event that both event *F* and event *R* occur—in other words, *C* is the event that you get a card that is *both* a red card and a face card. Then *C* contains these six outcomes:

$$\{JD, JH, QD, QH, KD, KH\}$$

Therefore, $\Pr(C) = N(C)/s = 6/52$.

There is a special name for the set that contains all of the elements that are in both of two other sets. It is called the **intersection**. For example, in this case set *C* is the intersection of set *F* and set *R*, because it contains the outcomes that are in both set *F* and in set *R*. The symbol of intersection is the symbol for union turned upside down: ∩. Now we can write:

C = set that contains the elements in both A and B;

$C = A$ intersect B; or

$C = A \cap B$

Here are some examples of intersections:

- If V is the set of vowels and C is the set of consonants, then $V \cap C$ is $\{y\}$.

- If A is the set of hands with five cards in sequence, and B is the set of hands with five cards of the same suit, then $A \cap B$ is the set of all straight flushes.

- If you flip a coin twice, and A is the event of getting a head on the first toss and B is the event of getting a head on the second toss, then $A \cap B$ is $\{HH\}$.

There is no general formula for the probability of the intersection of two events, but Chapter 5 discusses a formula that can sometimes be used. For now, we'll have to count the number of outcomes in the intersection and calculate the probability directly from that.

If two events can't happen together (in other words, they are mutually exclusive events), then their intersection is the empty set. For example, we can't get both a head and a tail on a single coin flip, so if H = event of getting a head, and T = event of getting a tail, then $\Pr(H$ and $T) = \Pr(H \cap T) = 0$. In general, if A and B are mutually exclusive, then $A \cap B = \emptyset$ and $\Pr(A \cap B) = 0$.

Now let us consider the possibility that we will get either a face card *or* a red card when we draw a card from the deck. If F is the event of getting a face card, and R is the event of getting a red card, then we want to know

$$\Pr(F \text{ or } R) = \Pr(F \cup R)$$

We can't use the simple formula $\Pr(F) + \Pr(R)$, because these two events can happen together. We can make a list of all the outcomes in $F \cup R$:

$$\{AH, 2H, 3H, 4H, 5H, 6H, 7H, 8H, 9H, 10H,$$
$$JH, QH, KH,$$
$$JC, QC, KC,$$
$$AD, 2D, 3D, 4D, 5D, 6D, 7D, 8D, 9D, 10D,$$
$$JD, QD, KD,$$
$$JS, QS, KS\}$$

Altogether there are 32 outcomes, so

$$\Pr(F \text{ or } C) = \Pr(F \cup C) = 32/52$$

AH	AD	AC	AS
2H	2D	2C	2S
3H	3D	3C	3S
4H	4D	4C	4S
5H	5D	5C	5S
6H	6D	6C	6S
7H	7D	7C	7S
8H	8D	8C	8S
9H	9D	9C	9S
10H	10D	10C	10S
JH	JD	JC	JS
QH	QD	QC	QS
KH	KD	KC	KS

Figure 4-3

Now we can figure out a general formula for the probability of $A \cup B$.

$$\text{Pr}(A \text{ or } B) = \text{Pr}(A) + \text{Pr}(B) - \text{Pr}(A \text{ and } B)$$

or, written mathematically:

$$\text{Pr}(A \cup B) = \text{Pr}(A) + \text{Pr}(B) - \text{Pr}(A \cap B)$$

This formula is good for any two events, whether or not they are mutually exclusive. Notice that, if A and B *are* mutually exclusive, then $\text{Pr}(A \text{ and } B) = 0$, so we get the same formula that we had before:

$$\text{Pr}(A \text{ or } B) = \text{Pr}(A) + \text{Pr}(B)$$

Examples: Determining the Probability of Union

PROBLEM   Suppose that, in the middle of a backgammon game, you want to know the probability that you will get a total of 8 or doubles on the dice.

SOLUTION   Let's say that $E1$ is the event of getting 8. Then:

$$E1 = \{(2,6), (3,5), (4,4), (5,3), (6,2)\}, \qquad \text{Pr}(E1) = 5/36$$

Let $E2$ be the event of getting a double. Then:

$$E2 = \{(1,1), (2,2), (3,3), (4,4), (5,5), (6,6)\}, \qquad \text{Pr}(E2) = 6/36$$

These two events are not disjoint, since you can get both 8 and a double (if the dice turn up (4,4)). Then ($E1 \cap E2$) is the event of getting (4,4), which, as we have learned, has a probability of 1/36. Now we can use our formula:

Pr(getting a double or 8)

$$= \Pr(E1 \text{ or } E2) = \Pr(E1) + \Pr(E2) - \Pr(E1 \text{ and } E2)$$
$$= 5/36 + 6/36 - 1/36$$
$$= 10/36$$

PROBLEM   What is the probability that you will get at least one 6 when you roll two dice?

SOLUTION   We'll call *E1* the event of getting a 6 on the first die and *E2* the event of getting a 6 on the second die. Then the event that you will get at least one 6 is $E1 \cup E2$. We know that $\Pr(E1) = \Pr(E2) = 1/6$. Since $E1 \cap E2$ is the event of gettting 6's on both dice, we know that the probability of that happening is 1/36. Therefore, we can use the formula:

$$\Pr(E1 \cup E2) = \Pr(E1) + \Pr(E2) - \Pr(E1 \cap E2)$$
$$= 1/6 + 1/6 - 1/36$$
$$= 11/36$$

So your chances of getting at least one 6 are 11/36.

(1 ,1)	(1 ,2)	(1 ,3)	(1 ,4)	(1 ,5)	(1 ,6)
(2 ,1)	(2 ,2)	(2 ,3)	(2 ,4)	(2 ,5)	(2 ,6)
(3 ,1)	(3 ,2)	(3 ,3)	(3 ,4)	(3 ,5)	(3 ,6)
(4 ,1)	(4 ,2)	(4 ,3)	(4 ,4)	(4 ,5)	(4 ,6)
(5 ,1)	(5 ,2)	(5 ,3)	(5 ,4)	(5 ,5)	(5 ,6)
(6 ,1)	(6 ,2)	(6 ,3)	(6 ,4)	(6 ,5)	(6 ,6)

Figure 4-4

YOU SHOULD REMEMBER

1. If there are *s* outcomes that are all equally likely, and if there are N(A) outcomes that are part of event A, then the probability of event A occurring is

$$\Pr(A) = \frac{N(A)}{s}$$

2. If A and B are two mutually exclusive events (that is, they cannot happen together), then the probability that either event A or event B will occur is

$$\Pr(A \text{ or } B) = \Pr(A) + \Pr(B)$$

In words: to calculate the probability that either one of two mutually exclusive events will occur, simply add the probabilities of the two events.

3. If A and B are not mutually exclusive, then

$$\Pr(A \text{ or } B) = \Pr(A) + \Pr(B) - \Pr(A \text{ and } B)$$

MULTIPLICATION PRINCIPLE

When we have a set of outcomes that are all equally likely, the probability that an event A will occur is given by

$$\Pr(A) = \frac{N(A)}{s}$$

where $N(A)$ is the number of outcomes in A and s is the total number of possible outcomes. This means that if we can determine these two numbers—$N(A)$ and s—we're done. We can then directly calculate the probability. An important part of probability, therefore, involves figuring out how to count the outcomes corresponding to a given event. If there are not too many outcomes, we can list them all; but for complicated situations the number of outcomes quickly becomes too large to list.

Suppose we want to choose a car that is one of these four colors; red, blue, green, or white. We are interested in three different body types: 4-door, 2-door, and wagon. How many different types of car do we need to consider? We can make a list:

red 4-door,   red 2-door,   red wagon
blue 4-door,   blue 2-door,   blue wagon
green 4-door, green 2-door, green wagon
white 4-door, white 2-door, white wagon

There are 12 possible types, since $12 = 4 \times 3$.

We can make a general statement of this principle. Suppose we are going to conduct two experiments. The first experiment can have any one of *a* possible outcomes, the second experiment can have any one of *b* possible outcomes, and let's suppose that any possible combination of the two results can occur. Then the total number of results of the two experiments is

$$a \times b$$

This rather obvious result is sometimes called the **multiplication principle**. Here are some examples:

- If you flip two coins, each flip has two possible outcomes, so the number of total outcomes is $2 \times 2 = 4$.

- Suppose you toss two dice. Since each die has six possible outcomes, the total number of possible outcomes from the two dice is $6 \times 6 = 36$.

- Suppose there are five candidates in the Republican primary for a particular office, and six candidates in the Democratic primary. Then the total number of possible general-election matchups is $5 \times 6 = 30$.

SAMPLING WITH REPLACEMENT

Now, suppose you have five sweaters in your drawer. Each morning you reach in and randomly select one sweater. In the evening you put the sweater back and mix the sweaters up again. How many different ways can you wear sweaters for the week?

We can use the same principle, only now there are more than two experiments. There are five possible sweaters for you to wear on Sunday. There are also five sweaters for you to wear on Monday, so there is a total of 25 possible different combinations of sweaters you can wear on Sunday and Monday. For each of these possibilities there are five more choices for Tuesday, so there are $25 \times 5 = 125$ possible sweater-wearing patterns for the first 3 days. In fact, for the first week there are

$$5 \times 5 \times 5 \times 5 \times 5 \times 5 \times 5 = 5^7 = 78,125$$

different ways of wearing the sweaters.

From this information we can answer another question: What is the probability that you will wear the same sweater every day? Since there are 78,125 possible outcomes, and only five outcomes in which you wear the same sweater every day, the probability of selecting the same sweater every day is

$$\frac{5}{78,125} = .000064$$

The sweater selection process described here is an example of what is called *sampling*. Sampling means choosing a few items from a larger group called the *population*. In this case the population consists of five sweaters. We are selecting a sample of one sweater on Sunday, one sweater on Monday, and so on, for a total of seven selections. This type of sampling is called **sampling with replacement**. It should be obvious why we use the words "with replacement," since we are replacing the sweater worn that day in the drawer each evening. (Later, we will discuss sampling without replacement.) In general, if you sample n times with replacement from a population of m objects, then there are m^n possible different ways to select the objects.

The key idea of sampling with replacement is that once an item has been selected it can still be selected again. Flipping a coin is an example of sampling with replacement. In this case the population is of size 2: heads and tails. The fact that you've selected heads once doesn't mean you can't select heads again the next time. Therefore if you flip a coin n times there are 2^n possible results.

Rolling a die is another example of sampling with replacement. In this case the population is of size 6. If you roll a 5 on the die once, nothing will prevent you from rolling a 5 the next time. Therefore, there are 6^n total possibilities if you roll a die n times.

Examples: Sampling with Replacement

PROBLEM   Suppose you have to take a 20-question multiple-choice exam in a subject you know absolutely nothing about. Each question has five choices. What is the probability that you will be able to get all of the answers right just by guessing?

SOLUTION   In this case we are sampling 20 times from a population of size 5, so the total number of possible ways of choosing the answers is 5^{20} = 9.5×10^{13}. There is only one possible outcome in which you have selected all of the right answers, so the probability of getting all of the answers right by pure guessing is $1/(9.5 \times 10^{13}) = 10^{-14}$ (approximately).

PROBLEM   Suppose you are trying to guess the license-plate number of a friend's car. (You haven't seen the car, but you do know that it doesn't have vanity license plates.) Assume that each license plate consists of three letters followed by three digits, such as DGM 235.

SOLUTION   First, calculate how many different possibilities there are for the three letters. That is the same as sampling 3 times with replacement from a population of 26 (since there are 26 letters in the alphabet), so there are $26^3 = 17,576$ ways of selecting the three letters. Since there are 10 possible digits, there are $10^3 = 1,000$ ways of selecting the three digits on the license plate. Each possible letter combination can be matched with each possible digit combination to give a valid license plate, so the total number of license plates is $17,576 \times 1,000 = 17,576,000$. Therefore, your chance of guessing the license plate correctly is $1/17,576,000 = 5.69 \times 10^{-8}$.

PROBLEM   You are visiting a 31-flavor ice-cream store. You intend to order a two-scoop cone, but you are having trouble deciding which flavors. How many possibilities are there?

SOLUTION   There are 31 possibilities for the top flavor and 31 choices for the bottom flavor, so there are $31 \times 31 = 961$ possible cones.

SAMPLING WITHOUT REPLACEMENT

Now, consider a different situation called **sampling without replacement**. Suppose that you have seven T-shirts. Each morning you reach into the drawer and randomly select one T-shirt to wear that day. However, this time, instead of putting the T-shirt back in the drawer in the evening, you put it in the bag of clothes to be washed. (The laundry gets done only once a week.) How many ways can you select the seven shirts for the week?

On Sunday you have seven choices. However, on Monday there are only six clean T-shirts left, so you have only six choices. Therefore, there are $7 \times 6 = 42$ possible ways of choosing the T-shirts that you will wear during the first 2 days. On Tuesday there are only five shirts left, so there are $7 \times 6 \times 5 = 210$ ways of choosing the shirts for the first 3 days. Continuing the process for the rest of the week, we can see that there are $7 \times 6 \times 5 \times 4 \times 3 \times 2 \times 1 = 5,040$ ways of selecting the T-shirts for the week.

We've already given a name to this quantity (see Chapter 3): $7 \times 6 \times 5 \times 4 \times 3 \times 2 \times 1$ is called 7 factorial and written as 7!. In this case we know for sure that you will wear each shirt exactly once during the week, so the only question is: In what order will you wear them? In general, if you have n objects, there are $n!$ different ways of putting them in order.

Examples: Sampling without Replacement

PROBLEM   How many different ways are there of shuffling a deck of 52 cards?

SOLUTION   There are 52 possibilities for the top card, 51 possibilities for the second card, and so on, so that altogether there are $52! = 8.07 \times 10^{67}$ ways of shuffling the deck.

PROBLEM   Suppose you have five different dinner menus to choose from for 5 days: hamburgers, hot dogs, pizza, macaroni, and tacos. In how many different orders can you arrange these five meals so that you don't repeat any meals during the 5 days?

SOLUTION   Since there are five meals, the number of different orderings is $5! = 120$.

PROBLEM   Suppose you have invited 20 people to a dinner party at your home. What is the probability that they will arrive at your house in alphabetical order?

SOLUTION   Since there are 20 guests, there are $20! = 2.43 \times 10^{18}$ different possible orders in which they can arrive. There is only one way of putting the guests in alphabetical order; the probability that they will arrive in alphabetical order is therefore $1/(2.43 \times 10^{18}) = 4.12 \times 10^{-19}$.

PROBLEM   An indecisive baseball manager decides to try out every possible batting order before deciding on the order that is best for the team. How many games will it take to test every possible order?

SOLUTION   There are nine players (assuming that this league does not have designated hitters), so there are $9! = 362,880$ different orders.

YOU SHOULD REMEMBER

1. The multiplication principle states that, if the first of two experiments can have *a* possible outcomes, the second experiment can have *b* possible outcomes, and any combination of these results can occur, then the total number of possible outcomes of the two experiments is

$$a \times b$$

2. In sampling with replacement, once an item has been selected it is returned to the population and can be chosen again.

3. In sampling without replacement, once an item has been selected it is not returned to the population and cannot be chosen again.

PERMUTATIONS

Suppose now that you have 10 T-shirts (and the T-shirts are still washed every week). How many different ways of selecting T-shirts are there during 7 days? Note that in this case you will not wear every shirt every week.

There are 10 choices for the shirt you wear on Sunday, then 9 choices for Monday, 8 choices for Tuesday, and so on down to 4 choices on Saturday. So the total number of choices is

$$10 \times 9 \times 8 \times 7 \times 6 \times 5 \times 4 = 604{,}800$$

We'd like a shorter way of writing that long expression, so we'll write it like this:

$$
\begin{aligned}
10 \times 9 &\times 8 \times 7 \times 6 \times 5 \times 4 \\
&= \frac{10 \times 9 \times 8 \times 7 \times 6 \times 5 \times 4 \times 3 \times 2 \times 1}{3 \times 2 \times 1} \\
&= \frac{10!}{3!}
\end{aligned}
$$

What we're doing is choosing a sample of size 7 without replacement from a population of size 10. It's obvious why we call this sampling without replacement, since this time we're *not* replacing the T-shirt in the drawer after it has once been selected. Sampling without replacement means that once an item has been selected it cannot be selected again.

In general, suppose you're going to select j objects without replacement from a population of n objects. Then there are

$$\frac{n!}{(n - j)!}$$

ways of selecting the objects. Each way of selecting the objects is called a **permutation** of the objects, so the formula $n!/(n - j)!$ gives the number of permutations of n things taken j at a time.

The number of permutations of n things taken j at a time is sometimes symbolized by the expression $_nP_j$.

Examples: Using Permutations

PROBLEM   The 31-flavor ice-cream store has introduced a new rule: you may not have the same flavor for both scoops of your cone. How many possibilities are there now?

SOLUTION   There are 31 choices for the top flavor, but now only 30 choices for the bottom flavor, so there are 31 × 30 = 930 possible cones. In this case we are sampling without replacement. Recall that we found there were 961 possibilities if we sample with replacement; now there are 31 fewer possibilities because we have eliminated the 31 cases with the same flavor on both top and bottom. Note that the order matters; we would count chocolate on top, vanilla on the bottom as one possibility and vanilla on top, chocolate on the bottom as another possibility. Because the order matters, we say there are 930 permutations of the 31 ice-cream flavors chosen two at a time.

PROBLEM   Suppose that you are attending an eight-horse race, and you are trying to guess the order of the top three finishers without knowing anything about the horses involved. What is the probability that you will guess right?

SOLUTION   Since three horses are to be chosen from the eight horses entered, this situation is equivalent to choosing a sample of size 3 without replacement from a population of size 8. Therefore, there are

$$\frac{8!}{(8-3)!} = \frac{8!}{5!} = 8 \times 7 \times 6 = 336$$

possible finishes. Your chance of randomly guessing the correct order is 1/336 = .003.

COMBINATIONS

Suppose you are in a card game in which you will be dealt a hand of 5 cards from a deck of 52 cards. There are 52 possibilities for the first card you will draw, then 51 possibilities for the second card, and so on. We can consider this an example of choosing a sample of size 5 without replacement from a population of 52 cards. (Once you've drawn a card, you can't draw that particular card again. If you do draw the same card again, something suspicious is going on.) Therefore, there are

$$\frac{52!}{(52-5)!} = \frac{52!}{47!}$$

$$= 52 \times 51 \times 50 \times 49 \times 48$$

$$= 311,875,200 \text{ ways of drawing the cards}$$

However, suppose you drew these cards:

5C, 8D, 6H, AH, AD

For practical purposes this hand means exactly the same as it would if you drew the cards like this:

8D, 5C, 6H, AH, AD

The second hand contains exactly the same cards, the only difference being that they were picked in a different order. In many card games, the order in which you draw the cards doesn't matter; it's only *which* cards you draw that matters. However, in the way we have been counting the hands, we have counted these two possibilities as separate hands, because they were drawn in different orders. There are, of course, a lot of different orders in which this hand could be drawn.

How many different orderings of these cards are there? We have already discussed this problem. Since there are 5 cards, there are 5! = 120 different ways of arranging the cards in different orders. In fact, for *every* possible hand there are 120 different orderings. Our formula 52!/(52 − 5)! has told us the total number of different orderings of all the hands, but in this case we are interested only in the total number of *different* hands, regardless of the ordering. Therefore, our formula gives us 120 times too many hands, so we have to divide by 120. Therefore:

(total number of distinct 5-card hands that can be drawn from a deck of 52 cards not counting different orderings)

$$= \frac{52!}{(52 - 5)! \, 5!}$$

$$= 2{,}598{,}960$$

In general, suppose we are going to select *j* items without replacement from a population of *n* items and we're interested only in the total number of selections, without regard to their order. Then the number of possibilities is given by the formula

$$\frac{n!}{(n - j)! \, j!}$$

The number of arrangements without regard to order is called the number of **combinations**. The formula $n!/[(n - j)! \, j!]$ is said to represent the number of combinations of *n* things taken *j* at a time. We will use the formula

$n!/[(n - j)!j!]$ frequently in probability and statistics. (Remember that we already used it in Chapter 3.) We'd like a shorter way of writing this formula, so we will symbolize it like this:

$$\binom{n}{j} = \frac{n!}{(n - j)!j!}$$

The number of combinations is sometimes symbolized by the expression $_nC_j$. Note that

$$_nC_j = \frac{_nP_j}{j!}$$

This expression is also called the *binomial coefficient* because it is used in a mathematical formula called the binomial theorem.

Here is an example of the formula. Suppose we have five letter blocks and we are going to select three of them. In how many ways can we make the selection? If the blocks have lettters A, B, C, D, E, then we can list all of the possible permutations:

ABC, ACB, BAC, BCA, CAB, CBA,
ABD, ADB, BAD, BDA, DAB, DBA,
ABE, AEB, BAE, BEA, EAB, EBA,
ACD, ADC, CAD, CDA, DAC, DCA,
ACE, AEC, CAE, CEA, EAC, ECA,
ADE, AED, DAE, DEA, EAD, EDA,
BCD, BDC, CBD, CDB, DBC, DCB,
BCE, BEC, CBE, CEB, EBC, ECB,
BDE, BED, DBE, DEB, EBD, EDB,
CDE, CED, DCE, DEC, ECD, EDC

There are 60 permutations in this list, a result that agrees with the formula:

$$\frac{5!}{(5 - 3)!} = \frac{5!}{2!} = 60$$

However, if we look closely at the list we can see that all of the arrangements in a particular row contain exactly the same letters, only arranged in different orders. Each of the 10 rows has different letters in it, so there are 10 different combinations of the letters. That agrees with the formula for combinations:

$$\frac{5!}{(5 - 3)!3!} = \frac{5!}{3!2!}$$

$$= 10$$

Examples: Calculating Combinations

PROBLEM   Previously we found there were 930 possible cones if we sample without replacement where the order matters from the 31-flavor ice-cream store. However, suppose we decide the order doesn't matter. In other words, we would count a cone with chocolate and vanilla as one possibility, regardless of which flavor is on top.

SOLUTION   We need to divide 930 by 2 to take care of the fact that each possibility is counted twice because of the different orderings. Therefore, there are $31 \times 30/2 = 465$ possible ways of choosing the ice cream without replacement where the order doesn't matter. Formally, we can say there are 465 combinations of the 31 flavors chosen two at a time. Using the formula:

$$\binom{31}{2} = \frac{31!}{29!2!} = \frac{31 \times 30}{2} = 465$$

PROBLEM   How many 13-card hands can be dealt from a deck of 52 cards?

SOLUTION   Using the formula for the number of combinations of 52 objects taken 13 at a time, we have

$$\frac{52!}{(52-13)!13!} = \frac{52!}{39!13!}$$

$$= 6.35 \times 10^{11}$$

Now we have all the tools we need to solve many probability problems. There's no sure-fire method that always works, because many problems contain subtle twists.

PROBLEM   Suppose you and your dream lover (whom you're desperately hoping to meet) are both in a group of 20 people, and five people are to be randomly selected to be on a committee. What is the probability that both you and your dream lover will be on the committee?

SOLUTION   The total number of ways of choosing the committee is

$$\binom{20}{5} = 15,504$$

Next, we need to calculate how many possibilities include both of you on the committee. If you've both been selected, then the three other committee members must be selected from the remaining 18 people, and there are

$$\binom{18}{3} = 816$$

ways of doing this. Therefore, the probability that you will both be selected is 816/15,504 = .053.

PROBLEM   Suppose 18 people are going to be randomly divided into two baseball teams. What is the probability that all 9 of the best players will be on the same team?

SOLUTION   There are $\binom{18}{9}$ = 48,620 ways of choosing the team that will bat first. Therefore, the chance of all the best players being on the team that bats first is 1/48,620. However, there is just as good a chance that all the best players will be on the team that bats second, so the chance they will be on the same team is 2/48,620 = 4×10^{-5}.

In the next three examples we will develop the formulas that are used in choosing samples. These results will play a crucial role later when we show why samples can give accurate pictures of populations.

PROBLEM   A group of nine letter blocks contains five letters from early in the alphabet (A, B, C, D, and E) and four letters from late in the alphabet (W, X, Y, and Z). Five blocks will be selected at random from this group. What is the probability that exactly three of the selected blocks will be early letters?

SOLUTION   Since we are choosing five blocks from the group of nine blocks, we can use the formula for combinations to find the number of possible ways of making the selection:

$$\binom{9}{5} = \frac{9 \times 8 \times 7 \times 6 \times 5 \times 4 \times 3 \times 2 \times 1}{4 \times 3 \times 2 \times 1 \times 5 \times 4 \times 3 \times 2 \times 1}$$
$$= \frac{9 \times 8 \times 7 \times 6}{4 \times 3 \times 2 \times 1} = 126$$

Fortunately, this number is small enough that we could list all of the possibilities and then count the number of outcomes with exactly three early letters. However, since most problems will involve far too many outcomes to list, we need a way to count those outcomes without having to list them all. We need to know: how many outcomes are there with three early letters (and two late letters)?

Let's ignore the late letters completely and ask: How many ways are there of choosing three early letters from among the five early letters? We can use the formula for combinations:

$$\binom{5}{3} = 10$$

Now, let's ignore the early letters, and ask: How many ways are there of choosing two late letters from among the four late letters? Again, use the formula for combinations:

$$\binom{4}{2} = 6$$

Now we come to the key insight, which will require you to think a bit before you believe it. Any of the 10 ways of choosing the early letters can be combined with any of the 6 ways of choosing the late letters. Therefore, we need to multiply to find that there are 60 possible ways of choosing three early letters and two late letters. In the future, you will need to accept this on faith, because there will be too many possibilities to list. However, here we can list all 60 possibilities:

	WX	WY	WZ	XY	XZ	YZ
ABC	ABCWX	ABCWY	ABCWZ	ABCXY	ABCXZ	ABCYZ
ABD	ABDWX	ABDWY	ABDWZ	ABDXY	ABDXZ	ABDYZ
ABE	ABEWX	ABEWY	ABEWZ	ABEXY	ABEXZ	ABEYZ
ACD	ACDWX	ACDWY	ACDWZ	ACDXY	ACDXZ	ACDYZ
ACE	ACEWX	ACEWY	ACEWZ	ACEXY	ACEXZ	ACEYZ
ADE	ADEWX	ADEWY	ADEWZ	ADEXY	ADEXZ	ADEYZ
BCD	BCDWX	BCDWY	BCDWZ	BCDXY	BCDXZ	BCDYZ
BCE	BCEWX	BCEWY	BCEWZ	BCEXY	BCEXZ	BCEYZ
BDE	BDEWX	BDEWY	BDEWZ	BDEXY	BDEXZ	BDEYZ
CDE	CDEWX	CDEWY	CDEWZ	CDEXY	CDEXZ	CDEYZ

Note how the possibilities are arranged; each column corresponds to 1 of the 6 ways of selecting the late letters, and each row corresponds to 1 of the 10 ways of selecting the early letters.

The probability of selecting three early letters and two late letters is therefore:

$$\frac{\binom{5}{3} \times \binom{4}{2}}{\binom{9}{5}} = \frac{10 \times 6}{126} = \frac{60}{126} = .476$$

By the same reasoning, we can figure out that the probability of drawing one early letter is given by the formula:

$$\frac{\binom{5}{1} \times \binom{4}{4}}{\binom{9}{5}} = \frac{5 \times 1}{126} = \frac{5}{126} = .040$$

The five outcomes with one early letter are AWXYZ, BWXYZ, CWXYZ, DWXYZ, EWXYZ.

The probability of drawing two early letters is given by the formula:

$$\frac{\binom{5}{2} \times \binom{4}{3}}{\binom{9}{5}} = \frac{10 \times 4}{126} = \frac{40}{126} = .317$$

The 40 outcomes with two early letters are:

	WXY	WXZ	WYZ	XYZ
AB	ABWXY	ABWXZ	ABWYZ	ABXYZ
AC	ACWXY	ACWXZ	ACWYZ	ACXYZ
AD	ADWXY	ADWXZ	ADWYZ	ADXYZ
AE	AEWXY	AEWXZ	AEWYZ	AEXYZ
BC	BCWXY	BCWXZ	BCWYZ	BCXYZ
BD	BDWXY	BDWXZ	BDWYZ	BDXYZ
BE	BEWXY	BEWXZ	BEWYZ	BEXYZ
CD	CDWXY	CDWXZ	CDWYZ	CDXYZ
CE	CEWXY	CEWXZ	CEWYZ	CEXYZ
DE	DEWXY	DEWXZ	DEWYZ	DEXYZ

Each column comes from one of the four ways of choosing three letters from among WXYZ; each row comes from one of the 10 ways of choosing two letters from among ABCDE.

The probability of drawing four early letters is given by the formula:

$$\frac{\binom{5}{4} \times \binom{4}{1}}{\binom{9}{5}} = \frac{5 \times 4}{126} = \frac{20}{126} = .159$$

The 20 outcomes with four early letters are:

ABCDW	ABCDX	ABCDY	ABCDZ
ABCEW	ABCEX	ABCEY	ABCEZ
ABDEW	ABDEX	ABDEY	ABDEZ
ACDEW	ACDEX	ACDEY	ACDEZ
BCDEW	BCDEX	BCDEY	BCDEZ

The probability of drawing five early letters is given by the formula:

$$\frac{\binom{5}{5} \times \binom{4}{0}}{\binom{9}{5}} = \frac{1 \times 1}{126} = \frac{1}{126} = .00794$$

We can summarize these results in a table:

Number of Early Letters in Hand	Number of Outcomes	Probability
1	5	.0397
2	40	.3175
3	60	.4762
4	20	.1587
5	1	.0079
Total	126	1.000

PROBLEM   You are drawing 10 cards at random from a 52-card deck. What is the probability that exactly 4 cards you draw will be hearts?

SOLUTION   First, calculate the total number of ways of selecting the cards, using the formulas for combinations:

$$\binom{52}{10} = \frac{52 \times 51 \times 50 \times 49 \times 48 \times 47 \times 46 \times 45 \times 44 \times 43}{10 \times 9 \times 8 \times 7 \times 6 \times 5 \times 4 \times 3 \times 2 \times 1} = 15,820,024,220$$

Now, determine how many ways there are of drawing 4 hearts from among the 13 hearts in the deck:

$$\binom{13}{4} = \frac{13 \times 12 \times 11 \times 10}{4 \times 3 \times 2 \times 1} = 715$$

If we draw 4 hearts among our 10 cards, then there must be 6 nonhearts. We can determine the number of ways of drawing 6 nonhearts from the 39 nonhearts in the deck:

$$\binom{39}{6} = 3,262,623$$

Now our reasoning is the same as in the previous example, except that there are so many possibilities that we cannot list them all. Each of the 715 ways of choosing the 4 hearts can be combined with each of the 3,262,623 ways of choosing the 6 nonhearts, so in total there are 715 × 3,262,623 = 2,332,775,445 total possible hands with 4 hearts and 6 nonhearts. Therefore, the probability of drawing 4 hearts is:

$$\frac{\binom{13}{4} \times \binom{39}{6}}{\binom{52}{10}} = 2{,}332{,}775{,}445/15{,}820{,}024{,}220 = .147$$

If we let X represent the number of hearts drawn in our 10 card hand, then Figure 4-5 shows the probability that X will equal i, for all values of i from 0 up to 10.

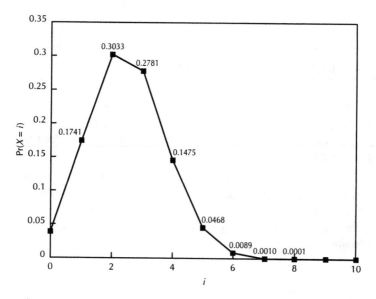

Figure 4-5

PROBLEM Suppose there are N people (M Republicans and $N-M$ Democrats) in the population. We will select a sample of n people (chosen randomly without replacement, and of course the order of selection of the people in the sample doesn't matter). What is the probability that exactly k people in the sample will be Republicans?

SOLUTION The reasoning is exactly the same as in the previous example, only now we have generalized the result.

There are $\binom{N}{n}$ ways of selecting the sample; there are $\binom{M}{k}$ ways of choosing the k Republicans in the sample, and there are $\binom{N-M}{n-k}$ ways of choosing the $n-k$ Democrats in the sample. Therefore, there are $\binom{M}{k} \times \binom{N-M}{n-k}$ ways of selecting the sample with k Republicans and $n-k$ Democrats, so the probability of this happening is given by:

$$\frac{\binom{M}{k}\binom{N-M}{n-k}}{\binom{N}{n}}$$

This is the general formula that applies to the selection of samples. We discuss it more in Chapters 7 and 12. However, this formula is cumbersome to work with, so we will also develop some easier shortcuts.

YOU SHOULD REMEMBER

1. The task of calculating probabilities requires calculating the number of possible outcomes that meet a specific condition. Two important concepts in counting outcomes are *permutations* and *combinations*.

2. Permutations: The number of permutations of n things taken *j* at a time is the number of different possible ways of choosing *j* objects from the group of n objects when each different way of ordering the chosen objects counts separately. The number of possibilities is

$$\frac{n!}{(n-j)!}$$

3. Combinations: The number of combinations of n things taken *j* at a time is the number of different possible ways of choosing *j* objects from the group of n objects when the order in which the objects are chosen doesn't matter. The number of possibilities is

$$\binom{n}{j} = \frac{n!}{j!(n-j)!}$$

KNOW THE CONCEPTS

DO YOU KNOW THE BASICS?

Test your understanding of Chapter 4 by answering the following questions:

1. When do you add the probabilities of two events?
2. If the probability of getting an order from firm *A* is .3, and the probability of getting an order from firm *B* is .4, does this mean there is a probability of .7 that you will get at least one order from firm *A* and/or firm *B*?
3. If there is a probability of .25 that firm *A* will build its new factory in your city, and a probability of .25 that the firm will build its new factory in the other city in your county, does this mean there is a probability of .5 that the firm will build its new factory in one of the cities in your county?
4. If you are checking a sample of electronic parts to determine how many are defective, should you choose the sample with or without replacement?
5. Why are there *n*! different ways of putting *n* objects in order?
6. If you are choosing 20 objects from a population of 80 objects, are there more permutations or more combinations?

TERMS FOR STUDY

classical approach to probability
combinations
complement
mutually exclusive events
intersection
multiplication principle
outcome
permutations
probability of an event

probability space
relative frequency view of probability
sampling with replacement
sampling without replacement
set
subjective view of probability
subset
union

PRACTICAL APPLICATION

COMPUTATIONAL PROBLEMS

1. What is the probability of getting a 7 or an 11 when two dice are rolled?
2. The town of Wethersfield, Connecticut, has an area of about 14 square miles. What is the probability that a meteorite thrown randomly at the earth will hit Wethersfield? (The surface area of the earth is about 200,000,000 square miles.)
3. What is the probability that a meteorite thrown randomly at the earth will hit an ocean?

4. If you flip a coin five times, what is the probability that you will get a head on either the first, second, or third toss?

5. If you roll a die three times, what is the probability that you will get a 1 on at least one of the three tosses?

6. Seventeenth-century Italian gamblers thought that the chances of getting 9 when they rolled three dice was equal to the chances of getting 10. Calculate these two probabilities to see whether they were right.

7. You're coach of a football team that has just scored a touchdown, so you are now down by 8 points in the fourth quarter. You expect to score one more touchdown in the game. Should you try for a 2-point conversion now, or should you kick a 1-point conversion now and then try a 2-point conversion after your next touchdown? Assume that you have a 100 percent chance of making the 1-point conversion but only a 50 percent chance of making the 2-point conversion.

8. If you toss a coin n times, what is the probability that you will get at least one head?

9. Assume that there are 21 one-hour slots for prime-time television shows during the week. Suppose that the networks select the time slots for their shows at random. If your two favorite shows are on different networks, what is the probability that they will conflict?

10. What is the probability that your two favorite shows will be televised on the same night?

11. What is the probability that your two favorite shows will both be televised on Monday?

12. Suppose your backgammon opponent has points 4, 5, and 6 covered, and you have two pieces on the bar. In order to get back on the board, you need to have both dice result in either 1, 2, or 3. What is the probability that both pieces will get back on the board?

13. Suppose you have a 40 percent chance of getting a job offer from your first-choice firm, a 40 percent chance of getting a job offer from your second-choice firm, and a 16 percent chance of getting a job offer from both firms. What is the probability that you will get a job offer from either firm?

14. Suppose 70 percent of the families in a certain town have children. Of these families, 30 percent have children under 6, and 60 percent have children 6 or over. How many families have children both over 6 and under 6?

15. A hostess has so many guests coming that she needs to use more than one set of plates. She has 22 guests coming, and she has 10 plates in one set and 12 plates in the other set. All of the people will sit at a round table. How many different ways can the plates be set at the table?

16. How many ways can all 32 chess pieces be arranged in a row? (Each color has 8 pawns, 2 rooks, 2 bishops, 2 knights, 1 king, and 1 queen.)

17. Suppose you suddenly find that you have to cook for yourself for 1 week. You have the following kinds of frozen dinners: 4 beef dinners, 2 chicken dinners, and 1 turkey dinner. How many ways can you arrange your menus for the week?

18. A choreographer is planning a dance routine consisting of four star dancers and an eight-member chorus line. The indifferent choreographer regards the four stars as indistinguishable, and likewise the chorus line members as indistinguishable. The dance will end with all 12 dancers in a row. In how many different ways can that row be arranged?

19. How many ways can the four aces be located in a deck of 52 cards?

20. Suppose you are getting dressed in a dark room. In your drawer you have four red socks, three blue socks, and two brown socks. If you randomly select two socks, what is the probability that you will get two socks that match?

21. If you have five pennies and four dimes in your pocket, and you reach in and pick two coins, what is the probability that you will get 20 cents?

22. Messrs. Smith, Jones, and Brown go to a garage sale. At the sale there are nine different bicycle horns. If Smith, Jones, and Brown each buy one horn, how many different possible purchases are there?

23. At a meeting, name tags are being made for four people named John, two people named Julie, and two people named Jane. In how many different ways can the name tags be distributed?

24. At a party five people have blue jackets, four people have brown jackets, and two people have red jackets. If, at the end of the party, each randomly selects a jacket of the correct color, in how many different ways can the jackets be mixed up?

25. If you have 10 large chairs and 5 small chairs to be arranged at a round table, how many different ways are there to arrange them?

26. If you are picking four cards from a standard deck, what is the probability that you will pick an ace, a 2, a 3, and a 4?

27. At a picnic the three indistinguishable Smith children and the two indistinguishable Jones children are sitting on a bench. In how many distinguishable ways can they be arranged on the bench?

28. If you are going to buy something from a vending machine that costs 55 cents, and you have five nickels and three dimes in your pocket, in how many different ways can you put change in the machine?

29. If you have 20 blue balls and 30 red balls in a box, and you randomly pull out 20 balls, what is the probability that they will all be blue?

30. What is $\binom{n}{0}$? Explain intuitively.

31. What is $\binom{n}{1}$? Explain intuitively.

32. What is $\binom{n}{n}$? Explain intuitively.

33. What is $\binom{n}{n-1}$? Explain intuitively.

34. Compare $\binom{n}{j}$ and $\binom{n}{n-j}$. Explain intuitively.

35. How many different possible five-letter words are there?

36. A combination for a lock consists of three numbers from 1 to 30. What is the probability that you could randomly guess the combination?

37. Have It Your Way Burgers, Inc., likes to give its customers a lot of choices, and it likes to keep all possible combinations on hand. Customers have their choice between these options: cheese/no cheese, onion/no onion, pickle/no pickle, well done/medium, ketchup/no ketchup, and there are six possible sizes: tiny, small, junior, medium, hefty, and jumbo. If any option can be selected with any other option, how many hamburgers will Have It Your Way have to keep on hand in order to have every single possibility available?

38. At a Chinese restaurant, you have a choice between five items in column A, six items in column B, and four items in column C. How many possible choices do you have?

39. If you roll three dice, what is the probability of getting at least two numbers the same?

40. Suppose you roll a die six times. What is the probability that no number will occur twice during the six rolls?

41. Suppose you roll a die n times. What is the probability that no number will occur twice during the n rolls?

42. If you roll three dice, what is the probability that they will all turn up the same?

43. If you roll five dice, what is the probability that they will all turn up the same?

44. If you roll a die n times, what is the probability that at least one 1 will turn up?

45. If you draw cards from a well-shuffled deck, what is the probability that you will draw all four kings before drawing the ace of spades?

46. What is the probability that you will draw all four kings before you draw a single ace?

47. Suppose you are playing a card game where you need to get either four in a row of the same suit, or four of a kind. You currently have 5D, 5C, 5H, 6H, and 7H, and you must discard one of these cards. Which card should you discard?

ANSWERS

KNOW THE CONCEPTS

1. You add the probabilities of two events when you want to find the probability that either one of the two events will occur, provided that the two events are disjoint.

2. No. The event of getting an order from firm *A* is not disjoint from the event of getting an order from firm *B*, so you cannot simply add the probabilities to find the probability of getting an order from at least one of the firms.

3. If you assume that the firm will not locate a factory in both cities (meaning that the event of the factory being located in your city is disjoint from the event of the factory being located in the other city), then it is correct to add the probabilities to determine that there is a .5 probability of the factory being located in one of the two cities.

4. You should sample without replacement because there is no need to check the same circuit more than once. However, if the population size is much larger than the sample size, it does not make very much difference whether the sample is chosen with or without replacement.

5. There are *n* possibilities for the first item, $n - 1$ possibilities for the second item, and so on.

6. There are more permutations, since each different ordering is counted separately when you are calculating permutations.

PRACTICAL APPLICATION

1. 2/9

2. .000 000 07

3. 3/4

4. 7/8

5. $1 - (5/6)^3 = .421$

6. Probability of getting 9 is .116; probability of getting 10 is .125.

7. If you go for the 2 points now, you have a 50 percent chance of succeeding, in which case you win the game. If you fail, then you can go for 2 points next time. If you succeed the next time, you will tie; otherwise you will lose. Therefore, if you go for the 2 points this time, you have a 50 percent chance of winning, a 25 percent chance of tying, and a 25 percent chance of losing. If you kick the extra point this time, then you have a 50 percent chance of winning and a 50 percent chance of losing.

8. $1 - (1/2)^n$

9. 1/21

10. $1/7$

11. $1/49$

12. $1/4$

13. Let A be the event of getting hired by the first firm and B be the event of getting hired by the second firm. Then $\Pr(A \text{ or } B) = \Pr(A \cup B) = .40 + .40 - .16 = .64$.

14. 20 percent

15. $\left(\dfrac{1}{22}\right)\left(\dfrac{22}{10}\right)$

16. $\dfrac{32!}{8! \ 8! \ 2! \ 2! \ 2! \ 2! \ 2! \ 2!}$

17. $\dfrac{7!}{4! \ 2! \ 1!} = 105$

18. $\dbinom{12}{4} = 495$

19. $\dfrac{52!}{48!} = 6{,}497{,}400$

20. $\dfrac{\dbinom{4}{2} + \dbinom{3}{2} + \dbinom{2}{2}}{\dbinom{9}{2}} = \dfrac{6 + 3 + 1}{36} = \dfrac{5}{18}$

21. $\dfrac{\dbinom{4}{2}}{\dbinom{9}{2}} = 1/6$

22. $9!/6! = 504$

23. $4! \ 2! \ 2! = 96$

24. $5! \ 4! \ 2! = 5{,}760$

25. $\dbinom{15}{10} = 3{,}003$

26. $256/270{,}725$

27. 10

28. 56

29. 2.1×10^{-14}

30. There is only one way to choose zero (that is, no) object from a group of n objects, so $\dbinom{n}{0}$ is 1.

31. There are n ways to choose one object from a group of n objects.

32. There is only one way to choose all n items.

33. There are n ways to select $n - 1$ objects, since you have n choices as to which object you will not select.

34. They are equal. Choosing j objects from the group is the same as *not* choosing $n - j$ objects.

35. $26^5 = 11,881,376$

36. $1/30^3 = 1/27,000$

37. $2^5 \times 6 = 192$

38. 120

39. 4/9

40. $6!/6^6 = 120/7776 = 5/324$

41. $n = 2{:}5/6; n = 3{:}5/9; n = 4{:}5/18; n = 5{:}5/54; n = 6{:}5/324; n > 6{:}0.$

42. 1/36

43. $1/6^4 = 1/1296$

44. $1 - (5/6)^n$

45. 1/5

46. 1/70

47. Discard either 5D or 5C. Then you can win with either 4H or 8H.

5
CONDITIONAL PROBABILITY

KEY TERMS

conditional probability the probability that a particular event will occur when it is given that another event has occurred

independent events two events that do not affect each other

With random events we're often totally in the dark about what will happen, as we have seen. However, it sometimes happens that we can get some information that sheds light on the issue by telling us whether a particular random event is more or less likely to occur.

For example, suppose we want to know the probability that a total of 8 will turn up when we roll two dice. Ordinarily, we know that the probability this will happen is 5/36. However, suppose we roll one of the dice first. Then we'll have a better idea about how likely we are to get 8. For example, suppose we get a 5 on the first die. Then, to get a total of 8, we need to roll a 3 on the second die, and we know that the probability of this happening is 1/6. Therefore, once we are given the fact that the first die turned up 5, our chances of rolling an 8 have improved from 5/36 to 1/6.

On the other hand, suppose that the first die comes up 1. Then we know that there is no way to roll an 8, no matter what happens with the second die. Therefore, the probability that we will roll an 8, given that we rolled a 1 on the first die, is 0.

Or suppose we're interested in the probability of flipping four heads in a row. We know that ordinarily the probability is 1/16. However, suppose that we've already flipped the coin twice and it turned up heads both times. Once that has happened, the probability of getting two more heads is 1/4. On the other hand, if we flip the coin twice and it comes up heads first and then tails, we know that there is no chance of getting four heads in a row.

Example: Calculating the Probability of a Royal Flush

PROBLEM   Suppose we're interested in the probability of getting a royal flush, that is, a hand consisting of A, K, Q, J, 10, all of the same suit.

Ordinarily this probability is only $4/2{,}598{,}960 = 1.54 \times 10^{-6}$. However, suppose we have already drawn the ace of hearts and the king of hearts, and we are going to draw three more cards.

SOLUTION   Then there are $\binom{50}{3} = 19{,}600$ ways of drawing the remaining three cards, so our probability of getting a royal flush has improved to $1/19{,}600 = 5.1 \times 10^{-5}$.

CALCULATING CONDITIONAL PROBABILITIES

All of these situations are examples of **conditional probability**. A conditional probability tells us the probability that a particular event will occur, *if* we already know that another specific event has occurred. In particular, suppose we know that event B has occurred and we want to know the probability that event A will occur. The conditional probability that event A will occur given that event B has occurred is written like this:

$$\Pr(A \mid B)$$

The vertical line means "given that."

Now we have to figure out how to calculate conditional probabilities. In ordinary circumstances the probability that event A will occur is $N(A)/s$, where s is the total number of outcomes and $N(A)$ is the number of events in A. However, we know that not all of these outcomes are possible. We know that event B has occurred, so only the outcomes in event B need to be considered. Therefore, the number of possibilities is $N(B)$. The next question is: How many of these remaining possibilities also have event A occurring? Ordinarily there are $N(A)$ ways for event A to occur, but not all of these are now possible. The outcomes that are in A but not in B cannot occur. Therefore, the number of possible outcomes in which event A can occur is equal to the number of outcomes that are in both event A and event B. But we already gave a name to that event:

$$A \text{ and } B = A \cap B$$

The event in which both A and B occur is called A intersect B. Therefore, the probability that event A will occur given that event B occurs is given by

$$\Pr(A \mid B) = \frac{N(A \cap B)}{N(B)}$$

We can rewrite this formula by dividing both the top and bottom by *s*:

$$\Pr(A \mid B) = \frac{N(A \cap B)/s}{N(B)/s}$$

$$= \frac{\Pr(A \cap B)}{\Pr(B)}$$

In words, the probability that event *A* will occur, given that event *B* occurs, is equal to the probability that both *A* and *B* will occur, divided by the probability that *B* will occur. (Note that this definition does not work when $\Pr(B) = 0$, because then we would be dividing by zero, which is no help at all.)

We can use this formula on the examples we discussed earlier.

Examples: Calculating Conditional Probabilities

PROBLEM   Let *A* = event of getting a total of 8 on a pair of dice.
Let *B* = event of getting a 5 on first roll; $\Pr(B) = 1/6$.
Calculate $\Pr(A \mid B)$.

SOLUTION $A \cap B$ = event of getting a 5 on first roll and 8 total.
$A \cap B$ can occur only if we roll (5,3), so $\Pr(A \cap B) = 1/36$. Thus

$$\Pr(A \mid B) = \frac{1/36}{1/6} = 6/36 = 1/6$$

(1 ,1)	(1 ,2)	(1 ,3)	(1 ,4)	(1 ,5)	(1 ,6)
(2 ,1)	(2 ,2)	(2 ,3)	(2 ,4)	(2 ,5)	(2 ,6)
(3 ,1)	(3 ,2)	(3 ,3)	(3 ,4)	(3 ,5)	(3 ,6)
(4 ,1)	(4 ,2)	(4 ,3)	(4 ,4)	(4 ,5)	(4 ,6)
(5 ,1)	(5 ,2)	(5 ,3)	(5 ,4)	(5 ,5)	(5 ,6)
(6 ,1)	(6 ,2)	(6 ,3)	(6 ,4)	(6 ,5)	(6 ,6)

Figure 5-1

PROBLEM   Let *A* be the event of getting four heads in a row.
Let *B* be the event of getting two heads in the first two rolls.
Calculate $\Pr(A \mid B)$.

SOLUTION $$\Pr(B) = 1/4$$

$$\Pr(A \cap B) = 1/16$$

$$\Pr(A \mid B) = \frac{1/16}{1/4} = 4/16 = 1/4$$

PROBLEM   Let A be the event of getting a royal flush.
Let B be the event of getting AH, KH on the first two cards.
Calculate $\Pr(A \mid B)$.

SOLUTION

$$\Pr(B) = \frac{\binom{50}{3}}{\binom{52}{5}}$$

$$= \frac{19{,}600}{2{,}598{,}960}$$

$$= 7.54 \times 10^{-3}$$

$A \cap B$ is the event of getting both KH, AH and a royal flush, so it means getting the hand AH, KH, QH, JH, 10H.

$$\Pr(A \cap B) = 1/2{,}598{,}960$$

Therefore,

$$\Pr(A \mid B) = \frac{1/2{,}598{,}960}{19{,}600/2{,}598{,}960}$$

$$= 1/19{,}600$$

PROBLEM   Let A be the event of getting a total of 8 on two dice.
Let B be the event of getting a 1 on the first roll.
Calculate $\Pr(A \mid B)$.

SOLUTION   In this case A and B can't happen together. (We used the term mutually exclusive for two events that cannot occur together.) Therefore, $A \cap B = \emptyset$, so $\Pr(A \cap B) = 0$; then $\Pr(A \mid B) = 0$. In general, when $\Pr(A \cap B)$ is zero, $\Pr(A \mid B)$ is zero. This result should be intuitively clear. It is like asking, "What is the probability that you will flip tails, given the fact that you have already done the flip and it came up heads?"

Another situation occurs if event A is a subset of event B. For example, suppose A is the event of getting a 1 on the dice and B is the event of rolling an odd number. Then A contains the outcome $\{1\}$, B contains the outcomes $\{1, 3, 5\}$, and $A \cap B$ contains the outcome $\{1\}$. Since $A \cap B = A$, then $\Pr(A \cap B) = \Pr(A)$, so $\Pr(A \mid B) = \Pr(A)/\Pr(B)$.

In general, if A is a subset of B, then

$$\Pr(A \mid B) = \frac{\Pr(A)}{\Pr(B)}$$

YOU SHOULD REMEMBER

The conditional probability that event A will occur, given that event B has occurred, is

$$\Pr(A \mid B) = \frac{\Pr(A \text{ and } B)}{\Pr(B)}$$

The vertical line, |, means "given that."

INDEPENDENT EVENTS

As we have seen, often knowing whether one event has occurred will help tell you whether another event is more likely or less likely. However, there are some cases in which knowing that one event has occurred does not give you a clue about whether another event will occur. For example, suppose you know that a family just had a baby girl. What is the probability that the same parents' next baby will be a girl? In this case, knowing about the last baby does not give you any information about the next baby.

Or suppose that you roll a 3 on the first roll of a die. What is the probability that you will roll a 5 on the next roll? Knowing that the first roll came up 3 does not help you know what will happen on the next roll. In this case, if A is the event of getting a 3 on the first roll and B is the event of getting a 5 on the second roll, then $\Pr(A) = 1/6$, $\Pr(B) = 1/6$, and $\Pr(A \mid B) = 1/6$, since the fact that B has occurred does not affect the probability that A will occur.

We will give a special name to this situation: we will say that these two events are **independent events**. That is a logical term, since two events are independent if they don't affect each other. Knowing that one of the events has occurred does not give you any information about whether the other event will occur.

The formal definition of independence is as follows:

Events A and B are independent if $\Pr(A \mid B) = \Pr(A)$

Here are some more examples of independent events:

- The probability that you will draw two pairs in a card game is not affected by the fact that you drew two pairs in a card game yesterday.
- The probability that you will roll a 4 on a die is not affected by the fact that you just flipped a head on a coin.

There is an interesting result that we can get from the formula for conditional probability. Note that, in general,

$$\Pr(A \mid B) = \frac{\Pr(A \cap B)}{\Pr(B)}$$

Also, if A and B are independent, then

$$\Pr(A \mid B) = \Pr(A)$$

Therefore, when A and B are independent, we can write:

$$\Pr(A) = \frac{\Pr(A \cap B)}{\Pr(B)}$$

Therefore:

$$\Pr(A \cap B) = \Pr(A)\,\Pr(B), \quad \text{or} \quad \Pr(A \text{ and } B) = \Pr(A)\,\Pr(B)$$

This result gives us a nice, simple rule to find the probability of $A \cap B$ if A and B are independent.

For example: You have vitally important data on your computer disk. You estimate there is a .01 probability that the system will fail on any day, causing a catastrophic loss of data. You realize you need a backup data system, which will also have a .01 probability of failing. What is the probability that both systems will fail?

If the two systems are independent, then you can simply multiply the probabilities:

Pr(main system fails and backup fails) =
= Pr(main system fails) × Pr(backup fails) = .01 × .01 = .0001

However, it is vital to remember the importance of the assumption of independence. Suppose your backup system is located next to the main system. In that case, a single bad event could disable both systems. Therefore, it would not be correct to treat the systems as independent, and the probability of a double failure is much greater than indicated above. The best strategy would be to have the backup system in a different building, so it would be more independent of the main system.

> # YOU SHOULD REMEMBER
>
> Two events, A and B, are independent if the probability that A will occur is not affected by whether or not B has occurred, and vice versa. In that case:
>
> $$\Pr(A \mid B) = \Pr(A)$$
>
> $$\Pr(A \text{ and } B) = \Pr(A) \times \Pr(B)$$
>
> In words: To calculate the probability that two independent events will both occur, simply multiply the probabilities of the two events.

Suppose 2 percent of the population has a particular disease. There is a test for the disease, but it is not perfectly accurate. In addition, 3.2 percent of the population tests positive for the disease. There is a 75 percent chance that a person with the disease will test positive. We need to calculate the probability that a person who does test positive really does have the disease.

Let $D+$ be the event of having the disease; $D-$ be the event of not having the disease; $T+$ be the event of testing positive; and $T-$ be the event of testing negative. Then we know:

$$\Pr(D+) = .020$$
$$\Pr(T+) = .032$$
$$\Pr(T+ \mid D+) = .750$$

We need to find $\Pr(D+ \mid T+)$. We can set up a table as follows:

	D+	D−	Total
T+			.032
T−			.968
Total	.020	.980	1.000

We can fill in the totals for the rows and columns with the information we're given, but we don't yet know the remaining numbers in the table. However, we can find

$$\Pr[(D+) \cap (T+)] = \Pr(D+ \text{ and } T+) = \Pr(T+ \mid D+) \Pr(D+) = .75 \times .02 = .0150$$

In other words, the fraction of the population that has the disease and will test positive is .015. Once we know this number, we can fill in its place in the table:

	D+	D–	Total
T+	.015		.032
T–			.968
Total	.020	.980	1.000

Since we know the totals for each row and column, we can now fill in the remaining numbers in the table:

	D+	D–	Total
T+	.015	.017	.032
T–	.005	.963	.968
Total	.020	.980	1.000

From the table we can see that $\Pr(D+\,|\,T+) = .015/.032 = .469$. Therefore, less than half the people that test positive actually have the disease.

It is important to remember that the way you state a conditional probability is important: the probability that you will test positive, given that you have the disease $[\Pr(T+\,|\,D+)]$, is not the same as the probability that you will have the disease, given that you tested positive $[\Pr(D+\,|\,T+)]$.

We can also solve this problem with a rule known as Bayes's rule, which states that for any two events A and B:

$$\Pr(B\,|\,A) = \frac{\Pr(A\,|\,B)\,\Pr(B)}{\Pr(A\,|\,B)\,\Pr(B) + \Pr(A\,|\,B^c)\,\Pr(B^c)}$$

where B^c means B complement; that is, the event that B does not occur.

To use Bayes's rule for our example, we need to find $\Pr(T+\,|\,D-) = \Pr(T+$ and $D-)/\Pr(D-) = .017/.980 = .01735$.

Now calculate:

$$\Pr(D+\,|\,T+) = \frac{\Pr(T+\,|\,D+)\,\Pr(D+)}{\Pr(T+\,|\,D+)\,\Pr(D+) + \Pr(T+\,|\,D-)\,\Pr(D-)}$$

$$\Pr(D+\,|\,T+) = \frac{.750 \times .020}{.750 \times .020 + .01735 \times .98} = .469$$

YOU SHOULD REMEMBER

Bayes's rule tells how to find the conditional probability Pr(B |A) provided that you know Pr(A | B):

$$Pr(B \mid A) = \frac{Pr(A \mid B) \, Pr(B)}{Pr(A \mid B) \, Pr(B) + Pr(A \mid B^c) \, Pr(B^c)}$$

KNOW THE CONCEPTS

DO YOU KNOW THE BASICS?

Test your understanding of Chapter 5 by answering the following questions:

1. What is the conditional probability Pr(A | B) if A and B are mutually exclusive events?

2. What is the conditional probability Pr(A | B) if A and B are independent events?

3. If the probability of getting an order from firm A is .3, and the probability of getting an order from firm B is .4, does this mean there is a probability of .12 that you will get an order from both firms?

4. If there is a probability of .25 that firm A will build its new factory in your city, and a probability of .25 that the firm will build its new factory in the other city in your county, does this mean there is a probability of .0625 that the firm will build factories in both cities?

5. If 50 percent of the poor families in a city are single-parent families, what fraction of the single-parent families in the city are poor? What other information do you think you might need to answer the first question?

TERMS FOR STUDY

Bayes's rule   independent events
conditional probability

PRACTICAL APPLICATION

COMPUTATIONAL PROBLEMS

1. Suppose you have two nickels in your pocket. You know that one is fair and one is two-headed. If you take one out, toss it, and get a head, what is the probability that it was the fair coin?

2. In blackjack, each player is given two cards to start with, and then tries to get a numerical total of 21 in the following way: 2's through 10's are worth their face value, face cards are worth 10, and an ace can be worth either 1 or 11, depending on the player's preference. The player can take more cards, trying to get as close to 21 as possible without going over (in which case the game is lost). Suppose you are dealt a 4 and a 9. If the dealer is dealing from a single deck of 52 cards, and the 4 and the 9 are the only cards that have been dealt from that deck that you know the value of, should you take another card? In other words, what is the probability that if you take another card you won't go over 21?

3. In some versions of blackjack, one of the two cards dealt each player is turned up. Suppose you can see that your fellow players have been dealt two aces, a 5, and a king. You again have a 4 and a 9. Should you draw another card?

4. What is the probability that you will get a royal flush in your next four cards if your first card is the ace of hearts? What is the probability if the next two cards you draw are the king of hearts and queen of hearts? What is the probability if the fourth card you draw is the jack of hearts?

5. What is the probability that you will draw a straight if you have already drawn a 2 and a 3? (A straight consists of five cards in sequence.)

6. If you flip a fair coin twice, and know that at least one head came up, what is the probability of two heads?

7. Suppose you remove all the diamonds from a 52-card deck, put the ace of diamonds back in, shuffle the remaining 40 cards, and pick a card at random. Given that the card you pick is an ace, what is the probability that it is the ace of diamonds?

8. Suppose you roll two dice and pick a card at random from a 52-card deck. Suppose further that the number rolled on the dice is the same as the number on the card. What is the probability that the number is 6?

9. What is the probability that you will win a lottery with a three-digit number, given that you have two digits of the winning number?

10. What is the probability of getting four aces in a poker hand, given that two of the cards in your hand are the ace and king of spades?

11. What is the probability of drawing a jack from a 52-card deck, given that you've drawn a face card?

12. Suppose you roll three dice. Given that one of the dice shows a 5, what is the probability of rolling 14?

13. Consider the experiment of tossing two coins. Let A be the event of getting a head on the first toss.

   (a) Can you name an event that is mutually exclusive but not independent of A?

(b) Can you name an event that is independent but not mutually exclusive of *A*?
(c) Can you name an event that is both mutually exclusive and independent of *A*?
(d) Can you name an event that is neither mutually exclusive nor independent of *A*?

14. Suppose that an election is being conducted with two candidates, Smith and Jones. Of the people in the city 2/3 support Jones, but 5/9 of the people in the country support Smith. Half of the people live in the country and half live in the city. If you randomly start talking with a voter who turns out to be a Jones supporter, what is the probability that that voter lives in the country?

15. Suppose that 5 percent of the people with blood type O are left-handed, 10 percent of those with other blood types are left-handed, and 40 percent of the people have blood type O. If you randomly select a left-handed person, what is the probability that he or she will have blood type O?

16. Suppose that 70 percent of the people with brown eyes have brown hair, 20 percent of the people with green eyes have brown hair, and 5 percent of the people with blue eyes have brown hair. Also, suppose 75 percent of the people have brown eyes, 20 percent have blue eyes, and 5 percent have green eyes. What is the probability that a randomly selected person with brown hair will also have green eyes?

ANSWERS

KNOW THE CONCEPTS

1. If *A* and *B* are mutually exclusive, then event *A* cannot occur if event *B* has occurred, so $\Pr(A \mid B)$ is 0.

2. If *A* and *B* are independent, then knowing that event *B* has occurred does not affect the probability that event *A* will occur, so $\Pr(A \mid B) = \Pr(A)$.

3. This reasoning is correct if the two events are independent; in other words, if the chance of getting an order from firm *B* is not affected by whether or not you get an order from firm *A*.

4. These two events are presumably mutually exclusive so they cannot be independent. Therefore it is not correct to multiply the two probabilities.

5. You are not given enough information to answer the question. You need to know what fraction of all families are poor and what fraction of all families are single-parent families. For example, if 15 percent of the families are poor and 30 percent of the families are single-parent families, then 25 percent of the single-parent families are poor.

PRACTICAL APPLICATION

1. Let A be the event that a head appeared and B be the event that the coin was fair.

$$Pr(B) = 1/2 \qquad Pr(A \mid B) = 1/2 \qquad Pr(A \mid B^c) = 1$$

$$Pr(B \mid A) = \frac{PR(A \mid B) \, Pr(B)}{Pr(A \mid B) \, Pr(B) + Pr(A \mid B^c) \, Pr(B^c)}$$

$$= \frac{(1/2)(1/2)}{(1/2)(1/2) + (1)(1/2)} = 1/3$$

2. $\dfrac{4 + 4 + 4 + 3 + 4 + 4 + 4 + 4}{50} = \dfrac{31}{50}$ yes

3. $\dfrac{2 + 4 + 4 + 3 + 3 + 4 + 4 + 4}{46} = \dfrac{28}{46} = \dfrac{14}{23}$ yes

4. $4! \left(\dfrac{1}{51 \times 50 \times 49 \times 48} \right) = \dfrac{1}{249,900}$

$2! \times \left(\dfrac{1}{49 \times 48} \right) = \dfrac{1}{1,176}$

$1/48$

5. There are two ways: A, 2, 3, 4, 5; 2, 3, 4, 5, 6.

$$2 \times 3! \left(\frac{4}{50} \times \frac{4}{49} \times \frac{4}{48} \right) = \frac{8}{1,225}$$

6. Let A be the event of getting at least one head. B is the event of getting two heads. Then

$$Pr(B \mid A) = \frac{Pr(A \cap B)}{Pr(A)} = 1/3$$

7. Let A = ace; B = AD;

$$Pr(B \mid A) = \frac{Pr(A \cap B)}{Pr(A)} = \frac{1/40}{4/40} = 1/4$$

10. $3! \left(\dfrac{1}{50} \times \dfrac{1}{49} \times \dfrac{1}{48} \right) = 1/19,600$

11. $\dfrac{1/13}{3/13} = 1/3$

12. A = one 5; B = 14; $Pr(A) = 25/72$; $Pr(A \cap B) = 1/36$; $Pr(B \mid A) = 2/25$.

13. (a) The event of getting tails on the first toss.
    (b) The event of getting heads on the second toss.
    (c) The null set Ø is both mutually exclusive and independent of every other event. However, there is no way for two events to be both mutually exclusive and independent if one of them is not the null set.
    (d) The event of getting heads on both tosses.

14. Let C mean that a person from the city is chosen, and C_0 that a person from the country is chosen. $\Pr(J|C) = 2/3$; $\Pr(J|C_0) = 5/9$; $\Pr(C) = \Pr(C_0) = 1/2$; $\Pr(C_0|J) = 2/5$.

15. Let A = left-handed; B = O blood type.

$$\Pr(A|B) = .05 \qquad \Pr(A|B^c) = .10$$

$$\Pr(B) = .40 \qquad \Pr(B|A) = \frac{(1/20)(2/5)}{(1/20)(2/5) + (1/10)(3/5)}$$

$$= 1/4$$

16. $\Pr(RH|BE) = 7/10 \qquad \Pr(RH|GE) = 1/5$

$\Pr(RH|BE) = 1/20 \qquad \Pr(RE) = 3/4$

$\Pr(BE) = 1/5 \qquad \Pr(GE) = 1/20$

$$\Pr(GE|RH) = \frac{(1/5)(1/20)}{(7/10)(3/4) + (1/5)(1/20) + (1/20)(1/5)}$$

$$= .018$$

6
RANDOM VARIABLES

KEY TERMS

expectation the average value that would appear if you observed a random variable many times; also called the *expected value* or *mean* and symbolized by μ

probability function for a discrete random variable, the probability function at a specific value is the probability that the random variable will have that value

random variable a variable whose value depends on the outcome of a random experiment

variance a measure of dispersion for a random variable

It often happens in probability that the events in which we're interested involve counting something or measuring something. For example, we have been interested in the number of times that "heads" appear when we flip a coin or the number that appears on a pair of dice. In these cases it is easier to talk about **random variables** than about probability spaces and events. If X is the number of heads that appear when you flip a coin three times, then X is a random variable. If Y is the number that appears when you toss one die, then Y is also a random variable. If W represents the number of times that the word *tennis* is used on the 11 o'clock news, then W is a random variable (which can take on values anywhere from zero during the dead of winter to perhaps 50 during Wimbledon). We'll use capital letters to stand for random variables, to avoid confusion with the ordinary variables used in algebra.

Suppose we roll a die and then write down the number Y that appears. This process is called *observing* (or *measuring*) the value of Y. If we roll the die 10 times, we have 10 observations of the random variable Y.

As we have seen, a random event is something which we don't know for sure will happen, but we can often calculate the probability that it will happen. By analogy, a random variable is a variable such that we're not sure what it will equal, but for which we can often calculate the probability that it will equal a particular value. (Formally, a random variable is a variable that takes on a specified value when a particular random event occurs, so it is a function from sets to numbers.)

Although we usually cannot tell exactly what the value of a random variable will be, we often can determine what its value will *not* be. For example, the tennis variable W cannot be $3\frac{1}{3}$ or π or any other rather unusual number; W must be a whole number. The number on the die (Y) must also be a whole number, but it can have only one of six possible values: 1, 2, 3, 4, 5, or 6. Random variables that can take on only isolated values are called **discrete random variables**. (We'll talk later about *continuous random variables*.)

Discrete random variables don't have to take just whole-number values. Suppose we roll two dice, and let T be the average of the two numbers that appear. Then T has the possible values 1, $1\frac{1}{2}$, 2, $2\frac{1}{2}$, 3, $3\frac{1}{2}$, 4, $4\frac{1}{2}$, 5, $5\frac{1}{2}$, and 6.

Examples: Two General Types of Random Variables with Frequent Practical Applications

PROBLEM   Suppose you are interested in a population of N people (or objects). Of these people, M have one particular characteristic and the other $N - M$ do not. If you do not know the value of M, how can you estimate it?

SOLUTION   Observe a random sample of n of these people, and let X be the number of people in the sample who have this characteristic. Then X is a random variable, since its value will depend on which people happen to be chosen to be part of your random sample.

PROBLEM   Suppose you are interested in a large population of people (or objects). Each person has a particular measurable attribute, such as height, income, age, or hours of television watched per week, in which you are interested. What is the value of this attribute, for one specific person in the population?

SOLUTION   Select one person at random, and let X represent the value of the attribute for the person you have chosen. Then X is a random variable since it depends on the distribution of that attribute among the entire population, as well as on the identity of the person you select.

PROBABILITY FUNCTIONS

Now let's figure out what we need to know about random variables. With an ordinary variable, about all we need to know is its value. However, with random variables the situation is much more complicated. First, we need to know which values are possible and which are impossible. For example, if a random variable X can never take the value 3/2, we can write

Probability that $X = 3/2$ is 0

We can write that in a shorter fashion:

$$\Pr(X = 3/2) = 0$$

Once we've made a list of all the possible values, the next thing we would like to know is: How likely is each of these different values? In the case of tossing one die, the situation is very simple: there are only six possible values, and they are all equally likely. We can make a list of these:

$$\Pr(Y = 1) = 1/6$$
$$\Pr(Y = 2) = 1/6$$
$$\Pr(Y = 3) = 1/6$$
$$\Pr(Y = 4) = 1/6$$
$$\Pr(Y = 5) = 1/6$$
$$\Pr(Y = 6) = 1/6$$

We can also make a list of the probabilities for the random variable X, defined as the number of heads turned up in three tosses of a coin:

$$\Pr(X = 0) = 1/8$$
$$\Pr(X = 1) = 3/8$$
$$\Pr(X = 2) = 3/8$$
$$\Pr(X = 3) = 1/8$$

In both of these cases we understand the process that is generating the random variable, so it is easy to calculate the probability of each possible value. In other circumstances, such as the tennis example, we cannot calculate the probabilities because we do not understand the process well enough. (However, later we will discuss ways to estimate the probabilities of these occurrences.)

To make things a little more convenient, we will define a **probability function** for a random variable. The value of the probability function for a particular number is just the probability that the random variable will equal that number. We'll use a small letter f to stand for the probability function. Then we can make this definition:

$$f(a) = \Pr(X = a)$$

(The probability function is also sometimes called the *probability mass function* or *probability density function*.)

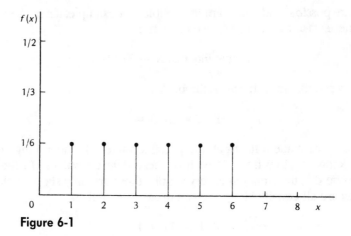

Figure 6-1

When we start talking about more than one random variable at a time, we will write the probability function as $f_x(a)$ to make it clear that f is the probability function for the random variable X.

Here is the probability function for the toss of one die:

$$f(1) = 1/6$$
$$f(2) = 1/6$$
$$f(3) = 1/6$$
$$f(4) = 1/6$$
$$f(5) = 1/6$$
$$f(6) = 1/6$$

(This function is graphed in Figure 6-1).

Here is the probability function for flipping a coin three times:

$$f(0) = 1/8$$
$$f(1) = 3/8$$
$$f(2) = 3/8$$
$$f(3) = 1/8$$

Figure 6-2 shows a graph of this function. (In both of these cases $f(a) = 0$ for all values of a except the ones that are listed.)

There is a close connection between the probability function of a random variable and the frequency histogram for the numbers in a sample. For example, suppose we roll a die 6,000 times and then make a frequency histogram showing the number of times that each possible result appears (Figure 6-3). The frequency histogram has approximately the same shape as the probability function.

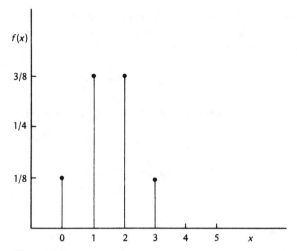

Figure 6-2

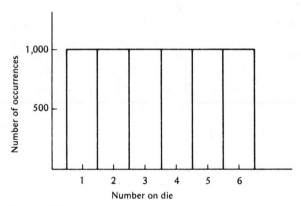

Figure 6-3

We can quickly establish two obvious properties that a probability function must satisfy:

$$f(a) \leq 1 \quad \text{for all possible values of } a$$
$$f(a) \geq 0 \quad \text{for all possible values of } a$$

These two statements point out that there is no such thing as a probability that is greater than 1 or less than 0.

Suppose we would like to know the probability that a random variable will have either one of two values. For example, what is the probability that X will equal 2 or 3? We can write it like this:

$$\Pr[(X = 2) \text{ or } (X = 3)]$$

We can rewrite that probability as the probability of a union of two events:

$$\Pr[(X = 2) \cup (X = 3)]$$

The two events $(X = 2)$ and $(X = 3)$ are clearly mutually exclusive. (Just try to make X equal both 2 and 3 at the same time!) Therefore, we can rewrite the probability:

$$\Pr[(X = 2) \text{ or } (X = 3)] = \Pr(X = 2) + \Pr(X = 3)$$
$$= f(2) + f(3)$$

In general, when we want to know the probability that X will equal either a or b, we can take the sum of $f(a) + f(b)$.

Now, suppose we make a list of all of the possible values and add up their probabilities. If the probabilities add up to less than 1 or more than 1, we will know that something is wrong. The probability must be exactly 1 that X will equal one of its possible values. Therefore, in order for a function f to be a valid probability function for a random variable, the sum of its possible values must be 1: If the possible values are $a_1, a_2, a_3, \ldots, a_n$, then

$$f(a_1) + f(a_2) + f(a_3) + \cdots + f(a_n) = 1$$

For example, when X represents the number of heads that appear on three coin tosses, we can add all of the probabilities:

$$1/8 + 3/8 + 3/8 + 1/8 = 8/8 = 1$$

to show that the sum does indeed equal 1.

It's nice to know what the probability is that X will take on a particular value, but sometimes we don't need to know that. Sometimes we'd like to know what the probability is that X will be less than or equal to a particular value a^*. For example, if you're playing blackjack and you've already drawn a 7 and an 8, then you're mainly interested in the probability that the value of the next card you draw will be less than or equal to 6.

To find the probability that the random variable Y in the case of rolling a die will be less than or equal to 3, we need to add $\Pr(Y = 1) + \Pr(Y = 2) + \Pr(Y = 3) = 1/2$. In general, if $a_1, a_2, a_3, \ldots, a^*$ are all of the possible values of Z that are less than or equal to a^*, then

$$\Pr(Z \le a^*) = f(a_1) + f(a_2) + f(a_3) + \cdots + f(a^*)$$

We will give a special name to the function that tells the probability that Z will be less than or equal to a particular value. We'll call that function the **cumulative distribution function**, and we'll represent it by a capital F:

$$F(a) = \Pr(Z \le a)$$

We can calculate the cumulative distribution function for the number of heads, X, appearing on three coin tosses:

$$F(a) = \begin{cases} 0 & \text{if } a < 0 \\ 1/8 & \text{if } 0 \le a < 1 \\ 1/2 & \text{if } 1 \le a < 2 \\ 7/8 & \text{if } 2 \le a < 3 \\ 1 & \text{if } 3 \le a \end{cases}$$

YOU SHOULD REMEMBER

1. A *random variable* is a variable whose value depends on the result of a random experiment.

2. The *probability function* of a discrete random variable is the function that gives the probability that the random variable will take on a particular value. For example, if $x_1, x_2, \ldots, x_m$ are the possible values of a random variable X and $f(x)$ is the probability function, then

$$f(x_1) = \Pr(X = x_1), f(x_2) = \Pr(X = x_2), \text{ and so on}$$

3. The *cumulative distribution function* $F(a)$ gives the probabilities that the random variable will be less than or equal to a particular value:

$$F(a) = \Pr(X \le a)$$

EXPECTATION

Once we know the probability function or the cumulative distribution function for a particular random variable, we know just about everything we might possibly need to know about it. However, many times we would like to summarize the information that we have about the random variable. We showed that we can summarize a group of numbers by a single number (the average). We will develop a similar concept for a random variable.

Consider the variable X (the number of heads appearing in three coin tosses.) Suppose we measure X eight million times. How many times are we likely to get no head, one head, two heads, or three heads? We can guess that we will probably observe the value $X = 0$ close to one million times, the value $X = 1$ close to three million times, the value $X = 2$ three million times, and the value $X = 3$ one million times. In reality, we would probably not get exactly these numbers, but let us assume for a moment that we performed this experiment eight million times and did get exactly these results. (In general, if you measure a random variable N times, then the number of times that you can expect to get the value a will be equal to $f(a) \times N$.)

Now, we have eight million numbers written down (one number for each of the eight million repetitions of our experiment.) Since we don't want to carry around all eight million numbers, we'd like to summarize all of these values by taking their average:

$$(0 + 0 + 0 + 0 + \cdots \qquad \text{(1 million zeros)}$$
$$+ 1 + 1 + 1 + 1 + \cdots \qquad \text{(3 million ones)}$$
$$+ 2 + 2 + 2 + 2 + \cdots \qquad \text{(3 million twos)}$$
$$+ 3 + 3 + 3 + 3 + \cdots)/8{,}000{,}000 \qquad \text{(1 million threes)}$$

We can rewrite that in a shorter fashion:

$$(1{,}000{,}000 \times 0$$
$$+ 3{,}000{,}000 \times 1$$
$$+ 3{,}000{,}000 \times 2$$
$$+ 1{,}000{,}000 \times 3)/8{,}000{,}000$$

Here's an even shorter version:

$$[1/8 \times 0] + [3/8 \times 1] + [3/8 \times 2] + [1/8 \times 3] = 1\tfrac{1}{2}$$

Therefore, the average of all the values is $1\tfrac{1}{2}$. We can also write that expression in terms of the probability function f:

$$\text{Average} = [f(0) \times 0] + [f(1) \times 1] + [f(2) \times 2] + [f(3) \times 3]$$

We will call this quantity the *expected value* of X, or the **expectation** of X. The expectation of a random variable tells us the average of all the values we would expect to get if we measured the random variable many times. We'll use the capital letter E to stand for expectation, and write this definition:

$$E(X) = \text{expectation of } X$$

The general formula for calculating an expectation is

$$E(X) = f(a_1)a_1 + f(a_2)a_2 + f(a_3)a_3 + \cdots + f(a_n)a_n$$

$$= \sum_{i=1}^{n} f(a_i)a_i$$

where $a_1, a_2, \ldots, a_n$ are all of the possible values of the random variable X.

The expectation of X is a weighted average of all the possible values of X. The weight for each value is equal to the probability that X will take on that value.

The expectation of X is also called the *mean* of X, or the mean of the distribution of X, and it is usually symbolized by the Greek letter μ (mu). Note that $E(X) = \mu$ is not itself a random variable. It is a regular constant number.

The expectation of a random variable does not itself have to be one of the possible values of the random variable. For example, we found that the expectation for the number of heads in three tosses is $1\frac{1}{2}$, but we strongly caution you against betting that the number of heads will ever come out to be $1\frac{1}{2}$ (unless you make the bet with us).

The expectation of the number (Y) that shows up on a die is easy to calculate:

$$E(Y) = [f(1) \times 1] + [f(2) \times 2] + [f(3) \times 3]$$
$$+ [f(4) \times 4] + [f(5) \times 5] + [f(6) \times 6]$$

$$= [1/6 \times 1] + [1/6 \times 2] + [1/6 \times 3]$$
$$+ [1/6 \times 4] + [1/6 \times 5] + [1/6 \times 6]$$

$$= 3\frac{1}{2}$$

There are two important properties of expectations that we can establish. If c is a constant number (in other words, not a random variable), then

$$E(cX) = c\,E(X)$$

For example, suppose you will win $100 multiplied by the number that appears when a die is tossed. If Y is the random variable representing the number on the die, and W represents your winnings, then $W = 100Y$. We can calculate:

$$E(W) = E(100Y) = 100E(Y) = 100 \times 3.5 = \$350$$

Therefore, your expected winnings from this game are $350.

Suppose we have two random variables X and Y, and we form a new random variable V which is equal to $V = X + Y$. In general, putting two random variables together creates a lot of complications, and we will not talk much about these problems until Chapter 9. However, one property we can establish is

$$E(X + Y) = E(X) + E(Y)$$

There will also be times when we would like to calculate the expectation of a particular function of a random variable.

For example, when X is the number of heads that appear on three coin tosses, we can calculate $E(X^2)$ by multiplying the square of each possible value by the corresponding probability:

$$E(X^2) = [0 \times 1/8] + [1 \times 3/8] + [4 \times 3/8] + [9 \times 1/8]$$

$$= 24/8 = 3$$

VARIANCE

Although the expectation value of a random variable tells us a lot about its behavior, it doesn't tell the whole story. For example, let's consider one simple "random variable" that really isn't a random variable at all. Suppose U represents the number of times that Wile E. Coyote, the cartoon character, will catch the Roadrunner. Then we know for sure that $U = 0$, so the probability function has the value 1 at $U = 0$ and the value 0 everywhere else. In this case we can easily see that $E(U) = 0$.

Let us consider another random variable, T, which represents the number on the first die you toss minus the number on the second die. Then $f(5) = 1/36$, $f(4) = 2/36$, $f(3) = 3/36$, $f(2) = 4/36$, $f(1) = 5/36$, $f(0) = 6/36$, $f(-1) = 5/36$, $f(-2) = 4/36$, $f(-3) = 3/36$, $f(-4) = 2/36$, $f(-5) = 1/36$. We can easily show that $E(T) = 0$. However, the behavior of the random variable T is much more unpredictable than the behavior of the random variable U.

To consider an even more extreme example, suppose S is a random variable that represents your profit on the following stock market transaction: You pay 1 million dollars for the stock, which has a 50 percent chance of doubling in value and a 50 percent chance of becoming worthless. Then $S = 1$ million with probability 1/2 and $S =$ minus 1 million with probability 1/2. $E(S) = 0$, but, as we can see, the actual value of S will never be close to $E(S)$. Therefore, in addition to the expectation, we need something that tells us how unpredictable a random variable is—in other words, we would like something that tells us whether the random variable is likely to be close to its expectation value.

You might suggest that we look at how far away from the mean each possible value of X is, and then find out what the expected value of the distance from $E(X)$ will be, like this:

(unpredictability measure) $= E[a - E(X)]$

$$= f(a_1)[a_1 - E(X)] + f(a_2)[a_2 - E(X)]$$
$$+ \cdots + f(a_n)[a_n - E(X)]$$

However, the problem with this measure is that some of the possible values of X will be less than $E(X)$ and others will be greater, so we will get a lot of positive and negative numbers that will cancel each other out.

We will take the *square* of the distance from each value a to $E(X)$, and then find the expectation of that quantity. The result is called the **variance** of the random variable X.

$$\text{Variance } (X) = f(a_1)[a_1 - E(X)]^2 + f(a_2)[a_2 - E(X)]^2 + \cdots + f(a_n)[a_n - E(X)]^2$$

$$= E[(X - E(X))^2]$$

The variance of X is usually written as $\text{Var}(X)$. The variance is also represented by the symbol σ^2; σ, you recall from Chapter 2, is the Greek lower-case letter sigma. The symbol σ itself stands for the square root of the variance, which is called the **standard deviation**.

$$(\text{standard deviation}) = \sigma = \sqrt{\text{Var}(X)}$$

For computational purposes, it is much easier to derive a shortcut formula for the variance:

$$\begin{aligned} \text{Var}(X) &= E[(X - E(X))^2] \\ &= E[X^2 - 2\,X\,E(X) + (E(X))^2] \\ &= E(X^2) - E[2\,X\,E(X)] + E[(E(X))^2] \\ &= E(X^2) - 2\,E(X)\,E(X) + (E(X))^2 \\ &= E(X^2) - (E(X))^2 \end{aligned}$$

Now we can calculate the variances of U, T, and S. It is clear that $E(U^2) = 0$, so $\text{Var}(U)$ also is 0. In general, if c is any constant, then $\text{Var}(c) = 0$.

We can find $E(T^2)$:

$$\begin{aligned} E(T^2) &= [25 \times 1/36] + [16 \times 2/36] + [9 \times 3/36] + [4 \times 4/36] \\ &\quad + [1 \times 5/36] + [0 \times 6/36] + [1 \times 5/36] + [4 \times 4/36] \\ &\quad + [9 \times 3/36] + [16 \times 2/36] + [25 \times 1/36] \end{aligned}$$

$$= 210/36$$

Since $E(T) = 0$, it follows that $\text{Var}(T) = 210/36$.

The variance of the stock market variable S is

$$\begin{aligned} \text{Var}(S) &= E(S^2) - (E(S))^2 = E(S^2) - 0^2 \\ &= 1/2 \times 1{,}000{,}000^2 + 1/2 \times (-1{,}000{,}000)^2 - 0 \\ &= 10^{12} \end{aligned}$$

Just as we suspected, the variance of S is much greater than the variances of the other two random variables.

We can calculate the variance of X, the number of heads in three coin tosses, since we have already found that $E(X^2) = 3$. Therefore,

$$\text{Var}(X) = 3 - (1\tfrac{1}{2})^2$$
$$= 3/4$$

To find the variance of Y, the number when a single die is tossed, set up the calculation as follows:

i	$\Pr(Y = i)$	$\Pr(Y = i)$	i^2	$i^2 \times \Pr(Y = i)$	$i^2 \times \Pr(Y = i)$
1	1/6	0.1667	1	1/6	0.1667
2	1/6	0.1667	4	4/6	0.6667
3	1/6	0.1667	9	9/6	1.5000
4	1/6	0.1667	16	16/6	2.6667
5	1/6	0.1667	25	25/6	4.1667
6	1/6	0.1667	36	36/6	6.0000
Total	6/6	1		91/6	15.1667

Therefore, $E(Y^2) = 15.1667$. Then:

$$\text{Var}(Y) = E(Y^2) - [E(Y)^2]$$
$$\text{Var}(Y) = 15.16667 - 3.5^2 = 2.91667$$

YOU SHOULD REMEMBER

1. The expectation of a random variable is the average value that would occur if you observed the random variable many times. It can be found from this formula:

$$E(X) = x_1 f(x_1) + x_2 f(x_2) + \cdots + x_m f(x_m)$$

where $f(x_i) = \Pr(X = x_i)$. The expectation of a random variable is also called the mean of its distribution, and it is symbolized by the Greek letter μ (mu).

2. The formula for the variance of a random variable is

$$\text{Var}(X) = E\{(X - \mu)^2\} = E(X^2) - (E(X))^2$$

A larger value of the variance means that there is a larger chance that the random variable is far away from its mean. The variance is symbolized by σ^2 (sigma squared).

BERNOULLI TRIALS

Now we'll discuss another simple random variable that illustrates all of these concepts. Let us consider a Mad Scientist who repeats a particular experiment each day. There is a probability p that the experiment will succeed on any particular trial; let's suppose that $p = 1/5$. (An experiment like this that can have only two possible results, success or failure, is called a **Bernoulli trial**.) Let Z be the random variable that is equal to the number of successes on a particular day. Then Z has only two possible values: 0 and 1. We can easily calculate the complete probability function:

$$f(0) = \Pr(Z = 0) = 1 - p$$

$$f(1) = \Pr(Z = 1) = p$$

$$f(a) = 0 \quad \text{for all other values of } a$$

Figure 6-4 shows a graph of this probability function for $p = 1/5$.

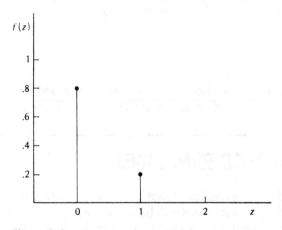

Figure 6-4

Next, we can calculate the cumulative distribution function:

$$
\begin{aligned}
F(a) &= 0 &&\text{if } a < 0 \\
F(a) &= 1 - p &&\text{if } 0 \le a < 1 \\
F(a) &= 1 &&\text{if } a \ge 1
\end{aligned}
$$

Figure 6-5 illustrates this function for $p = 1/5$.
  The expectation can be found from:

$$E(Z) = [0 \times (1 - p)] + [1 \times p] = p$$

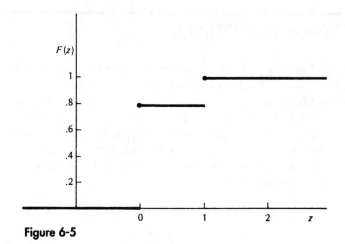

Figure 6-5

We can use the shortcut formula to calculate the variance:

$$E(Z^2) = [0^2(1 - p)] + [1^2 p] = p$$
$$\text{Var}(Z) = E(Z^2) - (E(Z))^2$$
$$= p - p^2$$
$$= p(1 - p)$$

Note that, if $p = 0$, we know for sure that $Z = 0$; and if $p = 1$, we know for sure that $Z = 1$. In either case, Var(Z) is 0, as we know must be the case.

YOU SHOULD REMEMBER

If you conduct an experiment that has a probability p of success, and Z is a random variable that will be 1 if the experiment succeeds and 0 if the experiment fails, then Z is a Bernoulli random variable:

$$E(Z) = p \qquad \text{Var}(Z) = p(1 - p)$$

VARIANCE OF A SUM

In this section we will develop two useful properties of variances. First, if c is a constant, we can show that

$$\text{Var}(cX) = c^2 \, \text{Var}(X)$$

Second, we would like a formula for the variance of a sum.

Suppose we toss two dice, letting X_1 be the number on the first die and X_2 be the number on the second die, and let $T_2 = X_1 + X_2$. We can calculate the expected value and variance of T_2:

i	# of outcomes	$Pr(T_2 = i)$	$i \times Pr(T_2 = i)$	$i \times Pr(T_2 = i)$
2	1	1/36	2/36	0.0556
3	2	2/36	6/36	0.1667
4	3	3/36	12/36	0.3333
5	4	4/36	20/36	0.5556
6	5	5/36	30/36	0.8333
7	6	6/36	42/36	1.1667
8	5	5/36	40/36	1.1111
9	4	4/36	36/36	1.0000
10	3	3/36	30/36	0.8333
11	2	2/36	22/36	0.6111
12	1	1/36	12/36	0.3333
Total	36	36/36	252/36	7

To find the variance:

i	# of outcomes	$Pr(T_2 = i)$	$i^2 \times Pr(T_2 = i)$	$i^2 \times Pr(T_2 = i)$
2	1	1/36	$4 \times 1/36$	0.1111
3	2	2/36	$9 \times 2/36$	0.5000
4	3	3/36	$16 \times 3/36$	1.3333
5	4	4/36	$25 \times 4/36$	2.7778
6	5	5/36	$36 \times 5/36$	5.0000
7	6	6/36	$49 \times 6/36$	8.1667
8	5	5/36	$64 \times 5/36$	8.8889
9	4	4/36	$81 \times 4/36$	9.0000
10	3	3/36	$100 \times 3/36$	8.3333
11	2	2/36	$121 \times 2/36$	6.7222
12	1	1/36	$144 \times 1/36$	4.0000
Total	36	36/36	1974/36	$E(T_2^2) = 54.8333$

We find $Var(T_2)$ is $54.8333 - 7^2 = 5.83333$.
Note that $5.83333 = 2.91667 + 2.91667$, so we are tempted to say:

$$Var(T_2) = Var(X_1 + X_2) = Var(X_1) + Var(X_2)$$

Unfortunately, we quickly realize this does not work as a general rule. For example, if X is the number on top, and B is the number on the bottom of a die, then $X + B$ always equals 7, so $\text{Var}(X + B) = 0$, which is not the same as $\text{Var}(X) + \text{Var}(B)$. The difference is that the random variables on the two different dice are **independent**—that is, knowing the value of one, the random variables provides you with no clue about the value of the other. On the other hand, the numbers on the top and bottom of a die are strongly related, so that if you know one of the numbers you also can tell what the other one is.

We can use the rule:

$$\text{Var}(X+Y) = \text{Var}(X) + \text{Var}(Y)$$

if X and Y are independent random variables. In Chapter 9 we will discuss a general formula for the variance of a sum that always works.

YOU SHOULD REMEMBER

1. $E(cX) = cE(X)$
2. $\text{Var}(cX) = c^2\text{Var}(X)$
   (In 1 and 2, c can be any constant; X is any random variable)
3. $E(X + Y) = E(X) + E(Y)$
   (In 3, X and Y can be any two random variables)
4. $\text{Var}(X + Y) = \text{Var}(X) + \text{Var}(Y)$
   (In 4, X and Y must be *independent* random variables)

RANDOM SAMPLES

Suppose we roll two dice. We'll let X_1 be the number that appears on the first die and X_2 be the number that appears on the second die. Then X_1 and X_2 are two random variables that are *independent and identically distributed*. They are independent because the dice don't affect each other at all, so knowing the value of X_1 does not tell us anything about X_2. They are identically distributed because they both have exactly the same probability function:

$$\Pr(X_1 = 1) = \Pr(X_2 = 1) = 1/6 \qquad \Pr(X_1 = 4) = \Pr(X_2 = 4) = 1/6$$

$$\Pr(X_1 = 2) = \Pr(X_2 = 2) = 1/6 \qquad \Pr(X_1 = 5) = \Pr(X_2 = 5) = 1/6$$

$$\Pr(X_1 = 3) = \Pr(X_2 = 3) = 1/6 \qquad \Pr(X_1 = 6) = \Pr(X_2 = 6) = 1/6$$

We will often deal with independent, identically distributed random variables. Suppose X is a random variable that we can observe n times, calling the first observation X_1, the second observation X_2, and so on, up to the last observation X_n. X could be a random variable involving dice, in which case we can measure X by rolling dice; or it could represent the height of a person selected at random from a specified population, in which case we can measure X n different times by selecting n people; or X might represent any of the many other possible types of random variables that we could have. In any case the sequence

$$X_1, X_2, X_3, \ldots, X_n$$

is called a **random sample** *of size n* taken from the distribution X.

In our case we are rolling two dice, so (X_1, X_2) form a random sample of size 2. We know that $E(X_1) = E(X_2) = 3.5$ and $\text{Var}(X_1) = \text{Var}(X_2) = 2.9167$. Now let's create a new random variable called T_2:

$$T_2 = X_1 + X_2$$

In other words, T_2 can be defined in this way: Roll two dice, and let T_2 be the sum of the numbers that appear on the dice. We already know enough about dice so that we can easily calculate the probability function of T_2:

$$f(2) = 1/36 \quad f(6) = 5/36 \quad f(10) = 3/36$$
$$f(3) = 2/36 \quad f(7) = 6/36 \quad f(11) = 2/36$$
$$f(4) = 3/36 \quad f(8) = 5/36 \quad f(12) = 1/36$$
$$f(5) = 4/36 \quad f(9) = 4/36$$

We can calculate the expectation and variance of T_2 by using the probability function:

$$
\begin{aligned}
E(T_2) =\ & 2 \times 1/36 + 3 \times 2/36 + 4 \times 3/36 \\
& + 5 \times 4/36 + 6 \times 5/36 + 7 \times 6/36 + 8 \times 5/36 \\
& + 9 \times 4/36 + 10 \times 3/36 + 11 \times 2/36 \\
& + 12 \times 1/36 = 252/36 = 7 \\
E(T_2^2) =\ & 4 \times 1/36 + 9 \times 2/36 + 16 \times 3/36 \\
& + 25 \times 4/36 + 36 \times 5/36 + 49 \times 6/36 \\
& + 64 \times 5/36 + 81 \times 4/36 + 100 \times 3/36 \\
& + 121 \times 2/36 + 144 \times 1/36 = 1974/36 = 54.8333 \\
\text{Var}(T_2) =\ & 54.8333 - 7^2 = 5.8333
\end{aligned}
$$

However, there is a much easier way. (Remember that in this book we will always try to find an easy way when we can.) We found formulas for the expectation and variance of the sum of two random variables, so we can calculate:

$$E(T_2) = E(X_1 + X_2) = E(X_1) + E(X_2) = 3.5 + 3.5 = 7$$

$$\text{Var}(T_2) = \text{Var}(X_1 + X_2) = \text{Var}(X_1) + \text{Var}(X_2) = 2.9167 + 2.9167 = 5.833$$

(Remember that the formula $E(X_1 + X_2) = E(X_1) + E(X_2)$ works for *any* two random variables, but the formula $\text{Var}(X_1 + X_2) = \text{Var}(X_1) + \text{Var}(X_2)$ works only because X_1 and X_2 are independent.) As you might have expected, the variance of $X_1 + X_2$ is greater than the variance of either X_1 or X_2 alone.

Now suppose we toss one die, and let Y_2 be a random variable equal to *twice* the number that appears ($Y_2 = 2X$). We can use this formula

$$E(cA) = cE(A) \quad \text{(where } A \text{ can be any random variable}$$
$$\text{and } c \text{ can be any constant)}$$

to tell us that $E(Y_2) = 2E(X) = 2 \times 3.5 = 7$. The expected value of Y_2 (which is double the number that appears on one die) is the same as the expected value of T_2 (the sum of the numbers that appear on two dice). However, the variance of Y_2 is different from the variance of T_2. We can use the formula

$$\text{Var}(cA) = c^2\text{Var}(A)$$

to find that

$$\text{Var}(Y_2) = \text{Var}(2X) = 2^2\text{Var}(X) = 4\text{Var}(X) = 11.667$$

Note that the variance of Y_2 is bigger than the variance of T_2. If we make a graph of the two probability functions, we can see why.

$$\text{Pr}(Y = 2) = 1/6 \qquad \text{Pr}(Y = 8) = 1/6$$
$$\text{Pr}(Y = 4) = 1/6 \qquad \text{Pr}(Y = 10) = 1/6$$
$$\text{Pr}(Y = 6) = 1/6 \qquad \text{Pr}(Y = 12) = 1/6$$

(See Figure 6-6.)

In each case the range of possible values is from 2 to 12. However, you can see that Y_2 is more likely than T_2 to have an extreme value such as 2 or 12. This is an example of a very important statistical property: If you add two independent random variables, the sum is relatively more likely to take on a value near the middle of its distribution than an extreme value. An extreme value will occur only if *both* random variables have extreme values in the same direction. For example, the random variable T_2 will take on the value 2 only if *both* X_1 and X_2 take on the value 1.

Now let's roll 100 dice, letting X_1 be the value that appears on the first dice, and so on, up to X_{100}. Then $X_1, X_2, \ldots, X_{100}$ are all independent and all identically

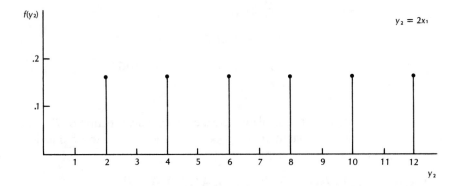

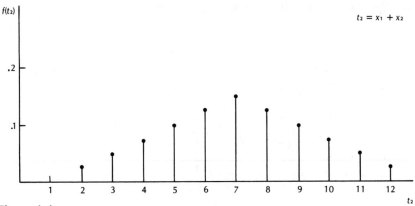

Figure 6-6

distributed, so they form a random sample of size 100 from the distribution of X. We'll let T_{100} be the sum of all the numbers:

$$T_{100} = X_1 + X_2 + \cdots + X_{100}$$

We can calculate $E(T_{100})$ and $\text{Var}(T_{100})$:

$$
\begin{aligned}
E(T_{100}) &= E(X_1 + X_2 + \cdots + X_{100}) \\
&= E(X_1) + E(X_2) + \cdots + E(X_{100}) \\
&= 3.5 + 3.5 + \cdots + 3.5 \\
&= 350
\end{aligned}
$$

$$\begin{aligned}
\text{Var}(T_{100}) &= \text{Var}(X_1 + X_2 + \cdots + X_{100}) \\
&= \text{Var}(X_1) + \text{Var}(X_2) + \cdots + \text{Var}(X_{100}) \\
&= 2.9167 + 2.9167 + \cdots + 2.9167 \\
&= 291.67
\end{aligned}$$

(It would be very tedious to calculate the exact probability function of T_{100}, but we will see in Chapter 8 that there is a good way to get a close approximation.)

• CALCULATING THE AVERAGE VALUE

Now suppose we are interested in the *average* number that appears when we roll some dice. We'll let A_2 represent the average of the two numbers that appear when we roll two dice, and we'll let A_{100} be the average of the numbers that appear when we roll 100 dice. Then

$$A_2 = \frac{T_2}{2} \quad \text{and} \quad A_{100} = \frac{T_{100}}{100}$$

We can calculate the expectation and variance:

$$E(A_2) = E\left(\frac{T_2}{2}\right) = \frac{1}{2} E(T_2) = \frac{1}{2} \times 7 = 3.5$$

$$E(A_{100}) = E\left(\frac{T_{100}}{100}\right) = \frac{1}{100} E(T_{100}) = \frac{1}{100} \times 350 = 3.5$$

$$\text{Var}(A_2) = \text{Var}\left(\frac{T_2}{2}\right) = \left(\frac{1}{2}\right)^2 \text{Var}(T_2) = \frac{1}{4} \text{Var}(T_2) = 1.4584$$

$$\text{Var}(A_{100}) = \text{Var}\left(\frac{T_{100}}{100}\right)$$

$$= \left(\frac{1}{100}\right)^2 \text{Var}(T_{100})$$

$$= 0.029167$$

Note that the expected value of each average is 3.5, which is also equal to $E(X)$. This is exactly what we would have expected. Remember that we initially said that the expected value of X was the average value we would expect to get if we observed X many, many times. However, we can see that the variance of A_2 is less than the variance of X, and the variance of A_{100} is much less than the variance of X. This means that, if we roll 100 dice, the average value will be very close to 3.5. If we roll one or two dice, then there is a reasonable chance that the average value will be as small as 1 or as large as 6. We can state this result as an important rule:

> The variance of the average value of a large number of independent, identically distributed random variables will be much less than the variance of any one of the random variables taken individually.

To show that this is true in general, we'll let $X_1, X_2, \ldots, X_n$ be a random sample of size n taken from a distribution of a random variable X that has mean μ and variance σ^2. In other words,

$$\mu = E(X_1) = E(X_2) = E(X_3) = \cdots = E(X_n)$$

and

$$\sigma^2 = \text{Var}(X_1) = \text{Var}(X_2) = \cdots = \text{Var}(X_n).$$

Then we'll let $\bar{x}$ represent the sample average for this sample:

$$\bar{x} = \frac{X_1 + X_2 + \cdots + X_n}{n}$$

Because $\bar{x}$ is formed by adding a bunch of random variables, $\bar{x}$ is itself a random variable. We can find its expectation and variance:

$$E(\bar{x}) = E\left(\frac{X_1 + X_2 + \cdots + X_n}{n}\right)$$

$$= \frac{1}{n} E(X_1 + X_2 + \cdots + X_n)$$

$$= \frac{1}{n} [E(X_1) + E(X_2) + \cdots + E(X_n)]$$

$$= \frac{1}{n} [\mu + \mu + \cdots + \mu]$$

$$= \frac{1}{n} (n\mu)$$

$$E(\bar{x}) = \mu$$

In words, the expected value of the sample average is equal to the mean of the distribution of X.

We can find the variance:

$$\text{Var}(\overline{x}) = \text{Var}\left(\frac{X_1 + X_2 + \cdots + X_n}{n}\right)$$

$$= \left(\frac{1}{n}\right)^2 \text{Var}(X_1 + X_2 + \cdots + X_n)$$

$$= \left(\frac{1}{n}\right)^2 [\text{Var}(X_1) + \text{Var}(X_2) + \cdots + \text{Var}(X_n)]$$

$$= \left(\frac{1}{n}\right)^2 (\sigma^2 + \sigma^2 + \cdots + \sigma^2)$$

$$= \left(\frac{1}{n}\right)^2 (n\sigma^2)$$

$$\text{Var}(\overline{x}) = \frac{\sigma^2}{n}$$

• THE LAW OF LARGE NUMBERS

The variance of the sample average is less than the variance of each random variable taken individually. The larger the sample size (n) becomes, the smaller the variance of $\overline{x}$ becomes. In other words, as the sample size increases, you can become more and more confident that the value of $\overline{x}$ will come close to the value of μ. This result is called the **law of large numbers**. We will see that it plays a crucial role in the theory of statistical inference.

YOU SHOULD REMEMBER

If you have n independent, identically distributed random variables $(X_1, X_2, \ldots, X_n)$, all chosen from a distribution with mean μ and variance σ^2, then these random variables are said to form a random sample of size n from the distribution of X. Let $\overline{x}$ be the average of all of these values. Then

$$E(\overline{x}) = \mu \qquad \text{Var}(\overline{x}) = \frac{\sigma^2}{n}$$

KNOW THE CONCEPTS

DO YOU KNOW THE BASICS?

Test your understanding of Chapter 6 by answering the following questions:

1. Why is the expected value of a random variable defined by the expression $\Sigma \, a_i \, f(a_i)$?

2. Why is the graph of a probability function for a discrete random variable like a frequency diagram?

3. Suppose X is a random variable representing the height of a person selected at random from a population. What is its mean and variance?

4. Suppose you took a sample of 30 people and calculated their average height (A). What is $E(A)$? How does the variance of A compare with the variance of X?

5. Let X be a random variable, and let $\bar{x}$ be the average of several independent random variables with the same distribution as X. Why is the variance of $\bar{x}$ smaller than the variance of X?

6. Why must the probabilities for a discrete random variable add up to 1?

7. What value of p makes the variance of a Bernoulli random variable as large as possible? Why?

8. If you roll a die several thousand times, the average of the numbers that appear will be close to 3.5 (according to the law of large numbers). If you find that the average after the first thousand rolls is 3.6, does this mean that you will be more likely to get values less than 3.5 in future rolls?

TERMS FOR STUDY

Bernoulli trial	probability function
cumulative distribution function	random sample
discrete random variable	random variable
expectation	standard deviation
law of large numbers	variance

PRACTICAL APPLICATION

COMPUTATIONAL PROBLEMS

For each exercise below, you are given the density function for a random variable X*. Calculate the expectation and variance of the random variable.*

1.

X	f(X)
1	.250
2	.250
3	.250
4	.250

2.

X	f(X)
10	.500
20	.250
30	.125
40	.125

3.

X	f(X)
−1	.500
1	.500

4.

X	f(X)
1	.100
2	.200
3	.300
4	.400

5.

X	f(X)
20	.100
25	.200
30	.400
35	.200
40	.100

6.

X	f(X)
−10	.990
1,000	.010

7.

X	f(X)
−100	.900
10,000	.100

8.

X	f(X)
−5,000	.100
0	.600
3,000	.300

9.

X	f(X)
−30	.200
0	.750
500	.050

10.

X	f(X)
1	.100
2	.100
3	.100
4	.100
5	.100
6	.100
7	.100
8	.100
9	.100
10	.100

ANSWERS

KNOW THE CONCEPTS

1. The expectation of a random variable is the average value that would appear if you observed the random variable many times. This average value will be a weighted average of all possible values of the random variable, with greater weight being attached to the possible values that have a greater probability of occurring.

2. If you observed the random variable many times and drew a frequency diagram for the results of those observations, the diagram would look like the density function.

3. The mean of X will be the average height of all persons in the population, and the variance of X will be the variance of the heights of all persons in the population.

4. The mean of A will be the average height of the persons in the population, but the variance of A will be much smaller than the variance of the population heights.

5. If you observe the random variable many times, you will get some values that are greater than the mean and some values less than the mean. The average of all these values is more likely to be close to the mean.

6. There is probability 1 that the random variable will take on one of its possible values, so the sum of the probabilities for all of the possible values must be 1.

7. The value $p = 1/2$. When p is closer to $1/2$, there is greater uncertainty as to whether any particular trial will be a success or a failure.

8. No. The probabilities for future rolls are not affected by past rolls.

PRACTICAL APPLICATION

The formulas you need to do Exercises 1–10 are as follows:

$$\text{expectation: } E(X) = \sum_{i=1}^{n} f(a_i)\, a_i$$

$$\text{variance:} \quad \text{Var}(X) = E(X^2) - (E(X))^2$$

The calculations for Exercise 1 are shown below. The same methods are used for Exercises 2–10.

1. $E(X) = [1 \times .25] + [2 \times .25] + [3 \times .25] + [4 \times .25] = 2.5$

 $E(X^2) = [1 \times .25] + [4 \times .25] + [9 \times .25] + [16 \times .25] = 7.5$

 $\text{Var}(X) = 7.5 - 2.5^2 = 1.25$

	$E(X)$	$\text{Var}(X)$
2.	18.75	110.938
3.	0	1
4.	3	1
5.	30	30
6.	0.1	10,099
7.	910	9,180,900
8.	400	5,040,000
9.	19	12,319
10.	5.5	8.25

7
THE BINOMIAL, POISSON, AND HYPERGEOMETRIC DISTRIBUTIONS

KEY TERMS

binomial distribution the discrete probability distribution that applies when an experiment is conducted *n* times with each trial having a probability *p* of success and each trial being independent of every other trial

hypergeometric distribution the discrete probability distribution that applies when a group of items is sampled without replacement

Poisson distribution the discrete probability distribution that gives the frequency of occurrence of certain types of random events; it can be used as an approximation for the binomial distribution

Several types of random variables are used so frequently that they have been given special names. One important random variable distribution is called the *binomial distribution*. We will also discuss two related distributions in this chapter: the *Poisson distribution* and the *hypergeometric distribution*.

THE BINOMIAL DISTRIBUTION

We'll return to the situation where a scientist conducts an experiment that has two possible results: success or failure. The probability of success for each trial is p, and the probability of failure is $1 - p$. If the experiment is conducted 10 times, how many trials do you think will result in success?

First we'll ask this question: If the scientist conducts the experiment two times, what is the probability that both trials will result in success? If A is the event of getting a success on the first trial and B is the event of getting a success on the second trial, then $\Pr(A) = p$ and $\Pr(B) = p$. The event of getting successes on both trials can be written as $A \cap B$ (A intersect B; see Chapter 4). We will make an important assumption: each trial is independent. This means that the chance of getting a success on any particular trial is not affected by the results of any of the other trials. If the trials are independent, then we can multiply the two probabilities:

$$\Pr(A \cap B) = \Pr(A \text{ and } B) = \Pr(A)\Pr(B) = p^2$$

Therefore, the probability of obtaining successes on both the first trial and the second trial is p^2. We can use the same reasoning to show that the probability of obtaining successes on all 10 trials is p^{10}. For example, if $p = .8$, then the probability of 10 successes is $.8^{10} = .107$. Even though the chance of success on any particular trial is good, the chance of 10 successes occurring in 10 trials is very slim.

We can also show that the probability of all 10 trials resulting in failure is $(1 - p)^{10}$. For example, if $p = .8$, then the chance of 10 failures in 10 trials is $.2^{10} = .0000001$.

What is the probability of obtaining one success and nine failures? This question is complicated, so let's first ask a simpler question: What is the probability that the first trial succeeds, and the other nine fail? That probability is $.8 \times .2^9 = 4.096 \times 10^{-7}$.

Make sure you understand the difference between these two questions:

- What is the probability that only the *first* trial succeeds?

- What is the probability that exactly 1 of the 10 trials succeeds?

To find the probability of exactly one trial succeeding, we need to add together the probability that only the first trial succeeds (4.096×10^{-7}), plus the probability that only the second trial succeeds (also 4.096×10^{-7}), and so on. There are 10 possibilities, so:

$$\Pr(\text{one success}) = 10 \times .8 \times .2^9 = 4.096 \times 10^{-6}$$

You can probably guess what comes next: we calculate the probability of exactly two successes, in a similar manner. The probability that the first two succeed (and the remaining eight fail) is

$$.8^2 \times .2^8 = 1.638 \times 10^{-6}$$

This is also the probability that any specified pattern of 2 successes and 8 failures will occur. The next question is: How many such patterns are there? In other words, how many ways are there of choosing 2 positions from among 10 possible positions? We have already seen the answer to that question. Use the formula for combinations:

$$\binom{10}{2} = 45$$

Therefore, the probability of exactly two successes is:

$$\binom{10}{2}\times.8^2\times.2^8 = 7.373\times10^{-5}$$

Now we can give a general formula for a special type of random variable distribution: the **binomial distribution**. The binomial distribution applies to any situation in which several independent trials are conducted, each of which can have one of two possible outcomes. We will call the two possible outcomes "success" and "failure," although in some cases these may be arbitrary designations. Suppose the scientist performs the experiment n times. Let X represent the number of successes. If the probability of success for each trial is p, then the probability that there will be i successes is

$$\Pr(X=i)=\binom{n}{i}p^i(1-p)^{n-i}$$

This formula gives the probability function for the binomial random variable. Formally, X is said to be a random variable having a binomial distribution with parameters n and p. Remember:

$$\text{The expression }\binom{n}{i}\text{ means }\frac{n!}{i!(n-i)!}$$

If n is large, it can be difficult to perform the calculations in the formula. In Chapter 8 we will see that it is possible to use another distribution called the *normal distribution* to calculate approximate values for the binomial distribution.

Here are the calculations for our case ($p = .8$ and $n = 10$):

i	$\Pr(X = i)$	Cumulative Probability (Pr X less than or equal to i)
0	1.024E-07	1.024E-07
1	4.096E-06	4.198E-06
2	0.00007	0.00008
3	0.00079	0.00086
4	0.00551	0.00637
5	0.02642	0.03279
6	0.08808	0.12087
7	0.20133	0.32220
8	0.30199	0.62419
9	0.26844	0.89263
10	0.10737	1.00000

(The table rounds the results to five decimal places, except for very small values that are given in exponential notation.)

We can see that the probability of getting exactly 6 successes in 10 trials is approximately .088.

This formula is a more general version of the formula we used in Chapter 3 to calculate the probability that a specified number of heads would result if we tossed a coin n times. The number of heads that will appear has a binomial distribution with $p = .5$.

• CALCULATING THE EXPECTATION AND VARIANCE OF A BINOMIAL RANDOM VARIABLE

We would like to calculate the expectation and variance for a binomial random variable. There is a clear intuitive explanation for the expectation. If, for example, you conduct an experiment 100 times where the probability of success for each trial is .75, you would expect, on average, to obtain 75 successes. In general, if X is a binomial random variable with parameters n and p, then the expected number of successes is np. We can easily show that this is the case. Let's suppose A_1 is a random variable that has only two possible values: A_1 will be 1 if trial 1 is a success, and A_1 will be 0 if trial 1 is a failure. Likewise, A_2 will be 1 if trial 2 is a success and 0 if trial 2 is a failure. We define $A_3, A_4, \ldots, A_n$ in the same way. Then

$$X = A_1 + A_2 + \cdots + A_n$$

The number of successes is just equal to the sum of all of the A's. We know from Chapter 6 that each A is a Bernoulli random variable, so

$$E(A_1) = p \quad \text{and} \quad \text{Var}(A_1) = p(1 - p)$$

Because each A is independent, we know that

$$E(X) = E(A_1) + E(A_2) + \cdots + E(A_n) = np$$

$$\text{Var}(X) = \text{Var}(A_1) + \text{Var}(A_2) + \cdots + \text{Var}(A_n) = np(1 - p)$$

Figure 7-1 shows a graph of a probability function for a sample binomial distribution $(n = 10, p = .8)$.

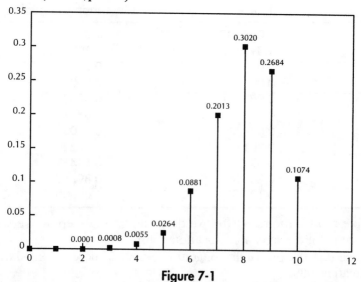

Figure 7-1

• *APPLICATIONS OF THE BINOMIAL DISTRIBUTION*
Some cases where the binomial distribution can be applied are as follows:

- the number of questions you can expect to guess right on a multiple-choice exam

- the number of casualty claims that an insurance company must pay

- the number of free throws that a basketball player will make during a season

Examples: Applying the Binomial Distribution

PROBLEM   Suppose you are taking a 20-question multiple-choice exam. Each question has four possible answers, so the probability is .25 that you can answer a question correctly by guessing. What is the probability that you can get at least 10 questions right by pure guessing?

SOLUTION   To solve this problem, we need to calculate the probabilities for a binomial distribution with $n = 20$ and $p = .25$:

k	$Pr(X = k)$	Cumulative Probability (Pr X less than or equal to k)
0	0.00317	0.00317
1	0.02114	0.02431
2	0.06695	0.09126
3	0.13390	0.22516
4	0.18969	0.41484
5	0.20233	0.61717
6	0.16861	0.78578
7	0.11241	0.89819
8	0.06089	0.95907
9	0.02706	0.98614
10	0.00992	0.99606
11	0.00301	0.99906
12	0.00075	0.99982
13	0.00015	0.99997
14	0.00003	1.00000
15	3.426E-06	1.00000
16	3.569E-07	1.00000
17	2.799E-08	1.00000
18	1.555E-09	1.00000
19	5.457E-11	1.00000
20	9.095E-13	1.00000

The table gives the probabilities rounded to five decimal places, except that exponential notation is used for very small values. By looking at the cumulative column, we can see that there is a probability of .98614 that you will get 9 or fewer questions right. Therefore, the probability of getting at least 10 right is $1 - .9861 = .01386$. It would seem in this case that studying is a better strategy than random guessing. (Of course, you can look at the bright side—there is only about a 9 percent probability that you will get fewer than 3 answers correct.)

PROBLEM Suppose you are running an airline that flies planes seating 200 people each. On average, 7 percent of the people who make reservations fail to show up for their flights. It seems wasteful to take only 200 reservations for each flight, because you know that then there will probably be some empty seats left. You decide to gamble by taking more than 200 reservations. If more than 200 people happen to show up, however, you will be in big trouble. Suppose you have decided that you are willing to run a 5 percent risk of having too many people show up for a flight. How many reservations can you accept while still keeping the probability of an overflow less than 5 percent?

SOLUTION We can regard each reservation as a "trial," and we can call each time a person with a reservation shows up for the flight a "success." (We are assuming that no people show up without making reservations.) If we let X be the number of people who show up for a particular flight, then X has a binomial distribution where n is the number of reservations and $p = .93$. Suppose you take 210 reservations. Then we can calculate these probabilities (rounded to three decimal places):

	Binomial Probability with $n = 210$ and $p = .93$
$\Pr(X = 201)$	.034
$\Pr(X = 202)$	.020
$\Pr(X = 203)$	.011
$\Pr(X = 204)$	.005
$\Pr(X = 205)$	.002
$\Pr(X = 206)$	.001

The probability that X will be greater than 206 is negligible. There will be an overflow if 201 or more people show up, so if we add up all of the probabilities in our table we can find that the probability of an overflow is about 7 percent. You want the overflow probability to be less than 5 percent, however, so you must take fewer than 210 reservations.

We can repeat the same calculations for 209 reservations; we find that the overflow probability is about 4 percent. Therefore, you should take 209 reservations for each flight since that is the largest number of reservations you can take and still have less than a 5 percent chance of an overflow.

YOU SHOULD REMEMBER

1. Suppose you conduct some form of "experiment" n times. The probability of success for any one trial is p. Let X be a random variable representing the number of successes that occur. Then X is said to have a binomial distribution with parameters n and p.

2. The probabilities for X are given by the formula

$$\Pr(X = i) = \binom{n}{i} p^i (1 - p)^{n-i}$$

3. The expectation and variance for X are given by these formulas:

$$E(X) = np \qquad \text{Var}(X) = np(1 - p)$$

• CALCULATING THE PROPORTION OF SUCCESSES

Often we will be interested not only in the number of successes in n trials but also the proportion of successes. If we let X represent the number of successes and P represent the proportion of successes, then $P = X/n$. We can find $E(P)$ and $\text{Var}(P)$ as follows:

$$E(P) = E\left(\frac{X}{n}\right) = \frac{1}{n} E(X) = \frac{1}{n} \times np = p$$

$$\text{Var}(P) = \text{Var}\left(\frac{X}{n}\right) = \left(\frac{1}{n}\right)^2 \text{Var}(X) = \frac{1}{n} \times np(1 - p) = \frac{p(1 - p)}{n}$$

The expected value of the proportion of successes is just equal to p, the probability of success. For example, if the probability that a machine will work correctly is 3/4, then you would expect that it would work 3/4 (75 percent) of the times you operate it.

* * * * *

We will now consider two more important types of random variable distributions: the Poisson distribution and the hypergeometric distribution.

THE POISSON DISTRIBUTION

Let X be the number of telephone calls that arrive at a particular office in an hour. X is a random variable, and it turns out that its probability function often looks like this:

$$f(k) = e^{-\lambda}\frac{\lambda^k}{k!} \quad (k = 0, 1, 2, 3, \ldots)$$

A random variable with this distribution is called a *Poisson random variable*. (The symbol λ is the Greek letter lambda, which is used as the parameter for the **Poisson distribution**, and e represents a special mathematical number equal to about 2.71828.)

As an example, suppose a study has determined that the number of phone calls arriving every hour at the office referred to above can be represented by a Poisson random variable with parameter $\lambda = 5$. Then we can calculate the probabilities for X:

k	Pr(X = k)(Probability of Getting Exactly k Phone Calls)	k	Pr(X = k)(Probability of Getting Exactly k Phone Calls)
0	.006	7	.104
1	.033	8	.065
2	.084	9	.036
3	.140	10	.018
4	.175	11	.008
5	.175	12	.003
6	.146		

(The probabilities are rounded to three decimal places)

Figure 7-2 shows a graph of a Poisson probability function.

Note that, in theory, there is an infinite number of possible values for X, but the probability that X will equal k becomes very small as k becomes large.

• *OTHER APPLICATIONS OF THE POISSON DISTRIBUTION*

Another important use of the Poisson distribution is to serve as an approximation for the binomial distribution. Suppose we have a binomial distribution such that n is very large and np is moderate (not very large and not very small). For example, suppose we have 500 students each of whom has a probability of .00002 of cutting a little finger on his or her test paper during finals. The calculation of the probability of i "successes" using the binomial density function becomes unmanageable. If we let $\lambda = np$, then the binomial density function can be approximated by the Poisson distribution:

$$P(X = i) = e^{-\lambda} \frac{\lambda^i}{i!}$$

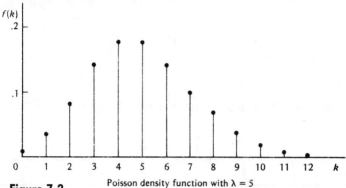

Figure 7-2   Poisson density function with $\lambda = 5$

In the example given above, $\lambda = np = 500 \times (.00002) = .01$, so the probability of two people getting paper cuts from their finals is

$$e^{-.01}(.01)^2(1/2) = 4.95 \times 10^{-5}$$

Other examples where a Poisson distribution is applicable include the following:

- the number of novae in our galaxy in a given decade
- the number of movies to gross over 25 million dollars in a year
- the number of Ph.D. students who don't finish their dissertations on time
- the number of people who have bought this book who bought it in New York City

• CALCULATING THE EXPECTATION AND VARIANCE OF A POISSON RANDOM VARIABLE

We can calculate that the expectation of a Poisson random variable with this probability function:

$$\Pr(X = i) = e^{-\lambda} \frac{\lambda^i}{i!}$$

is equal to λ. This result makes sense since we let $\lambda = np$ when the Poisson distribution is used as an approximation to the binomial distribution. We can also find that $\mathrm{Var}(X) = \lambda$. The Poisson distribution has the very peculiar property that its expectation is equal to its variance.

YOU SHOULD REMEMBER

1. The random variable X is said to have a Poisson distribution with parameter λ if its probability function is given by the formula

$$\Pr(X = k) = e^{-\lambda} \frac{\lambda^k}{k!}$$

where $e = 2.71828\ldots$

2. The expectation and variance are both equal to λ:

$$E(X) = \lambda, \quad \mathrm{Var}(X) = \lambda$$

THE HYPERGEOMETRIC DISTRIBUTION

Now we will introduce a discrete random variable distribution with an intimidatingly long name: the hypergeometric distribution. However, we have actually seen its probability function already (in Chapter 4), before we knew what it was called. The hypergeometric distribution applies to this situation:

- There are N objects in the population.

- The population is divided into two types: M objects of type A and N-M objects of type B.

- A random sample of size n is chosen from this population.

- Let X be a random variable equal to the number of type A objects in the sample. Then X has the hypergeometric distribution with parameters N, M, and n.

The probability function is:

$$\Pr(X = i) = \frac{\binom{M}{i}\binom{N-M}{n-i}}{\binom{N}{n}}$$

There are some obvious restrictions that apply to this formula:

$$0 \le M \le N; \quad 0 \le n \le N; \quad 0 \le i \le n; \quad 0 \le i \le M$$

We will not formally derive the formula for the expected value, but we can intuitively predict what it will be. Suppose a population of 1,000 parts contains 50 defective parts. If we select a random sample of 100 parts, how many would you expect to be defective? Since 5 percent of the population is defective, we would guess 5 percent of the sample, or 5 parts, are likely to be defective.

In general, the expectation for a hypergeometric random variable is given by this formula:

$$E(X) = \frac{nM}{N}$$

If we let $p = M/N$, the proportion of items in the population of type A, then the expectation is:

$$E(X) = np$$

That formula looks familiar; it suggests there is a similarity between the hypergeometric distribution and the binomial distribution. Suppose that we chose our sample of 100 parts with replacement: that is, once we had randomly selected 1 part and recorded whether or not it was defective, we put it back in the population so there was a chance of selecting that same part again. In that case, X (the number of defective parts) has a binomial distribution with parameters n and p. The binomial distribution works when we sample with replacement because each selection is independent.

The hypergeometric distribution applies when we sample without replacement. In that case the selections are not independent.

The variance of the hypergeometric distribution is given by the formula:

$$n \left(\frac{M}{N}\right) \left(1 - \frac{M}{N}\right) \left(\frac{N-n}{N-1}\right)$$

Since $p = M/N$, the hypergeometric variance is:

$$np(1-p) \left(\frac{N-n}{N-1}\right)$$

This is the same as the variance of the binomial distribution except for the factor $(N-n)/(N-1)$. If N is much larger than n, then the fraction $(N-n)/(N-1)$ will be close to 1, and the hypergeometric variance will be close to the binomial variance. In that case, it doesn't matter whether the sample is selected with or without replacement. Therefore, if N is much larger than n, a binomial distribution with parameters n and $p = M/N$ can be used as an approximation to the hypergeometric distribution with parameters N, M, and n. This is helpful because the binomial distribution is easier to deal with. (Later, we will see that the binomial distribution can be approximated by the normal distribution.)

YOU SHOULD REMEMBER

1. Consider a population of N objects consisting of M objects of type A and $N - M$ objects of type B. Choose n objects at random from the population, and let X represent the number of type A objects in the sample of n objects. Then X is a random variable having a hypergeometric distribution with parameters n, N, and M.

2. The probability function is

$$\Pr(X = i) = \frac{\binom{M}{i}\binom{N - M}{n - i}}{\binom{N}{n}}$$

3. The expectation and variance for X can be found from these formulas:

$$E(X) = \frac{nM}{N} \qquad \mathrm{Var}(X) = n\,\frac{M}{N}\left(1 - \frac{M}{N}\right)\frac{N - n}{N - 1}$$

KNOW THE CONCEPTS

DO YOU KNOW THE BASICS?

Test your understanding of Chapter 7 by answering the following questions:

1. Provide an intuitive explanation for the formula that gives the mean of the binomial distribution.

2. Why is the binomial distribution appropriate when a sample is chosen with replacement?

3. Suppose a worker attempts 80 identical tasks each day. Is it appropriate to regard the number of successful tasks as coming from a binomial distribution?

4. Is it appropriate to use the binomial distribution to represent the number of free throws made by a basketball team in a single game?

5. Are the binomial probabilities symmetric? In other words, is the probability of j successes the same as the probability of j failures?

6. Consider the airline reservation example in the text. Do you think you should try to make the probability of an overflow decrease almost to zero?

For questions 7–9, consider a hypergeometric distribution with population of size N, sample of size n, and M objects from the population having the desired characteristic. Compare that with a binomial distribution with n trials and probability of success M/N.

7. In what circumstance will the binomial distribution be appropriate? When will the hypergeometric distribution be appropriate?

8. When will the binomial distribution and the hypergeometric distribution give almost the same results?

9. Will the variance of the hypergeometric distribution be greater than or less than the variance of the binomial distribution? Explain.

TERMS FOR STUDY
binomial distribution   Poisson distribution
hypergeometric distribution

PRACTICAL APPLICATION

COMPUTATIONAL PROBLEMS

1. Suppose that 1,000 meteorites hit the earth each year. What is the probability that the town of Wethersfield, Connecticut will be struck by two meteorites in 11 years? (See Chapter 4, Exercise 2.)

2. What is the probability of getting three primes in five rolls of a die?

3. In tossing a fair coin, what is the probability of getting at least four heads in five tosses?

4. Pennsylvania has a daily lottery. A three-digit number is chosen every night. What is the probability of getting a number less than 100 more than five times in 1 week?

5. You're hunting Moby Dick. Each day you send out one small boat with harpooners from your ship. (You never catch Moby Dick, just because Moby Dick *is* Moby Dick.) The probability is 2/3 that the small boat will be sunk on any particular day. You plan to hunt Moby Dick for 4 days. What is the probability that you will lose three or more small boats?

6. Assume that you've been given a 100-question true/false exam on a subject that you know nothing about. If you guess randomly, what is the probability of getting at least 75 answers correct?

7. How many times must you toss a fair coin for the probability to be greater than 1/2 that you will get at least two heads?

8. Assume that 10 percent of the population is left-handed. If three people are chosen at random, what is the probability that at least one will be left-handed?

9. What is the probability that two of the next three Presidents of the United States will have been born on a Sunday?

10. Assume that 2/5 of the population have O+ blood type. If you randomly choose six people, what is the probability that four of them are O+?

11. Assume that 45 percent of the Smiths in the world are women. If you randomly run into three Smith siblings, what is the probability that at least two are sisters?

12. Suppose you are running an insurance company. You have N customers. There is a probability $p = .05$ that any particular customer will file a claim in a year, in which case you have to pay C = $1,000. You collect a premium of $50 per year from each customer.
    (a) What is the expected value for your profits each year?
    (b) Suppose you have $N = 20$ customers. What is the probability that your profits will be $2,000? What is the probability they will be $1,000? 0? − $1,000? − $2,000? − $3,000?

13. Calculate the mean and variance of a hypergeometric random variable with parameters $N = 1,000$, $M = 300$, and $n = 25$.

14. Suppose that you pull 15 balls out of a jar containing 30 white balls and 15 black balls. How many white balls would you expect to pull out on the average?

15. Let X be a random variable representing the number of times the word *platypus* is said on a given day. Assume X has a Poisson distribution with parameter $\lambda = 1/2$. What is $Pr(X > 1)$?

16. What is the maximum value for $Pr(X = n)$ if X is a Poisson random variable with parameter $\lambda > 0$?

17. If X is a Poisson random variable with parameter $\lambda = 10$, what is $Pr(1 \leq X \leq 3)$?

18. Given that 20 books in a shipment of 200 books to a bookstore contain misprints, and you buy 3 of the 200 books, what is the probability that one of your books will contain a misprint?

19. You're dressing in the dark because of a power failure. You have two black socks and six red socks in a drawer. You pull out three socks. What is the probability that two are black?

20. Let X be a Poisson random variable with parameter $\lambda = 3$, representing the number of people who use a given dictionary in a given library on a given day. If $F(a)$ is the cumulative distribution function, what is $F(4)$?

21. A dictionary has 300 pages. What is the probability that, if you look up five words at random, two of them will be on pages with page numbers ending in zero? (Assume that the two words you look up are on different pages.)

ANSWERS

KNOW THE CONCEPTS

1. On average you would expect the number of successes to be equal to the number of trials times the probability of success.

2. Consider it a "trial" each time you select one item from the population. Replace each item after it has been selected, and mix the population items before the next selection. Then each selection is independent, and the probabilities are the same for all selections, so the binomial distribution is appropriate.

3. The trials may not all be independent. For example, the worker may be more tired near the end of the day. Therefore, the binomial distribution might not be appropriate.

4. The probability of success will be different for each player, so the binomial distribution is not appropriate.

5. The probabilities are symmetric if $p = .5$; otherwise they are not.

6. You may reduce the probability of an overflow by reducing the number of reservations, but this will reduce your revenue. When you make your decision, you must balance the risk of overflow with the need to have the planes as full as possible.

7. The binomial distribution is appropriate if you select a sample of size n with replacement from the population described. The hypergeometric distribution is appropriate if you select a sample of size n without replacement.

8. If you select a sample without replacement from a population that is much larger than the sample, then the probabilities from the hypergeometric distribution will be very similar to the probabilities from the binomial distribution.

9. The variance of the hypergeometric distribution will be smaller by the factor $(N - n)/(N - 1)$. The variance associated with the sample will be slightly smaller if you select the sample without replacement.

PRACTICAL APPLICATION

1. The number of meteorites that hit Wethersfield will have a binomial distribution with $p = .000\ 000\ 07$ and $n = 11,000$. The probability of two hits is 2.96×10^{-7}. Interestingly enough, the town of Wethersfield *was* hit by two meteorites in an 11-year period.

2. $\binom{5}{3}\left(\frac{1}{2}\right)^3\left(\frac{1}{2}\right)^2 = \frac{5}{16}$

3. $\binom{5}{4}\left(\frac{1}{2}\right)^4\left(\frac{1}{2}\right) + \binom{5}{5}\left(\frac{1}{2}\right)^5\left(\frac{1}{2}\right)^0 = \frac{3}{16}$

4. $\binom{7}{6}\left(\frac{1}{10}\right)^6\left(\frac{9}{10}\right) + \binom{7}{7}\left(\frac{1}{10}\right)^7\left(\frac{9}{10}\right)^0 = \dfrac{1}{156{,}250}$

5. $\binom{4}{3}\left(\frac{2}{3}\right)^3\left(\frac{1}{3}\right) + \binom{4}{4}\left(\frac{2}{3}\right)^4\left(\frac{1}{3}\right)^0 = \dfrac{16}{27}$

6. $\sum\limits_{k=75}^{100} \binom{100}{k} \times \dfrac{1}{2^{100}}$

9. $\binom{3}{2}\left(\frac{1}{7}\right)^2\left(\frac{6}{7}\right) = \dfrac{18}{343}$

7. 4

8. $1 - (.9)^3 = .271$

10. $\binom{6}{4}\left(\frac{2}{5}\right)^4\left(\frac{3}{5}\right)^2 = \dfrac{432}{3125}$

11. $\binom{3}{2}\left(\frac{9}{20}\right)^2\left(\frac{11}{20}\right) + \binom{3}{3}\left(\frac{9}{20}\right)^3\left(\frac{11}{20}\right)^0 = .425$

12. Let X = number of claims. $E(X) = Np$ expected value of profits $= 50N - 1000Np = 0$.

Profits	Claims	Probability
2,000	–	0
1,000	0	0.3585
0	1	0.3774
–1,000	2	0.1887
–2,000	3	0.0596
–3,000	4	0.0133

13. Mean = 7.5   Variance = 5.12

14. Mean = 10

15. $\Pr(X > 1) = 1 - (3/2)e^{-1/2}$

16. $\Pr(X = n) = e^{-\lambda}\lambda^n/n!$, where $n \le \lambda < n + 1$

17. $e^{-10}(680/3)$

18. $\dfrac{\binom{20}{1}\binom{180}{2}}{\binom{200}{3}} = .245$

20. $e^{-3}(131/8)$

19. $\dfrac{\binom{2}{2}\binom{6}{1}}{\binom{8}{3}} = .107$

21. $\dfrac{\binom{30}{2}\binom{270}{3}}{\binom{300}{5}} = .072$

8

THE NORMAL DISTRIBUTION AND RELATED DISTRIBUTIONS

KEY TERMS

central limit theorem a theorem that states that the average of a large number of independent, identically distributed random variables will have a normal distribution

continuous random variable a random variable that can take on any real-number value within a certain range; it is characterized by a probability function curve such that the area under the curve between two numbers represents the probability that the random variable will be between those numbers

normal distribution the most important continuous random variable distribution; its probability function is bell-shaped; many real populations are distributed according to the normal distribution

THE BELL-SHAPED CURVE

Suppose you make a graph of the probabilities of the numbers of heads you will expect to see if you repeatedly flip a coin 15 times. (See Figure 8-1.) Or suppose you select 1,000 people off the street and make a frequency diagram of their heights (Figure 8-2).

These graphs look similar. This bell-shaped curve, called the *normal curve*, is the most important curve in statistics. There are many examples of quantities that are distributed according to the normal curve:

- the height, weight, or IQ of a population

- the results of the measurement of a physical quantity, such as the molecular weight of a chemical

- the total that appears if you toss many dice

- the number of weekly customers at many businesses

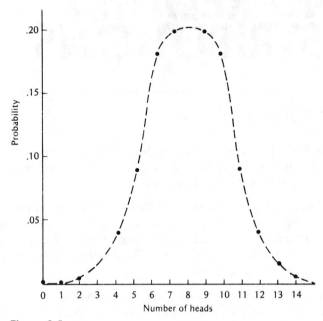

Figure 8-1

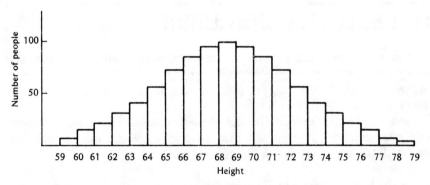

Figure 8-2

The normal distribution often applies in situations where extreme values are less likely to occur than moderate values. The normal distribution can also be used as an approximation to the binomial distribution, and it applies whenever you add together a bunch of identical, independently distributed random variables.

A normal random variable is not a discrete random variable, as were the random variables we discussed in Chapter 7. Instead, a normal random variable is an example of a different type of random variable called a **continuous random variable**.

CONTINUOUS RANDOM VARIABLES

Let us suppose that we randomly select a name from the phone book and then measure the height of the person selected. If H is the height in feet of the person, we can regard H as a random variable. However, it is different from the other random variables that we have discussed so far.

Suppose we try to list all of the possible values for H. There are clearly some values that are not possible. For example, H can never be less than 1/4 or greater than 9. However, we'll find that we can't list all of the possible values. The height might be 5 feet, or it might be 5.1 feet, or 5.00001 feet, or 5.000000001 feet. In fact, assuming that we can measure the height with perfect accuracy (oh, well, this is the theory—not the real world), there is an infinite number of possible values for the height. A discrete random variable cannot be used in a case like this where the result can be any number in a particular range. Instead, we need to use a *continuous random variable*.

Examples of continuous random variables include the following:

- the height above the floor at the point where a dart hits a dart board
- the length of time until a light bulb burns out
- the length of time until a radioactive atom decays
- the length of the life of a person

Discrete random variables are easier to understand intuitively. However, continuous random variables are usually easier to handle mathematically. If a discrete distribution has many possible values that are close together, then it can usually be approximated by a continuous distribution.

YOU SHOULD REMEMBER

1. The most important curve in statistics is the bell-shaped, or normal, curve, which shows the distribution of such quantities as the height of a population or the number of "heads" that appear if a coin is flipped many times.

2. A continuous random variable can take on any real-number value within a certain range. Examples include the length of time until an automobile battery "dies" and the height to which a beginning ballet student can leap.

CONTINUOUS CUMULATIVE DISTRIBUTION FUNCTIONS

Now we have to figure out how to describe the behavior of continuous random variables. There are many similarities between discrete random variables and continuous random variables, but there are also some important differences. We can estimate the probability that the person we select from the phone book will have a height less than 6 feet. Or we could calculate the probability that the person will have a height greater than 50 feet (which is, of course, zero). Therefore we can define a cumulative distribution function for a continuous random variable, just the same as we did for a discrete random variable. We'll use a capital letter, such as F, to stand for a cumulative distribution function, so we can make the definition

$$F(a) = \Pr(X \le a)$$

where X is the random variable we are discussing.

A continuous cumulative distribution function satisfies the same requirements that we found for a discrete cumulative distribution function:

1. $F(a)$ is always between 0 and 1.

2. As a becomes very large, $F(a)$ approaches 1.

3. As a becomes very small (approaches minus infinity), $F(a)$ approaches 0.

4. $F(a)$ is never decreasing.

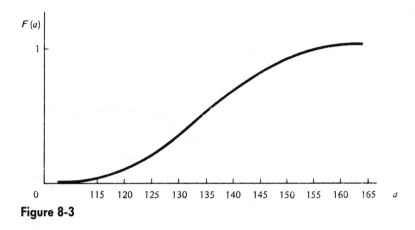

Figure 8-3

Two important practical properties are as follows:

If we want to find the probability that X will be greater than a particular value a, we can use the formula

$$\Pr(X > a) = 1 - \Pr(X < a) = 1 - F(a)$$

If we want to find the probability that X will be between two particular values b and c, we can use the formula

$$\Pr(b < X < c) = \Pr(X < c) - \Pr(X < b) = F(c) - F(b)$$

Warning: Do *not* use these two formulas for discrete random variables. They work only for continuous random variables.

Figure 8-3 shows a cumulative distribution function for the heights of a group of people.

One example of a continuous random variable is a uniform random variable, that is, a random variable that is equally likely to take on any value within a particular interval. For example, let's consider the random variable Y that has an equal chance of taking any value between 0 and 3. Then the probability that Y will be less than 1 is 1/3, the probability that Y will be between 1 and $1\frac{1}{2}$ is 1/6, and so on. If we make a graph of the cumulative distribution function for Y, it looks like the function shown in Figure 8-4.

CONTINUOUS PROBABILITY DENSITY FUNCTIONS

Now, let's find out the probability that Y will be exactly equal to 2. Any number from 0 to 3 has an equal chance of being selected, so if we let N be the number

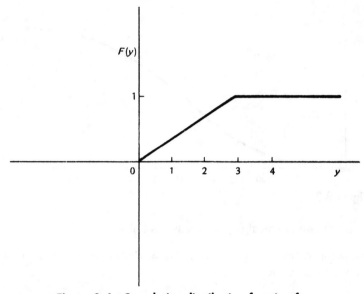

Figure 8-4. Cumulative distribution function for uniform continuous random variable

of numbers from 0 to 3, then $\Pr(Y = 2) = 1/N$. However, there is an infinite number of numbers between 0 and 3 (for example, 0.01, 0.011, 0.0111, 0.01111, and so on). This means that

$$\Pr(Y = 2) = \frac{1}{\infty} = 0$$

This property holds in general for continuous random variables: The probability that *any* continuous random variable will take on *any* specific precise value is zero!

Therefore, we can't define a probability function for a continuous random variable in the same way that we defined the probability function for a discrete random variable. We need to develop a new approach. To start with, remember that there is a connection between the probability function for a discrete random variable and the frequency diagram for a sample. So we'll start by drawing a frequency diagram for the weights of people in a particular sample. (See Figure 8-5.) Note that the height of each bar represents the number of people whose weights are between two specified values. For example, the height of the bar between 140 and 145 pounds is the number of people in the sample whose weights are between 140 and 145. We'll call the width of each bar Δx (Δ is the uppercase Greek letter delta). (In this case, $\Delta x = 5$.)

We can, by analogy, draw the graph of a function such that the height of the function in any given interval is equal to the probability that the random

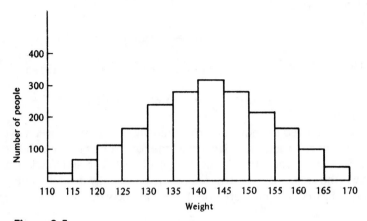

Figure 8-5

variable will have a value within that interval. It turns out to be more convenient to make the height of each bar equal to the probability of being in that interval divided by Δx, the width of the bar. (See Figure 8-6.)

Then:

$$\text{Height of bar between } a \text{ and } a + \Delta x = \frac{\Pr\left(a < X < a + \Delta x\right)}{\Delta x}$$

This function is a **density function**.

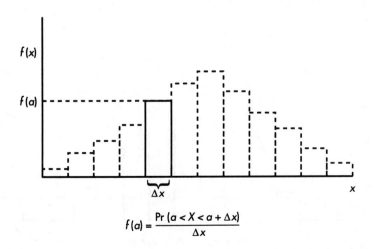

$$f(a) = \frac{\Pr\left(a < X < a + \Delta x\right)}{\Delta x}$$

Figure 8-6

And, therefore,

$$\Pr(a < X < a + \Delta x) = \text{ height of bar} \times \Delta x$$

We'll let $f(a)$ stand for the height of the bar from a to $a + \Delta x$, so

$$\Pr(a < X < a + \Delta x) = f(a) \, \Delta x$$

Suppose we need to know the probability that X will be between two values a and b. We need to add the heights of all the bars from a to b and then multiply by Δx. However, since $f(x)$ is the height of each bar and Δx is the width, $f(x) \, \Delta x$ is the area of the bar. Therefore, the probability that X will be between a and b is just equal to the area of all the bars between a and b. (See Figure 8-7.)

$$\Pr(\dot{a} < X < b) = \text{area of all of the rectangles between } a \text{ and } b$$

• DEFINITION OF THE DENSITY FUNCTION FOR A CONTINUOUS RANDOM VARIABLE

This is the basic defining feature of the density function for a continuous random variable: The area under the function between two values is the probability that the random variable will be between those two values. However, the bar diagram is only an approximate representation of the density function for a continuous random variable. We can get a better approximation of the true nature of the continuous random variable by making the bars narrower and narrower. When the bars become very, very narrow, the density function looks like a smooth curve. (See Figure 8-8.)

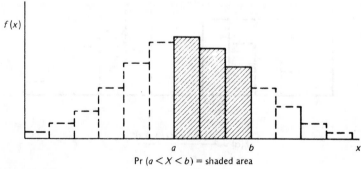

Pr $(a < X < b)$ = shaded area

Figure 8-7

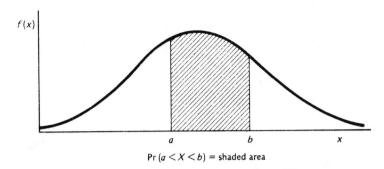

$$\Pr(a < X < b) = \text{shaded area}$$

Figure 8-8

We will make this definition: The function $f(x)$ is a density function for the random variable X if it satisfies the property that the area under the curve $y = f(x)$, to the left of the line $x = b$, to the right of the line $x = a$, and above the x axis, is equal to $\Pr(a < X < b)$. (Remember that capital letters represent random variables and small letters represent ordinary variables.)

We know that, if $F(x)$ is the cumulative distribution function, then

$$\text{Area under } f(x) \text{ from } a \text{ to } b = \Pr(a < X < b) = F(b) - F(a)$$

Now, suppose we look at the interval from minus infinity to plus infinity. We know that $F(+\infty) - F(-\infty) = \Pr(-\infty < X < \infty) = 1$, since the value of X must be somewhere between $-\infty$ and $+\infty$. (It doesn't have any other choice.) This means that

$$\text{Area under } f(x) \text{ from } -\infty \text{ to } +\infty = 1$$

In other words, the total area under the function $f(x)$ must be equal to 1. If $f(x)$ doesn't have this property, then it can't be a legitimate probability density function. We can show that this condition is met for the density function of the uniform variable Y. (See Figure 8-9.)

To calculate the probability that a continuous random variable will be between two numbers it is necessary to calculate the area under the density function between those two numbers. Calculating the area under a curve requires the use of a calculus technique called *integration*. However, even if you know calculus it is not possible to find simple formulas for the areas under many of the curves used in statistics, so it is necessary to look in a table to find the area. Later we will show how to use a table to find the area under the normal density function.

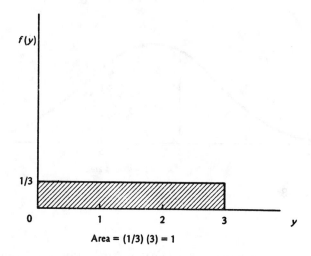

Area = (1/3) (3) = 1

Figure 8-9. Density function for uniform continuous random variable

• *EXPECTATION AND VARIANCE OF A CONTINUOUS RANDOM VARIABLE*

The expectation (or mean) of a continuous random variable is the average value that would appear if that random variable was observed many times. (We will not describe how to calculate the expectation of a continuous random variable in this book.) If X is a continuous random variable and we know $E(X)$ and $E(X^2)$, we can calculate the variance in the same way we did for a discrete random variable:

$$\text{Var}(X) = E(X^2) - (E(X))^2$$

YOU SHOULD REMEMBER

1. The density function for a continuous random variable is a function such that the area under the function and between two numbers is the probability that the random variable will be between these two numbers.

2. The function must always be positive, and the total area under the function must equal 1.

THE NORMAL DISTRIBUTION

• *PROPERTIES OF THE NORMAL DISTRIBUTION*

The normal distribution is an example of a continuous random variable distribution. In fact, there are many different normal distributions. A normal distribution can be identified by specifying two numbers: the mean and the variance (or standard deviation). The mean is located at the peak of the distribution. The variance determines the shape of the distribution—whether it will be spread out or whether most of the area will be concentrated near the peak. If X is a normal random variable with mean μ and variance σ^2, then the density function is given by this equation:

$$f(x) = \frac{1}{\sqrt{2\pi}\,\sigma}\, e^{-(1/2)[(x\ -\ \mu)/\sigma]^2}$$

where e is a special number approximately equal to 2.71828, and π is a special number about equal to 3.14159. As before, the standard deviation σ is the square root of the variance. σ is the distance from the mean to either of the points of inflection—the boundary points where the curvature changes from facing down to facing up. (See Figure 8-12.)

Figure 8-10 shows four normal distributions that have the same variance but different means.

Figure 8-11 shows four normal distributions that have the same mean but different variances. The area under each of these curves is 1, as it must be for a continuous probability distribution. The upper and lower ends of the curve are called the *tails* of the distribution. The tails never actually touch the axis. Therefore, the normal random variable could conceivably take on any value, including values far away from the mean. However, you can see that there is not very much area under the curve out in the tails, so the chance of the observed value appearing far from the mean is remote.

YOU SHOULD REMEMBER

1. The normal distribution is a common random variable distribution that arises in many situations where extreme values are less likely than moderate values.

2. The graph of the density function of a normal random variable forms a bell-shaped curve with its peak located at the mean (μ).

3. The variance (σ^2) determines the shape of the curve; a larger value of the variance means that the curve will be spread out more.

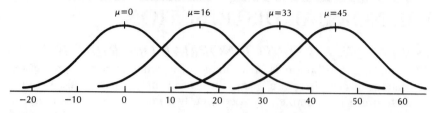

Figure 8-10

• *ADDITION PROPERTY OF NORMAL RANDOM VARIABLES*

An important property of a normal random variable is the addition property. If X is a normal random variable with mean μ and variance σ^2, and $Y = aX + b$, where a and b are two constants, then Y has a normal distribution with mean $a\mu + b$ and variance $a^2\sigma^2$.

Also, suppose X and Y are two independent random variables with normal distributions. (Two random variables are independent if they don't affect each other. We'll define exactly what independence means in Chapter 9.) Suppose that

$$E(X) = \mu_x, \quad \text{Var}(X) = \sigma_x^2, \quad E(Y) = \mu_y, \quad \text{and} \quad \text{Var}(Y) = \sigma_y^2.$$

If we form a new random variable by adding these two together, $V = X + Y$, then V will also have a normal distribution. (We already know that $E(V) = \mu_x + \mu_y$ and $\text{Var}(V) = \sigma_x^2 + \sigma_y^2$.)

For example, suppose you decide to enter the hamburger business by opening restaurants at two different locations. The number of hamburgers that you

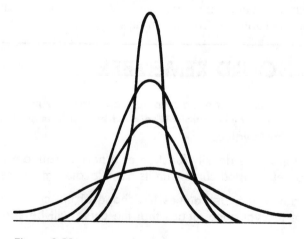

Figure 8-11

sell each day at the downtown location is given by a normal random variable with mean 200 and variance 1,600. The number of hamburgers sold at the suburban location has a normal distribution with mean 100 and variance 400. Then the total number of hamburgers that you will sell at both restaurants has a normal distribution with mean 300 and variance 2,000.

To find the probabilities associated with the normal distribution we must be able to calculate the area under the curve. Consider a normal random variable X with mean $\mu = 20$ and standard deviation $\sigma = 5$. (See Figure 8-12.) If we need to find the probability that X will be greater than 20, we need to find the area under the curve above 20. Because the distribution is symmetric, the total area under the curve above the mean is the same as the total area below the mean. Therefore, the probability that X will be greater than 20 is .5. Likewise, the probability that any normal random variable will be greater than its mean is equal to .5.

• *THE STANDARD NORMAL DISTRIBUTION*

In most cases when we need to know the area under a normal curve we will need to look in a table, or use a computer or calculator. It would be impossible to prepare a separate table for every normal distribution with every possible mean and every possible variance. Fortunately, we can find the results for any normal distribution by looking at a table for the normal distribution that has mean $\mu = 0$ and variance $\sigma^2 = 1$. This special normal distribution is called the **standard normal distribution**.

Suppose that Z is a random variable with a standard normal density function. Figure 8-13 shows a graph of the density function for Z. Suppose also that we need to know the probability that Z is between 0 and 1. Then we need to calculate the area under the curve between 0 and 1. (See Figure 8-14.)

Unfortunately, there is no simple formula that tells us what this area is. We have to look up the results in a table such as Table A3-1 at the back of the book.

The table gives the cumulative distribution function, which tells you the probability that Z will be less than a particular value, z. [The Greek uppercase letter Φ (phi) is often used to represent this function. $\Phi(a)$ equals $\Pr(Z < a)$.] The function satisfies this property: $\Phi(a) = 1 - \Phi(a)$. (See Figure 8-15).

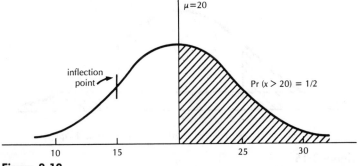

Figure 8-12

The probability that Z will be between any two numbers a and b can be found from the formula

$$\Pr(a < Z < b) = \Phi(b) - \Phi(a)$$

We have already figured out that $\Phi(0) = .5$. We can see from the table that $\Phi(1) = .8413$. Therefore, the probability that Z will be between 0 and 1 is $.8413 - .5000 = .3413$.

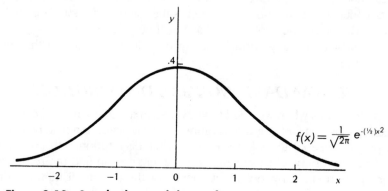

Figure 8-13. Standard normal density function: $\mu = 0$, $\sigma = 1$.

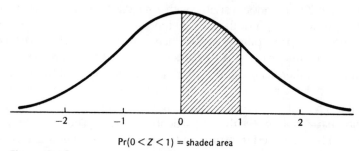

$\Pr(0 < Z < 1)$ = shaded area

Figure 8-14

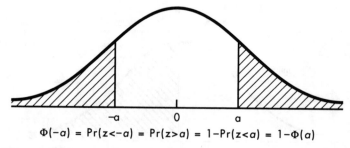

$\Phi(-a) = \Pr(z < -a) = \Pr(z > a) = 1 - \Pr(z < a) = 1 - \Phi(a)$

Figure 8-15

Because of the symmetry of the density function, we can see that there is also a .3413 probability that Z will be between –1 and 0. (See Figure 8-16.) We can add these two probabilities together:

$$Pr(-1 < Z < 0) + Pr(0 < Z < 1) = .3413 + .3413$$
$$Pr(-1 < Z < 1) = .6826$$

Therefore, there is a 68 percent chance that a standard normal random variable will be between -1 and 1. In other words, there is a 68 percent chance that the value of a standard normal random variable will be within one standard deviation of its mean. (In this case the mean is 0 and the standard deviation is 1.)

This particular property also holds for *any* normal random variable, regardless of its mean and standard deviation: there is a 68 percent chance that any normal random variable will be within one standard deviation of its mean. For example, if X is a normal random variable with mean 200 and standard deviation 30, then there is a 68 percent chance that X will be between 170 and 230.

We can also use the table to show that there is a 95 percent chance that Z will be between –1.96 and 1.96. In general, we can say that any normal random variable has a 95 percent chance of being less than about two standard deviations away from its mean. (See Figure 8-17.)

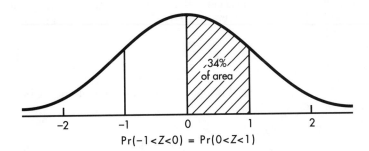

Pr(-1<Z<0) = Pr(0<Z<1)

Figure 8-16

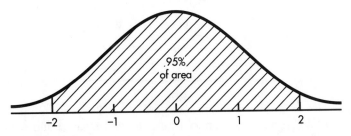

Figure 8-17

YOU SHOULD REMEMBER

1. Any normal random variable has a 68 percent chance of being within one standard deviation of its mean.

2. Any normal random variable has a 95 percent chance of being within two standard deviations of its mean.

It will often be helpful to know the probability that a standard normal random variable will be between a and $-a$, where a is a particular number. To make things more convenient, Table A3-2 at the back of the book lists these values. For example, the table shows that there is a .3830 probability that Z will be between -0.5 and 0.5.

The value of the standard normal random variable could conceivably be anything, since the density function never quite touches the axis. There is no number k such that $\Pr(Z > k) = 0$. However, you can see from Table A3-1 that there is only a .0002 probability that Z will be greater than 3.5. Larger values are even less likely, so we don't have to worry too much about the likelihood that Z might take on extreme values.

We can use the standard normal tables to find the probabilities for any normal random variable by using the following trick: Suppose Y has a normal distribution with mean 6 and variance 9, and we need to know the probability that Y will be between 5 and 8. We can create the random variable Z:

$$Z = \frac{Y - 6}{3}$$

which will have a normal distribution with mean 0 and variance 1 because of the addition property.

Now:

$$\Pr(5 < Y < 8) = \Pr\left(\frac{5-6}{3} < \frac{Y-6}{3} < \frac{8-6}{3}\right)$$

$$= P(-1/3 < Z < 2/3)$$

$$= \Phi(.6667) - \Phi(-.3333)$$

$$= .7486 - (1 - .6293)$$

$$= .7486 - .3707$$

$$= .3779$$

In general, if X is a normal random variable with mean μ and variance σ^2, then $(X - \mu)/\sigma$ is a standard normal random variable. For example, suppose we would like to know the probability that you will sell more than 230 hamburgers at your downtown store. Let X_1 represent the number of hamburgers. In this case $\mu = 200$, $\sigma^2 = 1{,}600$, and $\sigma = 40$. Let's create the standard normal random variable Z_1:

$$Z_1 = \frac{X_1 - 200}{40}$$

$$\Pr(X_1 > 230) = \Pr\left(\frac{X_1 - 200}{40} > \frac{230}{40}\right) = \Pr\left(Z > \frac{3}{4}\right)$$

If X_1 is greater than 230, then Z_1 is greater than $3/4$. Table A3-1 tells us that the probability of this occurring is $1 - \Pr(Z_1 < .75) = 1 - .7734 = .2266$.

Now, suppose we would like to know the probability that you will sell a total of more than 330 hamburgers at the two locations. Let X be the total number of hamburgers. Then $\mu = 300$, $\sigma^2 = 2{,}000$, and $\sigma = 44.72$. Set up the standard normal random variable:

$$Z = \frac{X - 300}{44.72}$$

$$\Pr(X > 330) = \Pr\left(\frac{X - 300}{44.72} > \frac{330 - 300}{44.72}\right) = \Pr(Z > .67)$$

If $X > 330$, then $Z > .67$, and the probability of this happening is $.2514$.

YOU SHOULD REMEMBER

1. A standard normal random variable is a normal random variable with mean 0 and variance 1. Table A3-1 in the back of the book can be used to calculate the probability that a standard normal random variable will be between two numbers.

2. The standard normal distribution can also be used to find the probabilities for any normal random variable. If X is a normal random variable with mean μ and variance σ^2, then $(X - \mu)/\sigma$ has a standard normal distribution.

3. The random variable $Z = (X - \mu)/\sigma$ has a standard normal distribution. The function $\Phi(a)$ is the cumulative distribution function of Z:

$$\Phi(a) = \Pr(Z < a)$$

This function satisfies these properties:

$$\Phi(0) = .5$$

$$\Phi(-a) = 1 - \Phi(a)$$

$$Pr(Z > b) = 1 - \Phi(b)$$

$$Pr(a < Z < b) = \Phi(b) - \Phi(a)$$

$$Pr(a < X < b) = \Phi\left(\frac{b - \mu}{\sigma}\right) - \Phi\left(\frac{a - \mu}{\sigma}\right)$$

4. If you multiply a normal random variable by a constant, then the result will still be a normal random variable. If you add together two independent normal random variables, then the result will still be a normal random variable.

5. In general, if X_1 has a normal distribution with mean μ_1 and variance σ_1^2, and X_2 has a normal distribution with mean μ_2 and variance σ_2^2, then this random variable:

$$c_1 X_1 + c_2 X_2 + c_3$$

will have a normal distribution with mean $(c_1\mu_1 + c_2\mu_2 + c_3)$ and variance $(c_1^2\sigma_1^2 + c_2^2\sigma_2^2)$, where c_1, c_2, and c_3 can be any constants.

CENTRAL LIMIT THEOREM

Now we will present an amazing property of the normal distribution. Suppose X is a random variable with any distribution. X could have a strange distribution that does not look at all like the normal distribution, and it can be either discrete or continuous. As usual, let $E(X) = \mu$ and $Var(X) = \sigma^2$. Take a random sample of size n from this distribution: $X_1, X_2, X_3, \ldots, X_n$. Let $\bar{x}$ equal the average of all of these numbers. In Chapter 6 we found that $E(\bar{x}) = \mu$ and $Var(\bar{x}) = \sigma^2/n$. Now for the amazing part:

As the sample size becomes very large, the distribution of $\bar{x}$ will be approximately the same as the normal distribution.

This result is called the **central limit theorem**. The astonishing feature of this theorem is that it applies to any random variable. (If X happens to have a normal distribution to begin with, then $\bar{x}$ will have a normal distribution for any sample size.)

We can establish two very simple corollaries. If we just add all of the x's together, the result will also have a normal distribution, this time with mean $n\mu$ and variance $n\sigma^2$. Or, if we calculate the random variable

$$Z = \frac{(\bar{x} - \mu)}{\sigma/\sqrt{n}} = \frac{\sqrt{n}\,(\bar{x} - \mu)}{\sigma}$$

then Z will have a standard normal distribution.

Since the binomial distribution can be thought of as the sum of a group of independent two-valued random variables, the central limit theorem says that the binomial probability function begins to look like the normal distribution as n becomes large. Note that this works even though the binomial distribution is a discrete distribution and the normal distribution is a continuous distribution. Table 8-1 illustrates what happens as n increases, and the binomial density function is illustrated in Figure 8-18, pages 168 and 169.

Table 8-1. Comparison of Normal and Binomial Probability Density Functions

	Pr(X = k)		p = .5					
	n = 6		n = 10		n = 20		n = 30	
k	Binomial	Normal	Binomial	Normal	Binomial	Normal	Binomial	Normal
1	.0938	.0859	.0098	.0103	—	.0001	—	—
2	.2344	.2334	.0439	.0417	.0002	.0003	—	—
3	.3125	.3257	.1172	.1134	.0011	.0013	—	—
4	.2344	.2334	.2051	.2066	.0046	.0049	—	—
5	.0938	.0859	.2461	.2523	.0148	.0146	.0001	.0002
6			.2051	.2066	.0370	.0360	.0006	.0007
7			.1172	.1134	.0739	.0725	.0019	.0020
8			.0439	.0417	.1201	.1196	.0055	.0056
9			.0098	.0103	.1602	.1614	.0133	.0132
10					.1762	.1784	.0280	.0275
11					.1602	.1614	.0509	.0501
12					.1201	.1196	.0806	.0799
13					.0739	.0725	.1115	.1116
14					.0370	.0360	.1354	.1363
15					.0148	.0146	.1445	.1457

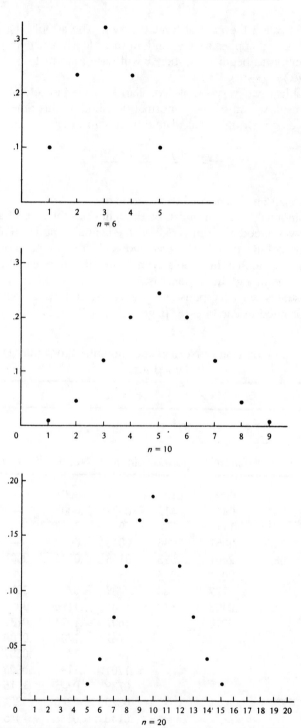

Figure 8-18 (continued)

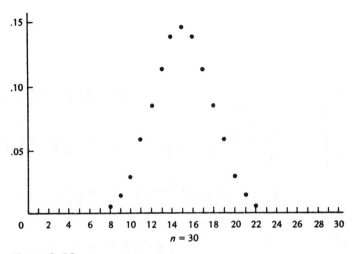

Figure 8–18

Of course, there are some limitations to the normal approximation. A normal random variable might be anywhere between minus infinity and plus infinity, whereas a binomial random variable must be between 0 and n. However, the probability that a normal random variable will take on such extreme values is very small, so there is no cause for worry. In practice, the normal approximation is all right when $n > 30$, provided that $np > 5$ and $n(1 - p) > 5$.

For another example, consider the probability function for the number that is the result of the role of a die. (See Table 8-2 and Figure 8-19.)

This function doesn't look at all like the normal distribution. However, if you roll two dice, the function for the result looks a bit more like the normal distribution, and if you increase the number of rolls, the function looks more and more like the normal density function.

YOU SHOULD REMEMBER

1. The central limit theorem states that the average of a large number of independent, identically distributed random variables will have a normal distribution.

2. This theorem applies to any random variable distribution.

Table 8-2

Number of Dice:	1		2		3		4		5	
Number on Dice	Number of Ways	Probability	Number of Ways	Probability	Number of Ways	Probability	Number of Ways	Probability	Number of Ways	Probability
1	1	.167								
2	1	.167	1	.028						
3	1	.167	2	.056	1	.005				
4	1	.167	3	.083	3	.014	1	.001		
5	1	.167	4	.111	6	.028	4	.003		∴
6	1	.167	5	.139	10	.046	10	.008	1	.001
7			6	.167	15	.069	20	.015	5	.002
8			5	.139	21	.097	35	.027	15	.005
9			4	.111	25	.116	56	.043	35	.009
10			3	.083	27	.125	80	.062	70	.016
11			2	.056	27	.125	104	.080	126	.026
12			1	.028	25	.116	125	.096	205	.039
13					21	.097	140	.108	305	.054
14					15	.069	146	.113	420	.069
15					10	.046	140	.108	540	.084
16					6	.028	125	.096	651	.095
17					3	.014	104	.080	735	.100
18					1	.005	80	.062	780	.100
19							56	.043	780	.095
20							35	.027	735	.084
21							20	.015	651	.069
22							10	.008	540	.054
23							4	.003	420	.039
24							1	.001	305	.026
25									205	.016
26									126	.009
27									70	.009
28									35	.005
29									15	.002
30									5	.001
									1	∴

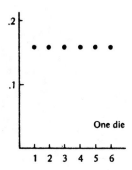

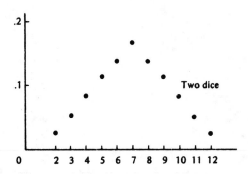

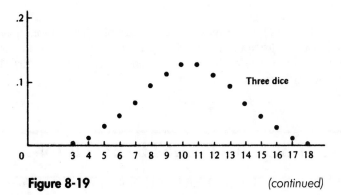

Figure 8-19   (continued)

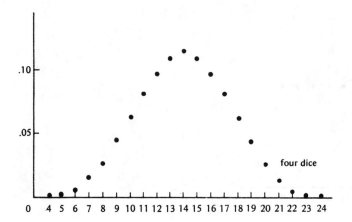

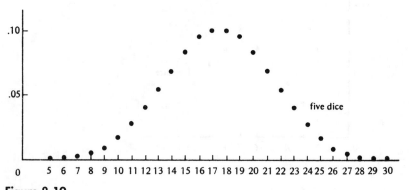

Figure 8-19

* * * * *

We will now discuss three more random variable distributions that are related to the normal distribution: the chi-square distribution, the *t* distribution, and the *F* distribution. These distributions seem rather esoteric at first, but they turn out to be important tools in statistics.

THE CHI-SQUARE DISTRIBUTION

Suppose Z is a standard normal random variable (that is, it has mean 0 and variance 1). Then suppose that $\chi = Z^2$. (χ is the Greek letter chi.) This means that χ will also be a continuous random variable. Its probability density function is shown in Figure 8-20.

The mean of this random variable is easy to calculate. Since Z has a standard normal distribution, by definition $E(Z) = 0$ and $\text{Var}(Z) = 1$. Since $\text{Var}(Z) = E(Z^2) - E(Z)^2 = E(Z^2)$, it follows that $E(Z^2) = E(\chi) = 1$. The variance of χ turns out to be 2.

The variable χ is, strictly speaking, called a **chi-square random variable** with one **degree of freedom**. Suppose we square many independent normal random variables and then take their sum, calling the result χ_n:

$$\chi_n = Z_1^2 + Z_2^2 + \cdots + Z_n^2$$

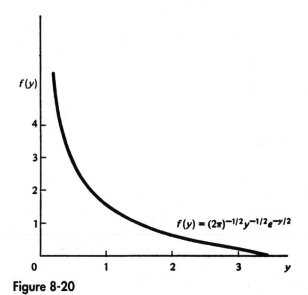

Figure 8-20

Then χ_n is said to have the chi-square distribution with n degrees of freedom. You're probably wondering why we use the term "degrees of freedom." You can think of it this way. Each of the normal random variables acts like a number that you can choose freely, so since you have n of these numbers, it's as if you have n different free choices that you can make.

Here are formulas for the expectation and variance:

$$E(\chi_n) = n$$
$$\text{Var}(\chi_n) = 2n$$

Figure 8-21 shows several different chi-square distributions with different degrees of freedom.

When the number of degrees of freedom is small, the density function is severely asymmetric. As the number of degrees of freedom increases, the den-

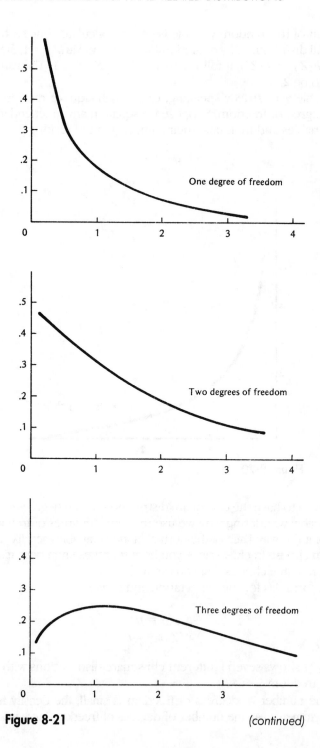

Figure 8-21 (continued)

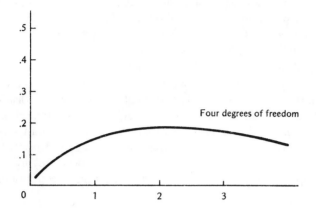

Figure 8-21

sity function gradually becomes more and more symmetric. As *n* becomes very large, the chi-square distribution begins to resemble a normal distribution. (This follows from the central limit theorem.)

Unfortunately, there is no simple expression for the cumulative distribution function for a chi-square random variable. The only way to calculate values of $F(y)$ is to use a computer. Table A3-3 lists some results.

The chi-square random variable is very important in statistical estimation. For example, the distribution of the sample variance of a random sample drawn from the normal distribution will be closely related to a chi-square distribution. (See Chapter 11.) Also, this distribution provides the basis for an important statistical test known as the chi-square test. (See Chapter 13.)

THE *t* DISTRIBUTION

Another important distribution related to the normal distribution is the *t* distribution. Suppose that Z and Y are independent random variables. Let Z be a standard normal random variable (mean 0, variance 1) and Y a chi-square variable with m degrees of freedom. Let us make the definition

$$T = \frac{Z}{\sqrt{Y/m}}$$

Then it is said that the variable T has the *t distribution with m degrees of freedom*. (The *t* distribution is sometimes called *Student's distribution*.)

The distribution is not just for students; Student was the pseudonym used by William Gossett when he first wrote about the distribution in 1908. He was a statistician employed by Guinness Brewing Company who was fearful of losing his job if he published under his own name.

Figure 8-22 shows a sample *t* density function.

The density function has a bell shape that is roughly similar to the standard normal distribution. Like the standard normal distribution, a *t* distribution is symmetric about its peak at 0. In general, though, the *t* distribution has thicker tails than the normal distribution. In other words, a *t* random variable has a higher chance of being far from 0 than does a standard normal random variable. However, as the number of degrees of freedom (*m*) increases, the *t* distribution approaches very close to the standard normal distribution.

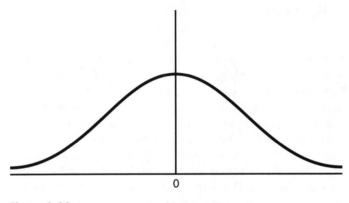

0

Figure 8-22

THE *F* DISTRIBUTION

Now we will mention one other distribution related to the chi-square distribution that has important uses in statistics. If X and Y are independent chi-square random variables with degrees of freedom m and n, respectively, then the random variable

$$F = \frac{X/m}{Y/n}$$

is said to have the *F* distribution *with m and n degrees of freedom*. Note that the order of m and n makes a big difference. Table A3-6 lists some values for the cumulative distribution function.

YOU SHOULD REMEMBER

1. The chi-square, t, and F distributions are distributions related to the normal distribution that are used extensively in statistical inference.

2. In order to identify a specific chi-square or t distribution you must specify the value of the *degrees of freedom*.

3. In order to identify a specific F distribution you must specify the degrees of freedom for the numerator and the degrees of freedom for the denominator.

KNOW THE CONCEPTS

DO YOU KNOW THE BASICS?

Test your understanding of Chapter 8 by answering the following questions:

1. Would it be appropriate to represent the weekly sales totals for an automobile dealer by a normal distribution?

2. If the heights of people in a population have a normal distribution with mean 66 inches and standard deviation 5 inches, is it possible that someone in the population has a height of 86 inches?

3. When can the normal distribution be used as an approximation to the binomial distribution? Why is that approximation appropriate? What are the advantages in making that approximation?

4. Can you explain intuitively why the central limit theorem is true?

5. If you add together two random variables that have normal distributions, will the result have a normal distribution?

6. Why must the area under the density function for a continuous random variable always be equal to 1?

7. Why is it impossible for the cumulative distribution function for a continuous random variable to ever by decreasing?

For questions 8 to 10, let Z be a standard normal random variable.

8. Is $\Pr(Z < 1)$ the same as the $\Pr(Z \leq 1)$?

9. Why is $\Pr(Z < -a) = \Pr(Z > a)$?

10. Why is $\Pr(Z > a) = 1 - \Pr(Z < a)$?

TERMS FOR STUDY

central limit theorem	degrees of freedom	standard normal
chi-square distribution	F distribution	distribution
continuous random variable	normal distribution	t distribution

PRACTICAL APPLICATION

COMPUTATIONAL PROBLEMS

1. Suppose the annual rainfall in a city has a normal distribution with mean 40 and standard deviation 5. What is the probability that the city will get less than 33 inches of rain next year? What is the probability that the city will get more than 38 inches of rain?

2. Suppose that the score that a student will get on an entrance exam is a random variable selected from a normal distribution with mean 550 and variance 900. If you need a score of 575 to get into a certain college, what is the probability that you will get in? If instead you need a score of 540, what is the probability that you will get in?

3. You're coach of a football team that faces a third-down situation with 4 yards needed for a first down. If you select a play involving a runoff tackle, the number of yards you will gain on the play is given by a normal random variable with mean 2.5 and standard deviation 1. What is the probability that you will make the first down if you run this play?

4. Consider the same situation as in Exercise 3. Another play you might run is a tricky end-around double reverse. The results of this play are given by a normal random variable with mean 3 and variance 6. What is the probability that you will make the first down if you run this play?

5. Suppose you are measuring the speed of light. The result of your measurement is given by a normal random variable whose mean is the true value and whose standard deviation is 5×10^9 centimeters per second. What is the probability that your measurement will be within 2×10^9 centimeters per second of the true value?

6. Suppose you are running a lemonade stand. The number of glasses of lemonade that you sell each day is given by a normal random variable with mean 15 and standard deviation 10. What is the probability that you will sell at least 120 glasses of lemonade in a week (7 days)? What is the probability that you will sell at least 100 glasses? Is it all right to represent a variable such as the number of glasses of lemonade as a normal random variable?

In Exercises 7–15, let X be a normal random variable with parameters μ and σ^2, density function $f_X(x)$, and cumulative distribution function $F_X(a)$. Use Table A3-1 to calculate $\Phi(x)$.

7. If $\mu = 0$ and $\sigma^2 = 100$, what is $\Pr(5 < X < 10)$?

8. If $\mu = -3$ and $\sigma^2 = 9$, and $F_X(a) = .6$, what is a?

9. If $\mu = 0$ and $F_X(5) = .8$, what is σ^2?

10. If $\mu = 3$, why can't $F_X(4) = .4$?

11. If $\mu = 73$ and $\sigma^2 = 81$, what is $Pr(|X| > 100)$?

12. If $\mu = 25$ and $\sigma^2 = 100$, what is $Pr(X = 25)$?

13. If $\mu = 1$ and $\sigma^2 = 64$, for what values of a is $.1 < F_X(a) < .3$?

14. If $f_X(X)$ takes a maximum value of 5 at $x = 10$, what are μ and σ^2?

15. If X is a normal random variable with mean μ_1 and standard deviation σ_1, and Y is a normal random variable with mean μ_2 and standard deviation σ_2, what is $Pr(Y > X)$?

16. Suppose you have your choice between two jobs. Your annual earnings from an industrial job will have a normal distribution with mean $15,000 and standard deviation $2,000. Your annual earnings from a traveling sales job will have a normal distribution with mean $12,000 and standard deviation $10,000. What is the probability that you would earn more from the traveling sales job?

17. Suppose you manage a bank where the amounts of daily deposits and daily withdrawals are given by independent random variables with normal distributions. For deposits, the mean is $12,000 and the standard deviation is $4,000; for withdrawals, the mean is $10,000 and the standard deviation is $5,000. For a particular day, calculate the probability of each of the following events:

(a) Deposits will be greater than $13,000.
(b) Withdrawals will be greater than $13,000.
(c) Withdrawals will greater than deposits.
(d) Withdrawals will exceed deposits by $5,000.

18. Suppose you have found that the heights of the adults in a town of 10,000 people are distributed according to a normal distribution with mean 5 feet 8 inches and standard deviation 5 inches.

(a) How many people are taller than 6'6"?
(b) How many people are between 6'0" and 6'6"?
(c) How many people are between 5'8" and 6'0"?
(d) How many people are between 5'4" and 5'8"?
(e) How many people are shorter than 5'4"?

19. Suppose that the lifetime of a particular electronic circuit has a normal distribution with mean 50,000 hours and standard deviation 8,000 hours.

(a) If you select one of these circuits at random, what is the probability that it will last less than 30,000 hours?
(b) What is the probability that your randomly selected circuit will last more than 55,000 hours?
(c) Suppose you buy two circuits. When the first circuit fails, you will replace it with the second circuit. Let X be a random variable representing the lifetime of the first circuit plus the lifetime of the second

circuit. What is the distribution of X? What is its mean? What is its variance?

(d) What is the probability that X will be greater than 80,000 hours?

20. Suppose you are a banquet manager. The number of people who arrive at your Chamber of Commerce banquet is given by a normal distribution with mean 65 and standard deviation 4. How many places may you set if the probability of an overflow must be less than 3 percent?

21. Consider a city of 100,000 families. If you took the common logarithms of the income for each family, you would find that the resulting numbers had a normal distribution with mean 4.3010 and standard deviation 0.3. (If you are not already familiar with logarithms, you may look ahead to the discussion in Chapter 15.) How many families in the city have incomes between $10,000 and $15,000? Between $15,000 and $20,000? Between $20,000 and $25,000? Between $25,000 and $30,000?

22. Suppose that the grade that a student receives in an individual course is not very accurate, since there are many random factors that could cause the grade to be higher or lower than the student's true abilities would indicate. Suppose that the grade is given by a random variable with a normal distribution with mean 3.5 and variance 1/16. What is the probability that the student's grade for an individual course will be between 3.4 and 3.6? Now, suppose a student takes 36 courses, whose grades all have the same normal distribution. What is the probability that the average grade for all of the courses will be between 3.4 and 3.6?

23. What is the probability that a chi-square random variable with 3 degrees of freedom will be greater than 4?

24. What is the probability that a chi-square random variable with 5 degrees of freedom will be less than 2?

25. What is the probability that a random variable with a t distribution with 5 degrees of freedom will be between 0 and 1?

26. What is the probability that a random variable with a t distribution with 10 degrees of freedom will be between -2 and -1?

27. Suppose someone tells you that he observed the value 16 as the result of a chi-square random variable with 12 degrees of freedom. Do you believe him?

28. If someone tells you that she observed the value 5 for a chi-square random variable with 16 degrees of freedom, do you believe her?

29. If someone tells you that he observed the value -2.3 for a random variable that was selected from a t distribution with 3 degrees of freedom, do you believe him?

30. If someone tells you that she observed the value 0.7 for a random variable selected from a t distribution with 12 degrees of freedom, do you believe her?

ANSWERS

KNOW THE CONCEPTS

1. The sales figures are likely to follow the normal distribution approximately. It would be necessary to observe the sales figures for several weeks and then draw a frequency diagram to see whether the normal distribution looked appropriate. Note that the normal distribution cannot exactly represent this situation since the normal distribution is a continuous distribution but the number of sales must always be a whole number.

2. It would be extremely unlikely for someone to have a height that is four standard deviations above the mean, but it is not impossible.

3. The normal distribution can be used as an approximation to the binomial distribution when the number of trials is large. The binomial distribution can be treated as the sum of n independent, identical Bernoulli random variables, so the central limit theorem applies. This approximation is advantageous because the calculations based on the normal distribution are easier than the calculations using the binomial distribution formula.

4. If you observe a random variable many times and take the average value, it is unlikely that this average will take on a value far from the mean of the distribution. For example, the average will only be far above the mean if most of the observed values are above the mean, which is unlikely. The normal distribution applies to many such situations where extreme values are less likely than moderate values.

5. The result will have a normal distribution if the two random variables are independent. If they are not independent, then you do not know for sure what the distribution of the sum will be.

6. The probability is 1 that the random variable will take on a value between minus infinity and plus infinity, so the area under the curve from minus infinity to plus infinity must be 1.

7. If $a < b$, and X is a continuous random variable, then $\Pr(X < a)$ must be less than (or equal to) $\Pr(X < b)$. For example, $\Pr(X < 2)$ must be less than or equal to $\Pr(X < 3)$.

8. Yes. If X is any continuous random variable and a is any number, then $\Pr(X < a) = \Pr(X \leq a)$. This property is not true for discrete random variables.

9. Because the standard normal curve is symmetric about zero, the area under the curve to the right of a must be equal to the area under the curve to the left of $-a$.

10. Since Z must be either greater than a or less than a, it follows that $\Pr(Z > a)$ must equal $1 - \Pr(Z < a)$.

PRACTICAL APPLICATION

1. .08; .66
2. .20; .63
3. .07
4. .34
5. .31
6. .28; .58. The normal distribution can represent this situation only approximately, since the number of glasses sold is a discrete random variable.
7. .15
8. -2.25
9. 35.43
10. $F(3) = F(\mu) = \Phi(0) = .5$, which is greater than $F(4) = .4$, thereby contradicting the fact that F must be an increasing function.
11. 0
12. 0
13. $-9.24 < a < -3.16$
14. Mean $= 10$ Variance $= .0064$
15. $1 - \Phi\left(\dfrac{\mu_1 - \mu_2}{\sqrt{\sigma_1^2 + \sigma_2^2}}\right)$
16. .39
17. (a) .4013   (b) .2743   (c) .3783   (d) .1379
18. (a) 227   (b) 1,892   (c) 2,881   (d) 2,881   (e) 2,119
19. (a) .0062   (b) .2676   (c) Normal distribution   Mean = 100,000
    Variance = 128,000,000   (d) .9616
20. 73
21. 17,850; 16,280; 12,550; 9,690
22. .3108; .9836
23. A bit less than .25
24. About .10
25. About .30
26. About .15
27. There is a probability of about .2 that this random variable might be as big as 16, so you can believe the person.
28. There is less than a .005 probability that this random variable will be less than 5, so you should not believe this claim.
29. Since there is only a .05 probability that this random variable will be smaller than -2.3, this value is implausible, but you cannot reject it with certainty.
30. Yes.

9
DISTRIBUTIONS WITH TWO RANDOM VARIABLES

KEY TERMS

correlation an indication of the degree of association between two quantities; its value is always between −1 and 1

covariance an indication of the degree of association between two quantities; it is related to the correlation but is not constrained to be between −1 and 1

independent random variables two random variables that do not affect each other; knowing the value of one of the random variables does not provide any information about the other variable

JOINT PROBABILITY FUNCTIONS

Suppose we have two discrete random variables X and Y, and we are interested in the probabilities that they will take on particular values. Just as in the case of one random variable, we can define a probability function and a cumulative distribution function.

Let's say that X has six possible values: X_1, X_2, X_3, X_4, X_5, and X_6. Also, we'll say that Y has six possible values: Y_1, Y_2, Y_3, Y_4, Y_5, Y_6. Now, let's conduct a random experiment in which we observe values for X and Y. The result of the experiment consists of two numbers: the observed value of X and the observed value of Y. Then there are 36 possible outcomes for the experiment. (In general, if there are m possible values for X and n possible values for Y, there will be mn possible results.) To characterize the experiment completely, we need to calculate the probability for each one of the 36 possible results. We can arrange our results in a table.

For example, suppose that X is the number that shows up on the top of a die when it is rolled and Y is the number on the bottom of the die. Then the probability table is as follows:

Y	X = 1	X = 2	X = 3	X = 4	X = 5	X = 6
1	0	0	0	0	0	1/6
2	0	0	0	0	1/6	0
3	0	0	0	1/6	0	0
4	0	0	1/6	0	0	0
5	0	1/6	0	0	0	0
6	1/6	0	0	0	0	0

(The table looks like this because the numbers on any two opposite faces of a die add up to 7.)

For another example, suppose that X and Z are the numbers that appear on two different dice. Then the probability table looks like this:

Z	X = 1	X = 2	X = 3	X = 4	X = 5	X = 6
1	1/36	1/36	1/36	1/36	1/36	1/36
2	1/36	1/36	1/36	1/36	1/36	1/36
3	1/36	1/36	1/36	1/36	1/36	1/36
4	1/36	1/36	1/36	1/36	1/36	1/36
5	1/36	1/36	1/36	1/36	1/36	1/36
6	1/36	1/36	1/36	1/36	1/36	1/36

(In this case, obviously, each of the 36 outcomes is equally likely.)

When there are many possible values for either X or Y, it will be too cumbersome to make a table. In general, though, we can define a **joint probability function** $f(x, y)$ like this:

$$f(x, y) = \Pr[(X = x) \text{ and } (Y = y)]$$
$$= \Pr[(X = x) \cap (Y = y)]$$

In this case f is a function of two variables. (An ordinary probability function is a function of only one variable.) Note that $f(x, y) = 0$ if (x, y) is not a possible result for X and Y. The sum of all of the possible values of the function $f(x, y)$ must be 1.

We can also define the **joint cumulative distribution function**:

$$F(x, y) = \Pr[(X \leq x) \text{ and } (Y \leq y)]$$
$$= \Pr[(X \leq x) \cap (Y \leq y)]$$

The joint cumulative distribution function for two continuous random variables can be defined in the same way. The joint density function for two continuous random variables can be pictured as follows. Imagine a little hill spread out across a flat plane with an x axis and a y axis marked on it. The total volume under the hill must be 1. Then the probability that the values of X and Y will be within any specified rectangle of the plane is equal to the volume of the hill above that rectangle.

MARGINAL DENSITY FUNCTIONS FOR INDIVIDUAL RANDOM VARIABLES

Suppose that we are interested in the value of X but don't care about the value of Y. (For example, most people don't really care what number shows up on the bottom of a die they have just tossed.) In that case, the joint probability function gives us far more information than we want. What we'd like to do is to figure out some way to derive the regular probability function for X alone. (We will write $f_X(x)$ to stand for the probability function of X.) Let's look at the joint probability function for X and Y shown above. We should be able to use this information to find the probability that X will equal 1. As you can see from the table, there are six possible outcomes of the experiment that have X equal to 1:

$$(X = 1, Y = 1); (X = 1, Y = 2); (X = 1, Y = 3);$$
$$(X = 1, Y = 4); (X = 1, Y = 5); (X = 1, Y = 6)$$

To get the probability that X will equal 1, we need to add all these probabilities. We get $0 + 0 + 0 + 0 + 0 + 1/6 = 1/6$.

Now, look at the table that gives us the joint density function for X and Z. Again we can get the probability that X will equal 1 by adding all of the terms in the first column: $1/36 + 1/36 + 1/36 + 1/36 + 1/36 + 1/36 = 1/6$. Likewise, by adding up each of the other columns in the tables we can find $\Pr(X = 2)$, $\Pr(X = 3)$, and so on.

In general, if there are n possible values for Y, then

$$\Pr(X = x) = \Pr[(X = x) \text{ and } (Y = y_1)]$$
$$+ \Pr[(X = x) \text{ and } (Y = y_2)] + \cdots$$
$$+ \Pr[(X = x) \text{ and } (Y = y_n)]$$

We can write this relation in terms of the joint probability function:

$$f_X(x) = f(x, y_1) + f(x, y_2) + \cdots + f(x, y_n)$$
$$= \sum_{i=1}^{n} f(x, y_i)$$

An individual probability function derived from a joint function in this way is sometimes called a **marginal probability function**.

We can also find the marginal probability function for y:

$$f_Y(y) = f(x_1, y) + f(x_2, y) + \cdots + f(x_m, y)$$

Here are the results for our random variables X, Y, and Z.

Y	X = 1	X = 2	X = 3	X = 4	X = 5	X = 6	$f_Y(y)$
1	0	0	0	0	0	1/6	1/6
2	0	0	0	0	1/6	0	1/6
3	0	0	0	1/6	0	0	1/6
4	0	0	1/6	0	0	0	1/6
5	0	1/6	0	0	0	0	1/6
6	1/6	0	0	0	0	0	1/6
$f_X(x)$:	1/6	1/6	1/6	1/6	1/6	1/6	

The row totals (in the far right column) give the marginal probability of Y, and the column totals (in the bottom row) give the marginal probability of X.

Z	X = 1	X = 2	X = 3	X = 4	X = 5	X = 6	$f_z(z)$
1	1/36	1/36	1/36	1/36	1/36	1/36	1/6
2	1/36	1/36	1/36	1/36	1/36	1/36	1/6
3	1/36	1/36	1/36	1/36	1/36	1/36	1/6
4	1/36	1/36	1/36	1/36	1/36	1/36	1/6
5	1/36	1/36	1/36	1/36	1/36	1/36	1/6
6	1/36	1/36	1/36	1/36	1/36	1/36	1/6
$f_X(x)$:	1/6	1/6	1/6	1/6	1/6	1/6	

Notice that the marginal probability functions for X, Y, and Z are all the same, even though the joint probability function for X and Y is much different from the joint probability function for X and Z.

With continuous random variables, it is necessary to use the calculus technique called *integration* to find a marginal probability function from a joint probability function.

CONDITIONAL PROBABILITY FUNCTIONS

Often we know the value of one of two random variables and would like to know the value of the other. Sometimes knowing the value of one random variable will help us when we try to guess the value of the other one. For example, if we know that the variable Y discussed above has the value of 4, then we know for sure that X has the value 3. However, knowing that Z (the number on another die) is 4 does not tell us anything about the value of X.

The **conditional probability function** for X tells us what the probability function for X is, given that Y has a specified value. Note that this concept is very much like the conditional probabilities that we discussed in Chapter 5. We will write "the conditional probability of X given that Y has the value of y^*" like this:

$$f(x \mid (Y = y^*)) \quad \text{or} \quad f(x \mid y^*)$$

(Remember that the vertical line, |, means "given that.") Then we can use the definition of conditional probability to find that

$$f(x \mid Y = y^*) = \frac{\Pr((X = x) \text{ and } (Y = y^*))}{\Pr(Y = y^*)}$$

$$= \frac{f(x, y^*)}{f_Y(y^*)}$$

In other words, the conditional probability function is equal to the joint probability function $f(x, y^*)$ divided by the marginal probability function of y. For example, suppose $Y = 3$. Then $f_Y(3) = 1/6$, so

$$f(x \mid Y = 3) = \begin{cases} \dfrac{0}{1/6} & \text{when } x = 1 \\[2ex] \dfrac{0}{1/6} & \text{when } x = 2 \\[2ex] \dfrac{0}{1/6} & \text{when } x = 3 \\[2ex] \dfrac{1/6}{1/6} & \text{when } x = 4 \\[2ex] \dfrac{0}{1/6} & \text{when } x = 5 \\[2ex] \dfrac{0}{1/6} & \text{when } x = 6 \end{cases}$$

YOU SHOULD REMEMBER

1. The *joint probability function* for two discrete random variables lists the probabilities for each pair of possible values for the two variables.

2. A *marginal probability function* is a probability function for a single random variable.

3. A *conditional probability function* is a probability function for one random variable, given that another random variable has a specified value.

4. Both the marginal probability function and the conditional probability functions can be calculated from the joint probability function.

INDEPENDENT RANDOM VARIABLES

It would be nice if we could tell whether two random variables affect each other. We will say that two random variables X and Y are **independent random variables** if knowing the value of Y does not tell us anything about the value of X, and vice versa. This means that the conditional probability for X, given Y, is equal to the regular marginal probability for X (since knowing the value of Y does not change any of the probabilities for X). Therefore, if X and Y are independent,

$$f(x \mid y) = f_X(x)$$

We know from the definition of conditional probability that

$$f(x \mid y) = \frac{f(x, y)}{f_Y(y)}$$

Therefore, when X and Y are independent,

$$f_X(x) = \frac{f(x, y)}{f_Y(y)}$$

$$f(x, y) = f_X(x) f_Y(y)$$

When the two variables are independent, the joint probability function can be found simply by multiplying together the two marginal probability functions. This means that independent random variables are much easier to deal with. In general, it is very difficult to reconstruct the joint probability function for two individual random variables if we are given their marginal probability functions. (In fact, it is impossible unless we have some additional information about how the two random variables are related.)

In the examples above, X, Y, and Z all have identical marginal probability functions. However, the joint probability for X and Y is much different from the joint probability function for X and Z, since X and Z are independent but X and Y are not.

Here is an example of the joint probability function for two independent random variables U and V:

V	U = 1	U = 2	U = 3	U = 4	$f_v(v)$
1	.02	.04	.06	.08	.2
2	.02	.04	.06	.08	.2
3	.06	.12	.18	.24	.6
$f_u(u)$	.1	.2	.3	.4	1

The marginal probability function for U can be found by adding each column, and the marginal probability function for V can be found by adding each row. You can see that for each entry in the table $f(u, v)$ is equal to $f_u(u)f_v(v)$, so the two random variables are independent.

Here is another useful result that works if two random variables are independent. If U and V are independent, then

$$E(UV) = E(U)E(V)$$

In other words, the expectation of their product is just equal to the product of their individual expectations. This is another reason why life is much simpler when we have independent random variables.

YOU SHOULD REMEMBER

1. Two random variables are independent if knowing the value of one of the variables does not provide any information about the value of the other variable.

2. Mathematically, two random variables are independent if their joint probability function can be found by multiplying the two marginal probability functions.

COVARIANCE AND CORRELATION

• *COVARIANCE*

Suppose X and Y are two random variables that are not independent. We would like to be able to measure how closely they are related. If X and Y are very closely related, learning the value of X tells you a lot about the value of Y. If they are only slightly related, then knowing the value of X helps a little, but not much, when we try to guess the value of Y.

The quantity that measures the degree of dependence between two random variables is called the **covariance**. The covariance of X and Y, written as Cov(X, Y), is defined as follows:

$$\text{Cov}(X, Y) = E[[X - E(X)][Y - E(Y)]]$$

Suppose that, when X is larger than $E(X)$, we know that Y is also larger than $E(Y)$. In that case, $E[[X - E(X)][Y - E(Y)]]$ will be positive. In general, when two random variables tend to move together, their covariance is positive. And if two random variables tend to move in the opposite direction (for example, if X tends to be big at the same time Y is small, and vice versa), their covariance is negative.

There is a shortcut formula for calculating covariances that is much easier to use than the defining formula:

$$E[[X - E(X)][Y - E(Y)]]$$
$$= E[XY - Y E(X) - X E(Y) + E(X) E(Y)]$$
$$= E(XY) - E[Y E(X)] - E[X E(Y)] + E(X) E(Y)$$
$$= E(XY) - E(X) E(Y) - E(X) E(Y) + E(X) E(Y)$$
$$\text{Cov}(X,Y) = E(XY) - E(X) E(Y)$$

We can calculate $E(XY)$ for X, the number on top of the die, and Y, the number on the bottom. There are only six possibilities with nonzero probabilities; each possibility has a probability of 1/6:

Probability	X	Y	XY	Prob × XY
1/6	1	6	6	6/6 = 1
1/6	2	5	10	10/6 = 5/3
1/6	3	4	12	12/6 = 2
1/6	4	3	12	12/6 = 2
1/6	5	2	10	10/6 = 5/3
1/6	6	1	6	6/6 = 1

$$E(XY) = \text{sum} = 9\tfrac{1}{3} = 9.333$$

Therefore,

$$\text{Cov}(X,Y) = 9.333 - 3.5 \times 3.5 = -2.917$$

The covariance is negative, because larger values of X are associated with smaller values of Y.

From the shortcut formula you can see immediately that $\text{Cov}(X,Y) = 0$ if X and Y are independent. (Use the result $E(XY) = E(X) E(Y)$ if X and Y are independent.) Therefore $\text{Cov}(X,Z) = 0$ when X and Z are the numbers on two different dice. [However, unfortunately it works out that just showing that $\text{Cov}(X,Y) = 0$ is not enough to ensure that X and Y are independent.]

• *CORRELATION*

If $\text{Cov}(X,Y)$ is nonzero, then we know that X and Y are not independent. However, the size of $\text{Cov}(X,Y)$ does not tell us much, because it depends mostly on the size of X and Y. We define a new quantity, called the **correlation**, that we can use directly to tell how strong the relation between X and Y is. We will write the correlation between X and Y as $r(X,Y)$ if the data comes from a sample. (The Greek letter rho (ρ) is often used to stand for correlation if you have population data.) The definition of the correlation coefficient is

$$r(X,Y) = \frac{\text{Cov}(X,Y)}{\sqrt{\text{Var}(X)\,\text{Var}(Y)}} = \frac{\text{Cov}(X,Y)}{\sigma_x \sigma_y}$$

The most important property of the correlation coefficient is that its value is always between -1 and 1. If X and Y are independent, then clearly their correlation is zero. If the correlation coefficient is positive, we know that, when X is big, Y is also likely to be big. They are then said to be **positively correlated**. X and Y are more closely related the closer the correlation coefficient is to 1. On the other hand, if the correlation is negative, then Y is more likely to be small when X is big. They are **negatively correlated**, and the negative relationship is stronger if the correlation coefficient is closer to -1.

Example: Determining the Correlation

PROBLEM   Suppose we flip three pennies and three nickels. Let U be the total number of heads that appear, and let V be the number of heads on the pennies. Calculate the joint probability table and the correlation.

SOLUTION   The joint probability table looks like this:

V	U = 0	U = 1	U = 2	U = 3	U = 4	U = 5	U = 6	f(v)
0	1/64	3/64	3/64	1/64	0	0	0	1/8
1	0	3/64	9/64	9/64	3/64	0	0	3/8
2	0	0	3/64	9/64	9/64	3/64	0	3/8
3	0	0	0	1/64	3/64	3/64	1/64	1/8
f(u)	1/64	6/64	15/64	20/64	15/64	6/64	1/64	1

We can calculate that $E(UV) = 5.25$. We already know that $E(U) = 3$ and $E(V) = 3/2$. Therefore, $\text{Cov}(U,V) = 5.25 - 3 \times 3/2 = .75$. We also have found that $\sigma_U = 1.225$, and $\sigma_V = .8660$. That means that $r(U,V) = .75/(1.225 \times .8660) = .7070$. The correlation is positive, because larger values of U are associated with larger values of V.

• PERFECTLY CORRELATED RANDOM VARIABLES

Suppose the joint density function for X and Y is as follows:

Y	X = 0	X = 1	X = 2	X = 3	X = 4	X = 5	X = 6	
0	1/7	0	0	0	0	0	0	1/7
4	0	1/7	0	0	0	0	0	1/7
8	0	0	1/7	0	0	0	0	1/7
12	0	0	0	1/7	0	0	0	1/7
16	0	0	0	0	1/7	0	0	1/7
20	0	0	0	0	0	1/7	0	1/7
24	0	0	0	0	0	0	1/7	1/7
	1/7	1/7	1/7	1/7	1/7	1/7	1/7	1

In this case it is clear that Y is always equal to $4X$. We can calculate that $E(X) = 3$, $\sigma_X = 2$, $E(Y) = 12$, $\sigma_Y = 8$, and $E(XY) = 52$. Therefore, $\text{Cov}(X,Y) = 52 - 3 \times 12 = 16$, and the correlation is $16/(2 \times 8) = 1$.

In general, if the correlation between two random variables is equal to 1, then there is a relationship between them of this general form:

$$Y = aX + b$$

where a and b are constants, and $a > 0$. This is called a linear relationship, because all of the points representing possible values of X and Y would be on a straight line with a positive slope if you drew their graph.

If we have the relationship $Y = aX + b$, but $a < 0$, then the correlation is -1. Again the points showing possible values would fit on a straight line, this time with a negative slope. For example, we can find the correlation between the top and bottom numbers of a die to be $-2.917/\sqrt{2.917 \times 2.917} = -1$.

There are often situations where we observe the outcomes of the random process but we don't know the nature of the relationship between the two variables. In that case, we turn to **regression analysis** to estimate the nature of the relationship; that topic is covered in Chapter 15.

VARIANCE OF A SUM

We can now derive a general formula for $\text{Var}(X + Y)$:

$$
\begin{aligned}
\text{Var}(X + Y) &= E[(X + Y)^2] - [E(X + Y)]^2 \\
&= E[X^2 + 2XY + Y^2] - [E(X) + E(Y)]^2 \\
&= E(X^2) + 2\,E(XY) + E(Y^2) \\
&\quad - (E(X))^2 - 2\,E(X)\,E(Y) - (E(Y))^2 \\
&= E(X^2) - (E(X))^2 + E(Y^2) - (E(Y))^2 \\
&\quad + 2\,[E(XY) - E(X)\,E(Y)] \\
&= \text{Var}(X) + \text{Var}(Y) + 2\,\text{Cov}(X,Y)
\end{aligned}
$$

For example, suppose you own stock in Worldwide Fastburgers, Inc. Your profit from the stock is a random variable (W) with mean 1,000 and variance 400. You would like to buy some more stock, and you are trying to decide between Have It Your Way Burgers, Inc., and Fun and Good Times Pizza, Inc. Both stocks also have profits that are random variables with mean 1,000 and variance 400. Which one should you choose? Let H represent your profit if you choose Have It Your Way Burgers, and let F represent your profit if you choose Fun and Good Times Pizza.

The expected value of your profit will be the same regardless of which one you choose. However, you would also like the variance of your profit to be as small as possible, since a smaller variance means that your stock holding is less risky. To calculate the total variance of your portfolio, you need to look at the covariance between Worldwide Burgers and the other two companies. Suppose that $\text{Cov}(W,H) = 380$. The covariance is positive, meaning that both firms will prosper if the hamburger market is strong, but both firms will suffer if the hamburger market is weak. Suppose also that $\text{Cov}(W,F) = -200$. The negative covariance means that the pizza market is booming when the hamburger market is in a slump, and vice versa.

If you buy the Have It Your Way Burger stock, then the total variance of your profit will be

$$
\begin{aligned}
\text{Var}(W + H) &= \text{Var}(W) + \text{Var}(H) + 2\,\text{Cov}(W,H) \\
&= \quad 400 \quad + \quad 400 \quad + \quad 2 \times 380 \\
&= 1{,}560
\end{aligned}
$$

If you buy the pizza stock, then the variance will be

$$
\begin{aligned}
\text{Var}(W + F) &= \text{Var}(W) + \text{Var}(F) + 2\,\text{Cov}(W,F) \\
&= \quad 400 \quad + \quad 400 \quad + 2 \times (-200) \\
&= 400
\end{aligned}
$$

Clearly, it is much less risky to buy the pizza stock. In general, you can lower the risk of your stock holdings by diversifying and buying stocks that have negative covariances with each other. In other words, don't put all your eggs in one basket.

Now consider X and Y, where X is the number on top of the die and Y is the number on the bottom. We have previously found $\text{Var}(X) = \text{Var}(Y) = 2.917$ and $\text{Cov}(X,Y) = -2.917$. Now, to find the variance of their sum:

$$
\text{Var}(X + Y) = \text{Var}(X) + \text{Var}(Y) + 2\text{Cov}(X,Y) = 2.917 + 2.917 - (2 \times 2.917) = 0
$$

This makes sense, because we know $X + Y = 7$ always, so there is no uncertainty about what $X + Y$ will be (even if there is lots of uncertainty about the value of X and the value of Y).

YOU SHOULD REMEMBER

1. The *covariance* of two random variables is a measure of their association. If the covariance is *positive*, then large values of one of the random variables are likely to be associated with large values of the other random variable. If the covariance is *negative*, then large values of one of the random variables are likely to be associated with small values of the other random variable. If the two random variables are *independent*, then the covariance is zero. The covariance can be calculated from this formula:

$$
\text{Cov}(X,Y) = E(XY) - E(X)\,E(Y)
$$

2. The correlation is a quantity similar to the covariance except that it is scaled so as always to be between -1 and 1. If the two random variables are independent, the correlation is zero. If there is a perfect linear relationship between the two random variables, the correlation is either -1 or 1, depending on whether the covariance is positive or negative.

3. If X and Y are two random variables, then the variance of their sum can be found from this formula:

$$
\text{Var}(X + Y) = \text{Var}(X) + \text{Var}(Y) + 2\,\text{Cov}(X,Y)
$$

KNOW THE CONCEPTS

DO YOU KNOW THE BASICS?

Test your understanding of Chapter 9 by answering the following questions:

1. What is the sum of all of the probabilities in a joint probability table?
2. How can you tell if two random variables are independent?
3. Do you think the number of new cars sold in a city is independent from the number of used cars sold?
4. Can you calculate the marginal density function for two random variables if you are given their joint density function?
5. Can you calculate the joint density function for two random variables if you are given their marginal density function?

TERMS FOR STUDY

conditional probability function
correlation
covariance
independent random variables
joint cumulative distribution function
joint probability function
marginal probability function

negatively correlated
perfectly correlated
positively correlated

PRACTICAL APPLICATION

COMPUTATIONAL PROBLEMS

1. Roll three dice (one red, one blue, and one green). Let X equal the sum of the numbers on the red and blue dice, and let Y equal the sum of the numbers on the blue and green dice. Find the joint distribution of X and Y.
2. Roll two dice (one red and one blue). Let X be the sum of the numbers on the two dice, and let Y be the number on the red die. Find the joint distribution of X and Y.

For Exercises 3–6 calculate the marginal probability functions for X *and* Y, *the covariance between* X *and* Y, *and the correlation between* X *and* Y. *In each case the table gives the joint probability function.*

3.

X	Y = 0	Y = 2	Y = 4	Y = 6
−2	.1	0	.1	.2
4	0	.1	0	.1
5	.1	0	.1	.2

4.

X	Y = 3	Y = 5	Y = 9
−2	.15	.05	0
−1	.05	0	.2
0	0	.1	0
1	.15	0	.2
2	0	.1	0

5.

X	Y = 1	Y = 4	Y = 5	Y = 6	Y = 9
0	.1	0	.1	0	0
5	0	.1	0	0	.15
9	0	.05	0	0	0
11	0	.3	0	.05	0
13	.15	0	0	0	0

6.

X	Y = −9	Y = −4	Y = −3	Y = −1	Y = 0
−3	0	.1	0	.05	.1
2	.05	0	0	0	0
4	0	0	.1	0	.05
5	0	0	0	.05	0
7	0	.25	.05	0	.2

7. Suppose the joint probability function for X and Y is given by the following table. Find $f_X(x)$ and $f_Y(y)$.

X	Y = 1	Y = 2	Y = 3	Y = 4
1	.2	.1	.1	0
2	0	0	0	0
3	0	.1	.1	.2
4	0	0	.1	.1

8. Roll two dice. Let X represent the sum of the numbers on the top of the dice, and let Y represent the sum of the numbers on the bottom of the dice. Determine the joint probability function.

9. In a lottery based on a randomly selected three-digit number, let X be the first two digits of the number and Y be the last two. Describe $f_{XY}(x, y)$.

10. Suppose the following table gives the joint probability function for two random variables X and Y:

X	Y = 7	Y = 10	Y = 13	Y = 14
1	.1	0	.1	0
2	0	0	0	.15
5	0	.25	0	.1
9	.1	.1	.1	0

Find the joint cumulative distribution function.

11. Why can't

$$g(x,y) = \begin{cases} .1 & x = 2, \ y = 3/2 \\ .3 & x = 1/2, y = 4 \\ .5 & x = 2, \ y = 1 \\ .2 & x = 4, \ y = 1 \\ 0 & \text{everywhere else} \end{cases}$$

be a joint probability function?

12. Calculate the covariance and correlation between X and Y when X represents the number on the top of a die and Y represents the number on the bottom of the die.

13. Calculate the covariance and correlation between X and Z when they represent the numbers that appear when two separate dice are tossed.

Exercises 14–16 refer to the random variables U *and* V *whose joint probability is given on page 189.*

14. Calculate the marginal probability functions for U and V.

15. Find the conditional probability function for U, given that $V = 2$.

16. Find the conditional probability function for V, given that $U = 3$.

ANSWERS

KNOW THE CONCEPTS

1. 1

2. If two random variables are independent, then their joint probability

function is found by multiplying their individual marginal probability functions.

3. Probably the two numbers are not independent, but you would need to collect some observations and then conduct a chi-square test (which will be discussed in Chapter 13) in order to tell for sure.

4. Yes, by addition.

5. In general, no. However, if you know that the two random variables are independent, then you can calculate the joint probability function by multiplying.

PRACTICAL APPLICATION

1.

X	Y=2	Y=3	Y=4	Y=5	Y=6	Y=7	Y=8	Y=9	Y=10	Y=11	Y=12
2	1/216	1/216	1/216	1/216	1/216	1/216	0	0	0	0	0
3	1/216	2/216	2/216	2/216	2/216	2/216	1/216	0	0	0	0
4	1/216	2/216	3/216	3/216	3/216	3/216	2/216	1/216	0	0	0
5	1/216	2/216	3/216	4/216	4/216	4/216	3/216	2/216	1/216	0	0
6	1/216	2/216	3/216	4/216	5/216	5/216	4/216	3/216	2/216	1/216	0
7	1/216	2/216	3/216	4/216	5/216	6/216	5/216	4/216	3/216	2/216	1/216
8	0	1/216	2/216	3/216	4/216	5/216	5/216	4/216	3/216	2/216	1/216
9	0	0	1/216	2/216	3/216	4/216	4/216	4/216	3/216	2/216	1/216
10	0	0	0	1/216	2/216	3/216	3/216	3/216	3/216	2/216	1/216
11	0	0	0	0	1/216	2/216	2/216	2/216	2/216	2/216	1/216
12	0	0	0	0	0	1/216	1/216	1/216	1/216	1/216	1/216

2.

X	Y = 1	Y = 2	Y = 3	Y = 4	Y = 5	Y = 6
2	1/36	0	0	0	0	0
3	1/36	1/36	0	0	0	0
4	1/36	1/36	1/36	0	0	0
5	1/36	1/36	1/36	1/36	0	0
6	1/36	1/36	1/36	1/36	1/36	0
7	1/36	1/36	1/36	1/36	1/36	1/36
8	0	1/36	1/36	1/36	1/36	1/36
9	0	0	1/36	1/36	1/36	1/36
10	0	0	0	1/36	1/36	1/36
11	0	0	0	0	1/36	1/36
12	0	0	0	0	0	1/36

3.

x:	−2	4	5
f(x):	.4	.2	.4

y:	0	2	4	6
f(y):	.2	.1	.2	.5

The covariance and correlation are 0.

4.

x:	−2	−1	0	1	2
f(x):	.2	.25	.1	.35	.1

y:	3	5	9
f(y):	.35	.25	.4

$Cov(X,Y) = .49$ $r(XY) = .139$

5.

x:	0	5	9	11	13
f(x):	.2	.25	.05	.35	.15

y:	1	4	5	6	9
f(y):	.25	.45	.1	.05	.15

$Cov(X,Y) = -2.5$ $r(X,Y) = -.214$

6.

x:	−3	2	4	5	7
f(x):	.25	.05	.15	.05	.50

y:	−9	−4	−3	−1	0
f(y):	.05	.35	.15	.1	.35

$Cov(X,Y) = -.17$ $r(X,Y) = -.018$

7.

x:	1	2	3	4
f(x):	.4	0	.4	.2

y:	1	2	3	4
f(y):	.2	.2	.3	.3

8.

Y	X = 2	X = 3	X = 4	X = 5	X = 6	X = 7	X = 8	X = 9	X = 10	X = 11	X = 12
2	0	0	0	0	0	0	0	0	0	0	1/36
3	0	0	0	0	0	0	0	0	0	2/36	0
4	0	0	0	0	0	0	0	0	3/36	0	0
5	0	0	0	0	0	0	0	4/36	0	0	0
6	0	0	0	0	0	0	5/36	0	0	0	0
7	0	0	0	0	0	6/36	0	0	0	0	0
8	0	0	0	0	5/36	0	0	0	0	0	0
9	0	0	0	4/36	0	0	0	0	0	0	0
10	0	0	3/36	0	0	0	0	0	0	0	0
11	0	2/36	0	0	0	0	0	0	0	0	0
12	1/36	0	0	0	0	0	0	0	0	0	0

9. For each possible value of X there will be 10 possible values for Y.

10.

	$y < 7$	$7 \leq y < 10$	$10 \leq y < 13$	$13 \leq y < 14$	$14 \leq y$
$x < 1$	0	0	0	0	0
$1 \leq x < 2$	0	.1	.1	.2	.2
$2 \leq x < 5$	0	0	.1	.2	.35
$5 \leq x < 9$	0	0	.35	.45	.7
$9 \leq x$	0	.2	.55	.75	1

11. Because the sum of the values is greater than 1.

12. $\text{Cov}(X,Y) = -2.917 \quad r(X,Y) = -1$

13. $\text{Cov}(X,Z) = 0 \quad r(X,Z) = 0$

10
STATISTICAL ESTIMATION

KEY TERMS

consistent estimator an estimator that tends to converge toward the true value as the sample size becomes larger

estimator a quantity based on observations of a sample whose value is taken as an indicator of the value of an unknown population parameter (for example, the sample average $\bar{x}$ is often used as an estimator of the unknown population mean μ)

maximum likelihood estimator an estimator with the following property: if the true value of the unknown parameter has this value, then the probability of obtaining the sample that was actually observed is maximized

statistical inference the process of using observations of a sample to estimate the properties of the population

unbiased estimator an estimator whose expected value is equal to the true value of the parameter it is used to estimate

Up to now, in most of the problems we have done we have known in advance what all of the probabilities were. For example, when we draw cards or toss coins, we can calculate all of the probabilities explicitly. However, in most real problems we don't know the probabilities in advance. Instead, we have to use the methods of **statistical inference** to estimate them.

Here are some examples of the use of statistical inference to estimate probabilities:

- Suppose that the distribution of the heights of all of the people in the country can be described by a normal distribution. However, we don't know in advance what the mean (μ) of the distribution is. Instead, we will have to estimate it.

- Suppose you are conducting a scientific experiment to measure the molecular weight of a chemical. On the average, you can expect that the result of the measurement will be the true value of the weight. However, any particular measurement is subject to random errors. It is often reasonable to suppose that the actual result of each measurement has a normal distribution whose mean is the true value of the quantity you are measuring.

• Suppose we know that the number of successes of an experiment can be characterized by a binomial distribution, but we don't know in advance the value of the probability-of-success parameter (p). We wish to use our observations of the results of the experiment to estimate the value of p.

• Suppose we know that a person's scores on two different kinds of tests are given by random variables with unknown correlation. We will try to estimate the correlations.

ESTIMATING THE MEAN

We will consider the general problem of trying to estimate the mean (μ) of a random variable X that has a normal distribution. An unknown population quantity such as μ is called a **parameter**. Let's say that we have n observations of the value of the random variable, which we can call $X_1, X_2, X_3, \ldots, X_n$. We need to make one more important assumption: we need to assume that each value of X is independent of all of the other values. (This process of drawing numbers is called taking a *random sample* of size n from this particular distribution.)

Intuitively, it is clear what our estimate for the mean should be:

$$(\text{estimate of mean}) = \bar{x} = \frac{X_1 + X_2 + X_3 + \cdots + X_n}{n}$$

This quantity is just the average value of all of the X's, which we called $\bar{x}$, or the sample average. The quantity $\bar{x}$ is an example of a **statistic**. A *statistic* is a particular function of the items in a random sample. When a statistic is used to estimate the value of some unknown quantity, it's called an **estimator**. In this case, $\bar{x}$ is used as an estimator for the quantity μ. Often a small hat sign ($\hat{\mu}$) is placed over a quantity to indicate that it is an estimator for a parameter. The statement "$\hat{\mu} = \bar{x}$" means, "We are using the sample average $\bar{x}$ as an estimator for the population mean μ." An **estimate** is the value of an estimator in a particular circumstance. If the sample we observe consists of the numbers 5, 7, 4, 10, 12, and 4, then $\bar{x} = \hat{\mu} = 7$ is an estimate for the population mean.

What we'll do now is show that the sample average $\bar{x}$ has some appealing properties when it is used to estimate the mean.

YOU SHOULD REMEMBER

1. *Statistical inference* is the process of using information from observations of a sample to estimate characteristics of the population from which the sample was drawn.

2. A *statistic* is a quantity calculated using observed values from a sample.

3. An estimator is a statistic that is used to estimate the value of an unknown population quantity. For example, the sample average $\bar{x} = (X_1 + \cdots + X_n)/n$ is used as an estimator for the unknown population mean. The unknown population value is called a parameter.

MAXIMUM LIKELIHOOD ESTIMATORS

We have already seen how the value of the population average μ can affect the probabilities associated with a sample average $\bar{x}$. For example, if $\mu = 10,000$, it's just not very likely that $\bar{x} = 7$. Thus if we don't know what the value of μ really is but we get a sample average $\bar{x} = 7$, we can probably rule out $\mu = 10,000$ as a possibility.

If we're trying to use $\bar{x}$ as a tool to figure out what μ really is, we should pick the value for the μ that is most likely to give us the $\bar{x}$ that we ended up with from our sample. If $\bar{x} = 7$, then we can reason as follows: if $\mu = 10,000$, we aren't likely to get $\bar{x} = 7$, so 10,000 is a bad estimate for μ. The chances of $\mu = 100$ and us getting $\bar{x} = 7$ are better, so 100 is a better estimate for μ. The value of μ that gives us the greatest probability of getting $\bar{x} = 7$ is $\mu = 7$. In this case, we say that $\bar{x}$ is a *maximum likelihood estimator* for μ, because its value corresponds to the value of μ that is most likely to give the sample average that we got.

The method of maximum likelihood can also be used for many other types of problems. In general, suppose that a is an unknown parameter in a particular probability distribution. Then in many cases we can calculate the **maximum likelihood estimator** for a. For example, we can show that the maximum likelihood estimator of the variance $(\hat{\sigma}^2)$ from a normal distribution is

$$\hat{\sigma}^2 = \frac{\left(X_1 - \bar{x}\right)^2 + \left(X_2 - \bar{x}\right)^2 + \cdots + \left(X_n - \bar{x}\right)^2}{n}$$

(We called this statistic s_1^2, the sample variance. See Chapter 2.)

If we are trying to estimate the probability of success p for a random variable with a binomial distribution, then the maximum likelihood estimator is also the obvious one:

$$\frac{(\text{number of successes})}{(\text{number of attempts})}$$

Another important property of maximum likelihood estimators is called the *invariance property*. Suppose that $\hat{a}$ is the maximum likelihood estimator for a parameter a, but we really want to know the maximum likelihood estimator of $\sqrt{a}$. If we were forced to make a guess, we would probably estimate that $\sqrt{a}$ is equal to $\sqrt{\hat{a}}$, and fortunately we would be right. For example, the maximum likelihood estimator of the standard deviation ($\hat{\sigma}$) is just the square root of the sample variance. In general, if $h(a)$ is any function of a parameter a, then the maximum likelihood estimator of $h(a)$ is just $h(\hat{a})$.

CONSISTENT ESTIMATORS

Another important property that we would like our estimators to have is the consistency property. You've undoubtedly been confused by inconsistent people, and you can be just as confused by inconsistent estimators. Here is what we mean by the consistency property for estimators. Suppose we are able to increase our sample size greatly and thereby acquire many more observations of the random variable X. In that case, do we know that the new value of $\bar{x}$ will be closer to the mean μ, or is there a chance that it might be farther away?

A **consistent estimator** is an estimator whose variance approaches 0 as the sample size increases. In this way, we can ensure that our estimator will be very close to its expected value (which is just the parameter that we're looking for if the estimator is unbiased) by choosing a large enough sample size.

Fortunately, $\bar{x}$ does satisfy the property of being a consistent estimator of the population mean.

UNBIASED ESTIMATORS

Another important question we might ask about an estimator is this: Do we expect that the result of this estimator will be the true value? An estimator is said to be *unbiased* if the expectation value of the estimator is equal to the true value of the parameter we are trying to estimate.

For example, we have already found that $E(\bar{x}) = \mu$ (see page 127) so the sample average is an **unbiased estimator** of the population mean μ. However, if we calculate the expectation of the sample variance s_1^2, we find that

$$E(s_1^2) = E\left[\frac{\sum_{i=1}^{n}(x_i - \bar{x})^2}{n}\right] = \frac{(n-1)\sigma^2}{n}$$

Since $E(s_1^2)$ does not equal σ^2, this means that s_1^2 is not an unbiased estimator of σ^2. We can calculate a new statistic:

$$s_2^2 = \frac{ns_1^2}{n-1} = \frac{\sum_{i=1}^{n}(x_i - \bar{x})^2}{n-1}$$

The expectation of this statistic is

$$E(s_2^2) = E\left(\frac{ns_1^2}{n-1}\right) = \frac{E(s_1^2)n}{n-1} = \sigma^2$$

Therefore, s_2^2 is an unbiased estimator of the variance. Note that s_2^2 is calculated in the same way as s_1^2 except that the sum of the distance of each x from $\bar{x}$ is divided by $n-1$ instead of by n. Now we know why we do so. We called s_1^2 the sample variance, version 1, and s_2^2 the sample variance, version 2. This illustrates a situation where it is not possible to find a single estimator that satisfies all of the desirable properties that we would like estimators to have.

In general, there can be many different unbiased estimators for the same parameter. When possible, we would like to select an estimator that has as small a variance as possible. Since an estimator is a statistic calculated from a random sample, it is a random variable whose variance we can calculate. (The distribution of an estimator is often called the **sampling distribution** of that estimator.)

For example, suppose two people are trying to estimate the value of an unknown population mean μ. One person randomly selects two items and uses the measure $A_2 = (X_1 + X_2)/2$. Since $E(A_2) = \mu$, it is an unbiased estimator. However, the other person randomly selects 100 items and uses their average: $A_{100} = (X_1 + \cdots + X_{100})/100$. Both estimators are unbiased, but A_{100} is clearly a much better estimator than A_2 because its variance is smaller ($\sigma^2/100$ as opposed to $\sigma^2/2$). The mere fact that an estimator is unbiased does not mean that it is the best one to use.

YOU SHOULD REMEMBER

1. There are several desirable characteristics that estimators can have if they provide good estimates of the population parameters.

2. A *maximum likelihood estimator* has the following property: if the true value of the unknown parameter has this value, then the probability of obtaining the sample that was actually observed is maximized.

3. A *consistent estimator* is an estimator whose value will converge to the true value as the sample size becomes very large.

4. An *unbiased estimator* is an estimator whose expected value is equal to the true value.

* * * * *

In this chapter we discussed ways of obtaining a single number that can be used to estimate the value of an unknown parameter. Such an estimate is called a **point estimate**. Often we will need to know whether the true value of the parameter is likely to be near the point estimate or whether there is a chance that the true value could be far from the point estimate. To answer that question we need to calculate an **interval estimate**, which we will discuss in the next chapter.

KNOW THE CONCEPTS

DO YOU KNOW THE BASICS?

Test your understanding of Chapter 10 by answering the following questions:

1. In statistical inference, what is the term for an unknown population quantity? What is the term for a known quantity calculated from the observations in a sample?

2. After you have calculated an estimate in a particular situation, do you know how accurate that estimate is?

3. Why is it desirable to use an estimator that has a small variance? Should you pick the estimator with the smallest possible variance?

4. Why does the formula for the sample variance (version 2) divide by $n - 1$ instead of by n?

5. Can you think of a situation where it would be best to use an estimator that is not unbiased?

6. What is the difference between point estimates and interval estimates?

TERMS FOR STUDY

consistent estimator

estimate

estimator

interval estimate

maximum likelihood estimator

parameter

point estimate

sampling distribution

statistic

statistical inference

unbiased estimator

PRACTICAL APPLICATION

COMPUTATIONAL PROBLEMS

1. Let X be a normal random variable with mean 7 and standard deviation 3. Suppose you collect 16 observations of X and then calculate the sample average $\bar{x}$. What is the probability that $\bar{x}$ will be between 7 and 7.5? Between 7.5 and 8? Between 8 and 8.5? Between 8.5 and 9? Between 9 and 9.5?

2. Suppose X_1, X_2, and X_3 are independent, identically distributed random variables with mean μ and variance σ^2. Let $V = 3/4\, X_1 + 1/8\, X_2 + 1/8\, X_3$. Calculate $E(V)$ and $\text{Var}(V)$. Is V an unbiased estimator of μ?

3. Let $W = 1/2\, X_1 + 1/2\, X_2 + 1/2\, X_3$ (X_1, X_2, and X_3 are defined in Exercise 2). Calculate $E(W)$ and $\text{Var}(W)$. Is W an unbiased estimator of μ?

4. Let X be a binomial random variable with parameters n and p. Let $\hat{p} = X/n$. Calculate $E(\hat{p})$. Is $\hat{p}$ an unbiased estimator of p?

5. What is $\text{Var}(\hat{p})$? (See Exercise 4.)

6. Suppose you conduct an experiment n times. Each trial has a probability p of success. Let X_i be equal to 1 if trial i is a success, and let X_i be equal to 0 if trial i is a failure. Let $A_2 = (X_1 + X_2)/2$. Calculate $E(A_2)$ and $\text{Var}(A_2)$. Do you think A_2 is a very good estimator for p?

7. Calculate $E(X_1)$ and $\text{Var}(X_1)$ (X_1 is defined in Exercise 6). Is X_1 a very good estimator for p?

8. Let X_1 and X_2 be two independent, identically distributed random variables drawn from a distribution with mean μ and variance σ^2. Calculate

$$E\left[\frac{(X_1 - \bar{x})^2 + (X_2 - \bar{x})^2}{n - 1}\right]$$

where $\bar{x} = (X_1 + X_2)/n$ and $n = 2$.

9. Let X be a random variable with mean μ and variance σ^2. Calculate $E(X^2)$. Let $X_1, X_2, \ldots, X_n$ be a random sample of size n drawn from this distribution. Let $\overline{x^2} = (X_1^2 + X_2^2 + \cdots + X_n^2)/n$. Calculate $E(\overline{x^2})$. Is $\overline{x^2}$ an unbiased estimator of μ^2?

ANSWERS

KNOW THE CONCEPTS

1. The unknown population quantity is called a parameter; the quantity calculated from the sample is called a *statistic*.

2. You do not know for sure how accurate the estimate is unless you have been able to check the entire population to learn the true value. When we discuss confidence intervals in the next chapter, we will learn how to determine the likelihood that the estimated value is close to the true value.

3. An estimator with a small variance is less likely to deviate randomly from the true value it is trying to estimate. However, that does not mean you should always choose the estimator with the smallest possible variance. You need to make sure that the estimator is likely to be near the true value. It is often desirable to choose the unbiased estimator that has the lowest variance.

4. Division by $n - 1$ produces an unbiased estimator.

5. An estimator with a small amount of bias but a small variance might be preferable to an unbiased estimator with a large variance.

6. A point estimate indicates a single value for the unknown parameter; an interval estimate expresses a range of likely values for the parameter.

PRACTICAL APPLICATION

1. Sample average $\bar{x}$ will have a normal distribution with mean 7 and standard deviation $3/\sqrt{16} = 3/4$. Then $z = 4/3(x - 7)$ will have a standard normal distribution. For $\bar{x}$ to be between 7 and 7.5, z will be between 0 and 0.67, which will occur with probability .249.

$$\Pr(7.5 < \bar{x} < 8.0) = .160$$
$$\Pr(8.0 < \bar{x} < 8.5) = .069$$
$$\Pr(8.5 < \bar{x} < 9.0) = .019$$
$$\Pr(9.0 < \bar{x} < 9.5) = .003$$

You can see that $\bar{x}$ will probably be near 7, which is the true value of the mean. Therefore, if you observe the value $\bar{x} = 7$ when you have taken observations from a distribution with unknown mean, you will be quite willing to estimate that the true value of the mean is 7. This is the concept behind maximum likelihood estimators.

2. $E(V) = \mu$, so V is an unbiased estimator of μ.
   $\text{Var}(V) = (38/64)\sigma^2$. Note that $\text{Var}(V) > \text{Var}(\bar{x})$, where
   $\bar{x} = (X_1 + X_2 + X_3)/3$. Therefore, $\bar{x}$ is a better estimator of the mean.

3. $E(W) = (3/2)\mu$, so W is not an unbiased estimator of μ.
   $\text{Var}(W) = (3/4)\sigma^2$.

4. $E(\hat{p}) = E(X/n) = E(X)/n = np/n = p$. Therefore, $\hat{p}$ is an unbiased estimator of p.

5. $\text{Var}(\hat{p}) = \text{Var}(X/n) = \text{Var}(X)/n^2 = np(1 - p)/n^2 = p(1 - p)/n$

6. $E(A_2) = E[(X_1 + X_2)/2] = E(X_1 + X_2)/2 = p$
   $\text{Var}(A_2) = p(1 - p)/2$

   A_2 is an unbiased estimator of p. However, if we let $\hat{p} = X/n$, where X is the total number of successes in n trials, then we know that $\hat{p}$ is a much better estimator of p. $\text{Var}(\hat{p}) = p(1 - p)/n$, which is smaller than $\text{Var}(A_2)$. It is obvious why A_2 is not a very good estimator: it is based only on the results of the first two trials, whereas $\hat{p}$ is based on the results of all n trials.

7. $E(X_1) = p$ and $\text{Var}(X_1) = p(1 - p)$. Although X_1 is an unbiased estimator of p, it is even worse than the estimator discussed in Exercise 6. It is not a good idea to conduct the experiment once and then estimate that $p = 1$ if the experiment was a success and that $p = 0$ if the experiment was a failure.

8. Since $n = 2$, $n - 1 = 1$. Therefore:

$$E[(X_1 - \bar{x})^2 + (X_2 - \bar{x})^2]$$

$$= E\left[\left(X_1 - \frac{X_1}{2} - \frac{X_2}{2}\right)^2 + \left(X_2 - \frac{X_1}{2} - \frac{X_2}{2}\right)^2\right]$$

$$= E\left[\left(\frac{X_1}{2} - \frac{X_2}{2}\right)^2 + \left(\frac{X_2}{2} - \frac{X_1}{2}\right)^2\right]$$

$$= E\left[2\left(\frac{X_1}{2} - \frac{X_2}{2}\right)^2\right]$$

$$= 2E\left(\frac{X_1^2}{4} - \frac{X_1 X_2}{2} + \frac{X_2^2}{4}\right)$$

$$= 2\left[\frac{E(X_1^2)}{4} - \frac{E(X_1 X_2)}{2} + \frac{E(X_2^2)}{4}\right]$$

$$= 2\left[\frac{E(X^2)}{2} - \frac{E(X_1)E(X_2)}{2}\right]$$

$$= E(X^2) - [E(X)]^2$$

$$= \text{Var}(X)$$

Note: $E(X_1 X_2) = E(X_1) E(X_2)$ because X_1 and X_2 are independent. Also, $E(X_1) = E(X_2) = E(X)$ because they have identical distributions. This calculation shows (for $n = 2$) that the sample variance formed by dividing by $n - 1$ is an unbiased estimator of the variance.

9. Since $\text{Var}(X) = \sigma^2 = E(X^2) - \mu^2$, it follows that $E(x^2) = \sigma^2 + \mu^2$. You can also show that $E(x^2) = \sigma^2 + \mu^2$, so x^2 is not an unbiased estimator of μ^2.

11
CONFIDENCE INTERVALS

KEY TERMS

confidence interval an interval based on observations of a sample and so constructed that there is a specified probability that the interval contains the unknown true value of a parameter (for example, it is common to calculate confidence intervals that have a 95 percent chance of containing the true value)

confidence level the degree of confidence associated with a confidence interval; the probability that the interval contains the true value of the parameter

In Chapter 10 we discussed how to use observations of a random variable to get information about an unknown parameter for the distribution that generates that variable. However, we still have to face one important question: Is such an estimate likely to be very close to the true value?

For example, suppose you are trying to estimate the fraction of days that it rains in Florida. You naturally would use the estimator

$$\frac{\text{(number of days you've been in Florida when it rained)}}{\text{(number of days you've been in Florida)}}$$

However, if you've been in Florida only one day and it rained that day, your estimate that it rains in Florida every day is not likely to be very accurate. If you've spent 10 years in Florida, you'll be able to estimate the fraction of rainy days much more accurately.

The value of a statistic that is used as an estimator depends on the values of a group of random variables, meaning that the estimator is itself a random variable. It would help if we could figure out what the distribution of the estimator looks like. For example, let's suppose we are using the sample average $\bar{x}$ to estimate the value of the mean of a normal random variable.

$$\bar{x} = \frac{X_1 + X_2 + \cdots + X_n}{n}$$

Each of these X's has a normal distribution, so because of the addition property for normal random variables $\bar{x}$ must also have a normal distribution ($\bar{x}$ is found by adding up a bunch of normal random variables). We have already found (see pages 127–128) that $E(\bar{x}) = \mu$ and $\text{Var}(\bar{x}) = \sigma^2/n$.

CALCULATING CONFIDENCE INTERVALS FOR THE MEAN WHEN THE VARIANCE IS KNOWN

Now that we know the distribution of $\bar{x}$, we can be more precise about how good our estimate is. We know that the true value of μ is likely to be close to $\bar{x}$, but how close is close? Is $\bar{x}$ likely to be 1 unit away from μ? Or is it likely to be 50 units away? We'd like to know the probability that the distance from $\bar{x}$ to μ will be less than some specific value c. In other words, we want to know the probability that the true value of μ is between $(\bar{x} - c)$ and $(\bar{x} + c)$.

Obviously, the probability depends a lot on the value of c that we choose. (See Figure 11-1.) If we choose a very large value of c, we can be almost certain that the true value of μ will be in the interval. For example, we could set c to be infinity. Then the probability that μ will be in the interval is 100 percent, since obviously μ must be between $(\bar{x} - \text{infinity})$ and $(\bar{x} + \text{infinity})$. However, an interval that wide is not very useful. If we make the interval narrower by choosing a smaller value for c, we can be more precise about the true value of μ. However, when we make the interval narrower, there is a greater chance that μ won't even be contained in the interval.

The normal procedure in statistics works like this. First, we choose the probability that we want—in other words, we set in advance the probability that μ will be in the interval. Often this probability is set at 95 percent. Then we calculate how wide the interval must be so that there is a 95 percent chance that it contains the true value. This type of interval is called a **confidence interval**, and .95 is the **confidence level**.

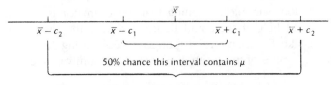

Figure 11-1

Now we need to calculate the value of c that satisfies the equation

$$\Pr(\bar{x} - c < \mu < \bar{x} + c) = .95$$

or

$$\Pr(-c < \bar{x} - \mu < c) = .95$$

Once we know the value of c, we know how wide the confidence interval must be. This means that our job is to calculate the value of c. Let's create a new random variable called Z:

$$Z = \frac{\bar{x} - \mu}{\sqrt{\sigma^2/n}}$$

$$= \sqrt{n}\,\frac{\bar{x} - \mu}{\sigma}$$

Using the properties we established for normal random variables, we know that Z has a standard normal distribution (mean 0 and variance 1; see Chapter 8). We can rewrite the equation:

$$\Pr\left(\frac{-c\sqrt{n}}{\sigma} < \sqrt{n}\,\frac{\bar{x} - \mu}{\sigma} < \frac{c\sqrt{n}}{\sigma}\right) = .95$$

$$\Pr\left(\frac{-c\sqrt{n}}{\sigma} < Z < \frac{c\sqrt{n}}{\sigma}\right) = .95$$

This problem calls for the standard normal probability table. Let's define a as follows: $a = c\sqrt{n}/\sigma$. Then

$$\Pr(-a < Z < a) = .95$$

Now we need to look in Table A3-2 until we find a value of a that satisfies this equation. Scanning down the columns, we can see that in this case the correct value for a is 1.96.

Now we can find the value for c:

$$c = \frac{1.96\sigma}{\sqrt{n}}$$

Thus we know how wide the confidence interval should be. There is a 95 percent chance that the interval from $\bar{x} - 1.96\sigma/\sqrt{n}$ to $\bar{x} + 1.96\sigma/\sqrt{n}$ will contain the true value of μ.

Two features of this result are appealing to common sense. First, the confidence interval is wider (that is, more uncertain) if σ is bigger. If the variance of each individual observation is bigger, then it will be harder for us to pin down the true value of μ. Second, the confidence interval is smaller if n is bigger. This means that as we take more and more observations we will be able to predict the true value of μ more accurately.

We can be even more cautious if we want to. Suppose that we want to be 99 percent sure that our confidence interval contains the true value of μ. Then we will have to settle for a wider, less precise interval. Or, if we wanted to be less careful, we could have calculated a smaller confidence interval that had a lower probability of containing the true value.

The confidence intervals that we talk about here and later on are symmetric in the following way: the probability that the left end point of the interval is greater than our parameter (μ here) is the same as the probability that the right end point of the interval is less than the parameter. In theory you could construct an asymmetric confidence interval, but symmetric confidence intervals are the narrowest ones possible, and thus more precise.

General Procedure for Calculating Confidence Intervals for the Mean When You Have *n* Observations from a Normal Distribution with Known Standard Deviation σ

1. Decide on the confidence interval you want. If you are more cautious, pick a higher level (.95 is one of the most common levels).

2. Look up the value of *a* in Table A3-2. If CL is the confidence level, then

$$\Pr(-a < Z < a) = CL$$

3. Calculate $\bar{x}$ and $a\sigma/\sqrt{n}$.

The confidence interval is from $\bar{x} - a\sigma/\sqrt{n}$ to $\bar{x} + a\sigma/\sqrt{n}$.

YOU SHOULD REMEMBER

1. When we calculate an estimate for an unknown population parameter, we need to know how accurate that estimate is likely to be.

2. It is valuable to calculate a *confidence interval*—that is, an interval constructed so that there is a fixed probability that the interval will contain the unknown value of the population parameter.

3. This fixed probability is known as the *confidence level*, and it is often set at 95 percent.

4. A narrow confidence interval is better because it means that you are able to make a more precise estimate of the true value of the parameter.

5. In general, the confidence interval will become narrower as the number of observations increases.

CALCULATING CONFIDENCE INTERVALS BY USING THE *t* DISTRIBUTION

There is one major difficulty with calculating confidence intervals in the way just described. Often we don't know the true value of σ^2. Our first guess might be that we could use the sample variance to estimate the value of σ^2. It turns out that, if the sample size (n) is large enough (for example, if $n > 30$), we can use the same confidence interval formula as in the preceding section, with the sample variance s_1^2 used in place of σ^2. However, for small samples we need to develop a new method.

Remember that our initial confidence interval calculation was based on the fact that

$$Z = \frac{\sqrt{n}(\bar{x} - \mu)}{\sigma}$$

has a standard normal distribution. Let's create a new random variable that we'll call T:

$$T = \frac{\sqrt{n}(\bar{x} - \mu)}{s_2}$$

Note that T is exactly the same as Z except that the known value $s_2 = \sqrt{s_2^2}$ has been substituted in place of the unknown value of σ. We would expect the distribution of T to be very similar to a standard normal distribution.

In Chapter 8, we found that the t distribution is similar to the normal distribution. Our statistic T will have a t distribution with $n - 1$ degrees of freedom.

Now we can calculate the width of the confidence interval. We need to find the value of c that satisfies the equation

$$\Pr(\bar{x} - c < \mu < \bar{x} + c) = .95$$

Using our definition of the statistic T, we can rewrite the equation like this:

$$\Pr\left(\frac{-c\sqrt{n}}{s_2} < T < \frac{c\sqrt{n}}{s_2}\right) = .95$$

We define a to be equal to $c\sqrt{n}/s_2$.

Now we have to look in a table of the t distribution. We need to find a value of a such that

$$\Pr(-a < T < a) = .95$$

For example, if $n - 1 = 8$, we can see from Table A3-5 that $a = 2.306$. Once we find a, we can find c from the formula

$$c = s_2 \frac{a}{\sqrt{n}}$$

Therefore, the 95 percent confidence interval for μ is from $\bar{x} - s_2 a/\sqrt{n}$ to $\bar{x} + s_2 a/\sqrt{n}$.

Example: Finding the Confidence Interval by Using the t Distribution

PROBLEM   We have been given a list of 750 major cities around the world and would like to estimate the average population of these cities.

SOLUTION   Choose a random sample of 20 of these cities. To choose the cities, use a computer random-number generation program. Here is one example of 20 randomly selected world cities:

Population	City
473,800	Bangui, Central African Republic
8,243,400	Bombay, India
550,000	Bucaramanga, Colombia
160,800	Cádiz, Spain
1,170,000	Cheng-chou, China
1,279,200	Curitiba, Brazil
4,884,200	Delhi, India
923,300	Goiânia, Brazil
1,144,000	Ibadan, Nigeria
1,056,400	Kitakyushu, Japan
346,900	Miami, USA
285,000	Nouakchott, Mauritania
1,287,600	Recife, Brazil
818,300	Sakai, Japan
484,700	Scarborough, Canada
112,300	Thunder Bay, Canada
348,000	Toulouse, France
336,500	Tucson, USA
566,000	Vilnius, Lithuania
266,500	Wiesbaden, Germany

To calculate a 95 percent confidence interval for the mean, we first need to calculate $\bar{x} = 1,236,845$ and $s_2 = 1,939,221.09740$. Since $n = 20$, we need to look in Table A3-5 for $20 - 1 = 19$ degrees of freedom, to find $a = 2.093$. Therefore, the confidence interval is $1,236,845 \pm 2.093 \times 1,939,221.09740/\sqrt{20}$, or 329,272–2,144,417.

This interval is so wide that it is not very useful. The problem is that there is a very large variance among the population of cities; there are some very large cities and a lot of smaller cities. The city populations do not follow the normal distribution, but we know from the central limit theorem that $\bar{x}$ will have a normal distribution anyway. As you can guess, the way to find a narrower confidence interval is to take a larger sample.

Note: It is correct to say, "There is a 95 percent chance that the interval $\bar{x} - s_2 a/\sqrt{n}$ to $\bar{x} + s_2 a/\sqrt{n}$ contains the true value of μ." The two end points of the interval are random variables. However, once you have performed the calculation, it is not correct to say, "There is a 95 percent chance that the interval 329,272 to 2,144,417 contains the true value of μ." Since μ is not a random variable, it does not make sense to talk about the probability that it takes on values in a certain range.

Procedure to Calculate Confidence Intervals Using the *t* Distribution When You Have *n* Observations of the Random Variable *X*

(Use for small samples when σ is unknown.)

1. Decide on the confidence level CL (.95 is one of the most common levels).

2. Calculate $\bar{x}$:

$$\bar{x} = \frac{x_1 + x_2 + \cdots + x_n}{n}$$

3. Calculate s_2:

$$s_2 = \sqrt{\frac{(x_1 - \bar{x})^2 + (x_2 - \bar{x})^2 + \cdots + (x_n - \bar{x})^2}{n - 1}}$$

$$= \sqrt{\frac{n}{n - 1}(\overline{x^2} - \bar{x}^2)}$$

4. Look up the value of *a* in Table A3-5.

$$\Pr(-a < T < a) = CL$$

where *T* has a *t* distribution with *n* − 1 degrees of freedom. (Note that, if *n* is larger than 30, the *t* distribution is almost the same as the standard normal distribution.)

The confidence interval for μ is from $\bar{x} - s_2 a/\sqrt{n}$ to

$$\bar{x} + s_2 a/\sqrt{n}.$$

CALCULATING THE CONFIDENCE INTERVAL FOR THE VARIANCE

Suppose you have a sample taken from an approximately normal population whose variance you would like to estimate. We know how to calculate the sample variance s_1^2. Now we would like to determine a confidence interval for

σ^2. We have found that

$$Y^2 = \frac{ns_1^2}{\sigma^2}$$

is a chi-square random variable with $n - 1$ degrees of freedom. For any two positive numbers a and b we know that

$$\Pr(a < Y^2 < b) = \Pr\left(a < \frac{ns_1^2}{\sigma^2} < b\right)$$

$$= \Pr\left(\frac{ns_1^2}{b} < \sigma^2 < \frac{ns_1^2}{a}\right)$$

For example, here are the heights (in centimeters) of 25 students selected at random from Millard Fillmore Elementary School:

$$135, 139, 128, 143, 122, 123, 142, 135, 140,$$
$$141, 115, 133, 128, 137, 142, 128, 135,$$
$$142, 129, 133, 141, 137, 125, 127, 138$$

We can find that s_1^2 is 54.17. We will construct a 90 percent confidence interval for the variance. We need to find two numbers a and b such that

$$\Pr(Y^2 < a) = .05 \quad \text{and} \quad \Pr(Y^2 < b) = .95$$

Looking in Table A3–3 for a chi-square distribution with $25 - 1 = 24$ degrees of freedom, we find $a = 13.85$ and $b = 36.4$. Then the confidence interval is from $ns_1^2/b = 37.20$ to $ns_1^2/a = 97.78$.

Procedure to Calculate the Confidence Interval for the Variance

1. Decide on the confidence level (CL).

2. Calculate $\bar{x}$, $\overline{x^2}$, and $s_1^2 = \overline{x^2} - \bar{x}^2$.

3. Use Table A3-3 to find a and b such that, if Y^2 is a chi-square random variable with $n - 1$ degrees of freedom, then

$$\Pr(Y^2 < a) = \frac{1 - CL}{2} \quad \text{and} \quad \Pr(Y^2 < b) = \frac{1 + CL}{2}$$

For example, if CL $= .90$, then $\Pr(Y^2 < a) = .05$ and $\Pr(Y^2 < b) = .95$.

The confidence interval for σ^2 is from ns_1^2/b to ns_1^2/a.

CALCULATING THE CONFIDENCE INTERVAL FOR THE DIFFERENCE BETWEEN TWO MEANS

Often a situation will arise where you want to compare two populations with respect to some random variable. Here are some examples:

- average income in two different cities

- sales revenue of two companies

- number of readers of two newspapers

In all of these examples it would be of interest to estimate the difference of the means, $\mu_a - \mu_b$, of the two random variables X_a (mean μ_a, variance σ_a^2) and X_b (mean μ_b, variance σ_b^2). This is done by calculating the difference of the sample means, $\bar{x}_a - \bar{x}_b$. As before, the question of accuracy arises. How close is $\bar{x}_a - \bar{x}_b$ to the unknown value of $\mu_a - \mu_b$? Again we turn to confidence intervals.

Let X_a have sample mean $\bar{x}_a$ and sample size n_a; and similarly for X_b, $\bar{x}_b$, and n_b. If X_a and X_b are normal random variables, then so is $\bar{x}_a - \bar{x}_b$, which has mean $\mu_a - \mu_b$ and variance $(\sigma_a^2/n_a + \sigma_b^2/n_b)$. As above, we create a new random variable Z:

$$Z = \frac{(\bar{x}_a - \bar{x}_b) - (\mu_a - \mu_b)}{\sqrt{\sigma_a^2/n_a + \sigma_b^2/n_b}}$$

which has a standard normal distribution. Then look in the Appendix: Table A3-2 to find the value of a such that

$$\Pr(-a < Z < a) = \text{CL}$$

where CL is the confidence level. Then calculate c:

$$c = a\sqrt{\sigma_a^2/n_a + \sigma_b^2/n_b}$$

The confidence interval for $\mu_a - \mu_b$ is then:

$$(\bar{x}_a - \bar{x}_b) \pm c$$

For example, here is a set of sales figures (in thousands) for a given newspaper in two nearby towns over a period of days:

Town A: 25, 13, 14, 19, 23, 30, 35, 29, 28, 17, 17, 16, 13, 18, 20

Town B: 10, 12, 15, 13, 7, 6, 11, 5, 9, 14, 15, 18, 17, 16, 12, 12, 10, 11, 13, 14

Assume that $\sigma_a^2 = 40$ and $\sigma_b^2 = 14$. Then we can calculate that $\bar{x}_a = 21.13$, $n_a = 15$, $\bar{x}_b = 12.00$, $n_b = 20$, and $\bar{x}_a - \bar{x}_b = 9.13$. Using a 99 percent confidence interval, we find that $a = 2.58$, $c = 2.58 \sqrt{40/15 + 14/20} = 4.73$, and the confidence interval is from 4.40 to 13.86.

We then have the following procedure (similar to the one used to calculate the confidence interval for the mean of one variable):

Procedure to Calculate the Confidence Interval for the Difference Between Two Means (σ_a, σ_b known)

1. Decide on the confidence level you want (such as 95 percent).

2. Look up the value of a in Table A3-2.

3. Calculate $\bar{x}_a - \bar{x}_b$.

4. Calculate $c = a\sqrt{\sigma_a^2/n_a + \sigma_b^2/n_b}$

The confidence interval is $(\bar{x}_a - \bar{x}_b) \pm c$.

Again, if the sample sizes are large enough, the sample variances can be substituted for σ_a^2 and σ_b^2 when the latter are unknown. For small samples we turn to the t distribution. Here, however, we must make two more assumptions, namely, that $\sigma_a^2 = \sigma_b^2$, and that the samples are selected independently. (With small sample sizes, it would be very easy to lose the randomness essential to good statistics.)

Remember that we're not assuming that we know σ_a^2 or σ_b^2. We are just assuming that they are equal. If you do know them, good. If not, we can approximate them by the *pooled estimate* s_p^2:

$$s_p^2 = \frac{(n_a - 1)s_a^2 + (n_b - 1)s_b^2}{(n_a - 1) + (n_b - 1)}$$

We will use s_a^2 to mean the sample variance, version 2, of sample a:

$$s_a^2 = \frac{(\overline{x_a^2} - \bar{x}_a^2)n_a}{n_a - 1}$$

Likewise, s_b^2 is the sample variance, version 2, for sample b. Also, s_p^2 is a weighted average of s_a^2 and s_b^2, so that, if n_a is much larger than n_b, then s_p^2 is closer to s_a^2 than s_b^2, and vice versa.

Now, in the above formula for large samples, if $\sigma_a^2 = \sigma_b^2$, we have

$$Z = \frac{(\bar{x}_a - \bar{x}_b) - (\mu_a - \mu_b)}{\sqrt{\sigma_a^2/n_a + \sigma_b^2/n_b}}$$

$$= \frac{(\bar{x}_a - \bar{x}_b) - (\mu_a - \mu_b)}{\sqrt{\sigma_a^2(1/n_a + 1/n_b)}}$$

We substitute in our pooled estimator s_p^2 in the place of σ_a^2 and create a new t statistic:

$$T = \frac{(\bar{x}_a - \bar{x}_b) - (\mu_a - \mu_b)}{\sqrt{\sigma_p^2(1/n_a + 1/n_b)}}$$

This t statistic has $(n_a - 1) + (n_b - 1) = n_a + n_b - 2$ degrees of freedom. If we know what the value of σ_a^2 is, we insert it for s_p^2 (no need to estimate if we've got the real thing).

We find the value of a from Table A3-5 so that

$$\Pr(-a < T < a) = \text{CL}$$

where CL is the confidence level and T has a t distribution with $n_a + n_b - 2$ degrees of freedom. Then we calculate c:

$$c = a\sqrt{s_p^2\left(\frac{1}{n_a} + \frac{1}{n_b}\right)}$$

Then the confidence interval for $\mu_a - \mu_b$ is

$$(\bar{x}_a - \bar{x}_b) \pm c$$

For example, here are the sales figures (in thousands) for two different brands of kites for randomly selected months:

Brand A: 15, 20, 33, 27

Brand B: 23, 42, 39

Assume that $\sigma_a^2 = \sigma_b^2$. We can calculate that $\bar{x}_a = 23.75$, $\bar{x}_b = 34.67$, $\bar{x}_a - \bar{x}_b = -10.92$, $s_a^2 = 62.25$, $s_b^2 = 104.3$, and $s_p^2 = 79.08$. Using a 95 percent confidence interval for $n_a + n_b - 2 = 5$ degrees of freedom, we find that $a = 2.571, c = 2.571 \sqrt{79.08(1/4 + 1/3)} = 17.46$, and the confidence interval is from -28.38 to 6.54.

Procedure to Calculate the Confidence Interval for the Difference between Population Means When We Have Small Samples with Equal Population Variances

1. Decide on the confidence interval (such as 95 percent).

2. Calculate $\bar{x}_a - \bar{x}_b$.

3. Calculate s_p^2:

$$s_p^2 = \frac{(n_a - 1)s_a^2 + (n_b - 1)s_b^2}{n_a + n_b - 2}$$

4. Look up the value of a in Table A3-5 for $n_a + n_b - 2$ degrees of freedom.

5. Calculate c:

$$c = a \sqrt{s_p^2 \left(\frac{1}{n_a} + \frac{1}{n_b} \right)}$$

The confidence interval is $(\bar{x}_a - \bar{x}_b) \pm c$.

KNOW THE CONCEPTS

DO YOU KNOW THE BASICS?

Test your understanding of Chapter 11 by answering the following questions:

1. Why shouldn't you calculate a confidence interval with a very high confidence level, such as 99.99 percent?
2. Why is a confidence interval more valuable than a single point estimate?
3. Name three ways to make a confidence interval narrower.
4. When should you calculate a confidence interval using the t distribution instead of the normal distribution?
5. If you double the number of observations, will you be able to cut the width of the confidence interval in half?

TERMS FOR STUDY
confidence interval   confidence level

PRACTICAL APPLICATION

COMPUTATIONAL PROBLEMS
Each of the following lists of numbers represents weekly net revenue figures for one company. Assume that for each company the net revenue figures have a normal distribution with unknown mean and unknown variance. Calculate a 95 percent confidence interval for the true value of the mean.

1.	23	12	1	6	4
	16	28	14	6	28
	15	18	6	2	14
	19	11	15	20	20
2.	5	48	6	49	0
	11	44	47	35	13
	10	4	50	7	22
	1	40	4	4	26
	8	18	35	23	12
3.	12	1	22	8	24
	25	22	21	13	12
	18	8	0	16	20
	11				
4.	10	20	19	30	17
	28	26	26	29	28
	6	4			
5.	23	8	18	29	14
	20	1	19	19	28
	18	16	30	4	19

    For each of the following exercises you are given the total of a list of observations, the average of these observations, and the standard deviation (s_2) of these observations. Calculate a 95 percent confidence interval and a 99 percent confidence interval for the unknown value of the mean of each distribution.

6. Total: 523.114
   Average: 52.311
   s_2: 8.813

7. Total: 1,571.322
   Average: 104.755
   s_2: 3.002

8. Total: 1,150.295
   Average: 127.811
   s_2: 4.611

9. Total: 4,528.186
   Average: 283.012
   s_2: 8.386

10. Total: 2,957.990
    Average: 147.899
    s_2: 1.688

ANSWERS

KNOW THE CONCEPTS

1. An interval with that high a confidence level generally is too wide to be useful.

2. A point estimate contains no information about how accurate the estimate is likely to be.

3. More observations; smaller value of the population variance; lower confidence level.

4. The t distribution should be used when the population variance is unknown and the number of observations is less than 30.

5. No. The width of the confidence interval is proportional to $1/\sqrt{n}$.

PRACTICAL APPLICATION

Calculations are shown below for Exercise 1 and Exercise 6. Similar methods can be used for Exercises 2–5 and 7–10, respectively.

1. $\bar{x} = (23 + 12 + 1 + 6 + 4 + 16 + 28 + 14 + 6 + 28$
    $+ 15 + 18 + 6 + 2 + 14 + 19 + 11 + 15 + 20 + 20)/20$
    $= 13.9$

    $\overline{x^2} = (23^2 + 12^2 + 1^2 + 6^2 + 4^2 + 16^2 + 28^2 + 14^2 + 6^2$
    $+ 28^2 + 15^2 + 18^2 + 6^2 + 2^2 + 14^2 + 19^2 + 11^2$
    $+ 15^2 + 20^2 + 20^2)/20$
    $= 253.7$

    $s_2 = \sqrt{253.7 - 13.9^2} \times \sqrt{19/20} = 7.98$

From Table A3-5 we can find that $a = 2.093$ for a t distribution with 19 degrees of freedom. Therefore the .950 confidence interval is

$$13.9 \pm 2.093 \times 7.98/\sqrt{20} \quad \text{or} \quad 10.165–17.635$$

2. Average = 20.880 s_2 = 17.405
   .950 confidence interval = 13.695–28.065

3. Average = 14.562 s_2 = 7.763
   .950 confidence interval = 10.427–18.698
4. Average = 20.250 s_2 = 9.255
   .950 confidence interval = 14.369–26.131
5. Average = 17.733 s_2 = 8.455
   .950 confidence interval = 13.050–22.416

95% Confidence Intervals	99% Confidence Intervals
6. 52.311 ± 2.262 × 8.813/$\sqrt{10}$	52.311 ± 3.250 × 8.813/$\sqrt{10}$
46.007– 58.615	43.254– 61.368
7. 103.092–106.417	102.447–107.063
8. 124.266–131.355	122.654–132.968
9. 278.544–287.479	276.834–289.190
10. 147.110–148.689	146.819–148.979

12
POLLS AND SAMPLING

KEY TERMS

cluster sampling a sampling method where the population is divided into clusters; some clusters are selected at random, and then some members of the chosen clusters are selected at random to make up the sample

stratified sampling a sampling method where the population is divided into strata that are as much alike as possible

OPINION POLLS

How can we find out how many people in a population have a particular characteristic? For example, we might want to know how many voters support our favorite presidential candidate. Or we might be interested in some general characteristics of people in a certain state—for example, how many are children, how many live in cities, or how many are employed.

One way to find the answers to these questions is to check with everybody. This method will be very accurate. Since we're asking *everybody*, we can get a detailed view of the entire population. This method is used sometimes. An election is held every 4 years in the United States to find out the presidential preferences of all voters. Every 10 years a census is held to obtain detailed information about all of the people in the country.

However, there are disadvantages to the ask-everybody method. The main one is that this method is very expensive. Elections and censuses are costly. Also, there will often be times when we would like to obtain information about the population but we can't wait until the next census or the next election.

Another possible method is to ask some people who are part of a sample. If the sample is representative of the entire population, we can use the characteristics of the people in the sample to estimate the characteristics of the people in the population. For example, opinion poll interviewers typically check with about 1,000 people in an attempt to estimate the opinions of all 250 million people in the country.

Are these results likely to be accurate? At first glance you might be suspicious. The poll takers are talking to only about 1 out of every 250,000 people. You probably think it is unlikely that each person in the poll has the same opinions as his or her 250,000 closest neighbors. However, polls do seem to be fairly accurate. The election results predicted by polls are usually close to the actual results (with a few notable exceptions).

Now we will look at the theory that explains why these results tend to be accurate. We'll let N stand for the number of people in the population we are investigating. We'll assume that M of these people favor our candidate. Our goal is to estimate M/N—the fraction of people who support our candidate. Let $p = M/N$. We'll ask a sample of n people. X will represent the number of people in the sample who favor our candidate. If our poll is any good, X/n will be close to M/N. Let $\hat{p} = X/n$.

For example, suppose that we are trying to estimate the preferences of people in a town with a population of 30,000. We will ask all the people in a sample of 500 which candidate they support. Suppose that in reality there are 16,500 people who support our candidate (16,500 is 55 percent of the total population). If it turns out that 270 people (54 percent) in the sample support our candidate, then the sample did a good job of representing the entire population. On the other hand, if 330 people (66 percent) in the sample are on our side, then the sample is quite unrepresentative and our poll will give us very misleading results.

• *METHODS FOR SELECTING A SAMPLE*

It makes a big difference how the sample is selected, since the value of X will depend on exactly who is in the sample. We need to figure out a good system for choosing the sample so that it will be representative of the population as a whole. We can't just start asking our friends, since our friends might be more likely to be on our side. We can't select just one neighborhood and interview everyone there, since people in one particular neighborhood are not likely to be representative of the diverse characteristics of all of the people in the town.

There are subtle problems with other systems as well. We can't just mail out a lot of postcards and ask people to return them, since the people on our side might be more likely to take the trouble to send the cards back. We might decide that we can make our sample representative by deciding in advance that we want our sample to contain certain quotas of people with particular characteristics. For example, we might decide that we want our sample to contain 50 percent women, 15 percent minorities, and 0.5 percent veterinarians. However, that method still doesn't answer the question about how to select the sample, since it doesn't tell us which minorities or which women or which veterinarians to include in the sample. We obviously can't set in advance a quota for the number of people in the sample who favor our candidate, since we don't know what that fraction is until after we've taken the sample.

It turns out that the best system for selecting the sample is to have no system at all—in other words, select the sample completely at random. We should design the sampling system so that each person has an equal chance of being selected. Not only that, we should design the system so that every single possible sample that we might conceive of has an equal chance of being the sample that we actually choose.

How are we going to do this? One way is to write everyone's name on a slip of paper, put the papers in little capsules, and then put all of the capsules in a large drum. If we mix up the capsules very thoroughly and then pull n capsules out of the drum, we will have a random sample of n people. However, even this approach is difficult if N is very large. For one thing, we would need a very large drum. Another problem is that it is difficult to mix the capsules well if there are a large number of them. If the capsules are not mixed well, then the people whose names were put in last have a greater chance of being selected and we will no longer have a pure random sample.

Something akin to this actually did happen with the draft lottery in the United States in 1970. People were assigned draft numbers on the basis of their birth dates, with a low number indicating a greater chance of being inducted. The 366 dates were put into capsules, mixed, and drawn and assigned lottery numbers 1, 2, etc. Apparently, the capsules were not mixed very well—people born in December had lottery numbers that averaged 121.5, which is pretty far away from the average of #1–366 = 183.5. Steps were taken with the 1971 draft lottery to make the results more random by drawing both the date and the lottery number from drums, after mixing them more thoroughly.

An easier method is to give everyone in the population a number. Then we can select a bunch of random numbers and interview the people whose numbers we've chosen. How do we select the random numbers? It's not as easy as it sounds. We can't just start making up numbers, since it is hard for a person to make up a long string of numbers without falling into some sort of pattern. (Try it yourself sometime.) We could use a die if we needed only numbers from 1 to 6, but to get a larger group of numbers we need a better system.

In the old (precomputer) days, the best way was to use a table of random digits. A random-digit table is a table that has been created by someone whose job it is to create random numbers. The numbers have been tested to make sure that they pass certain tests of randomness. Nowadays, you can have a computer generate the random numbers. Most computer systems have built-in random-number generators. The numbers that they generate are not true random numbers, since they are generated according to a fixed rule. However, the rule is unpredictable enough that for all practical purposes the numbers seem to have been selected totally at random.

YOU SHOULD REMEMBER

1. It is expensive to survey everyone in the population, so samples are selected.

2. The proportion of people in a sample with a particular characteristic is used to estimate the proportion of people in the population with the same characteristic.

3. The best method for choosing a sample is to select the sample at random, that is, in such a way that every possible sample is equally likely to be chosen.

• *USING THE BINOMIAL DISTRIBUTION*

We will choose the sample without replacement, since we don't want to run the risk of annoying someone by calling them twice. We already know the probability distribution to use: the hypergeometric distribution. We know:

$$\Pr\left(X = i\right) = \frac{\binom{M}{i}\binom{N-M}{n-i}}{\binom{N}{n}}$$

where $p = M/N$; $E(X) = np$, $\text{Var}(X) = np(1-p)(N-n)/(N-1)$.

The hypergeometric calculations are very cumbersome to deal with. To make the calculations easier, let's change our sampling procedure slightly. We'll still select 1,000 names from the drum. However, this time after we select each name we'll put it back in the drum and mix all of the names again. That means that a person who has been selected once in the sample might be selected again, and would thus be interviewed twice. (That person probably would be annoyed, which is one reason why this method is not used in practice.) There is even a minuscule chance that we will select the same person 1,000 times.

Once again we'll let X represent the number of people in the sample who favor our candidate. Let's call it a "success" if a particular person in the sample favors our candidate. Then we know exactly what the distribution of X looks like—it has a binomial distribution with parameters n and $p = M/N$.

It turns out that if the population size is much larger than the sample size we can use the binomial distribution even if we sample without replacement. For example, let's suppose that, in a population of 200 million, half of the people are on our side and half are on the other side. Then there is a probability of 1/2 that the first person we select will support our candidate. Suppose that the first person we select does support our candidate. If we sample with replacement, there is a probability of .5 that the second person we select will also be on our side. If we sample without replacement, there is a probability of

$$\frac{99,999,999}{199,999,999} = .4999999975$$

that the second person will support our candidate. So it does make a slight difference whether we sample with replacement or without replacement—but not much. Therefore, we can approximate the hypergeometric distribution by a binomial distribution.

CONFIDENCE INTERVAL FOR PROPORTION

• *USING THE NORMAL DISTRIBUTION*

The binomial distribution itself is rather cumbersome, so while we're making approximations we may as well go all the way and approximate the binomial distribution with a normal distribution. We know that this works when n becomes large. (See Chapter 8.)

Let $\hat{p}$ be the sample proportion X/n, where X is the number of people in the sample that support our candidate. Since X can be approximated by a normal distribution with mean np and variance $np(1 - p)$, it follows that $\hat{p}$ has a normal distribution with mean p and variance $p(1 - p)/n$.

Figures 12-1 and 12-2 illustrate what happens as we increase the sample size from 100 to 1,000. In both cases the population proportion is $p = .28$. The curve illustrates the density function for $\hat{p}$; note how the density function becomes taller and narrower for larger sample sizes. This illustrates that the probabilities for $\hat{p}$ become concentrated closer to the value of p as n becomes larger.

In reality, we will usually not know the true value of p, but we can calculate a confidence interval for it.

Let c represent half of the width of the confidence interval, as in the previous chapter. The confidence interval will be centered around $\hat{p}$. It is

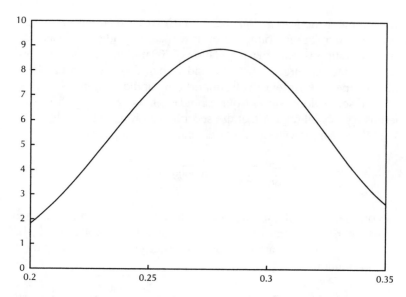

Figure 12-1. Choosing a random sample from a large population with the population proportion equal to 0.28 and the sample size equal to 100

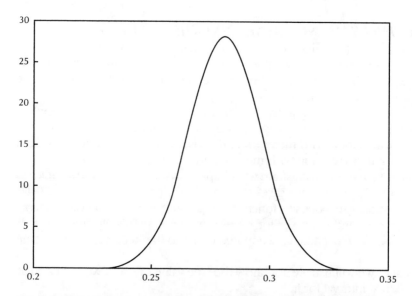

Figure 12-2. Choosing a random sample from a large population with the population proportion equal to 0.28 and the sample size equal to 1000

defined by this equation:

$$\Pr(\hat{p} - c < p < \hat{p} + c) = .95$$

Subtract p from all three parts of the inequality:

$$\Pr(-c < p - \hat{p} < c) = .95$$

Multiply all three parts by -1:

$$\Pr(c < \hat{p} - p < -c) = .95$$

Reverse the order:

$$\Pr(-c < \hat{p} - p < c) = .95$$

We have a normal random variable ($\hat{p}$) minus its mean (p); as you might guess, the next step is to divide by its standard deviation $\sqrt{p(1-p)/n}$.

$$\Pr\left[\frac{-c}{\sqrt{\frac{p(1-p)}{n}}} < \frac{\hat{p}-p}{\sqrt{\frac{p(1-p)}{n}}} < \frac{c}{\sqrt{\frac{p(1-p)}{n}}}\right] = .95$$

The middle expression can be replaced by the standard normal random variable Z; then make this definition:

$$a = \frac{-c}{\sqrt{\frac{p(1-p)}{n}}}$$

Then:

$$\Pr(-a < Z < a) = .95$$

From the table at the back of the book, we see that $a = 1.96$ (we would use a different value of a if the confidence level were different from .95). Therefore:

$$c = 1.96\sqrt{\frac{p(1-p)}{n}}$$

The population proportion p is unknown, but we can substitute the sample proportion $\hat{p}$ in its place. Therefore, the 95 percent **confidence interval for the proportion p** is:

$$\hat{p} - 1.96\sqrt{\frac{\hat{p}(1 - \hat{p})}{n}} \quad \text{to} \quad \hat{p} + 1.96\sqrt{\frac{\hat{p}(1 - \hat{p})}{n}}$$

For example, let $\hat{p} = .62$ and $n = 200$. The confidence interval is from

$$.62 - 1.96\sqrt{\frac{.62 \times .38}{200}} \quad \text{to} \quad .62 + 1.96\sqrt{\frac{.62 \times .38}{200}}$$

or from

$$.552 \text{ to } .687$$

General Procedure for Confidence Interval for Proportion

Let n be the number of items in the sample, X be the number of items in the sample with the characteristic you're interested in, $\hat{p} = X/n$ be the proportion of people in the sample with that characteristic, and p be the unknown proportion of people in the population with that characteristic.

1. Determine the confidence level you want.

2. Look up the value of a in Table A3-2. (If the confidence level is 95 percent, then $a = 1.96$.)

3. Calculate $c = a\sqrt{\hat{p}(1 - \hat{p})/n}$.

The confidence interval is from $\hat{p} - c$ to $\hat{p} + c$.

One minor concern about our analysis is that our approximation might break down if the population is not much larger than the sample. Recall that we can approximate the hypergeometric distribution with parameters N, M, and n by the binomial distribution with parameters $p = M/N$ and n, provided N is much larger than n. Fortunately, in many realistic cases, the population is much larger than the sample, so the approximation works just fine. To analyze situations where the approximation does not work, we need to consider the formula we saw in Chapter 7 that gives the variance of the hypergeometric distribution:

$$n\left(\frac{M}{N}\right)\left(1 - \frac{M}{N}\right)\left(\frac{N - n}{N - 1}\right) = np(1 - p)\left(\frac{N - n}{N - 1}\right)$$

As long as N is much larger than n, then the fraction $(N - n)/(N - 1)$ will be close to 1, and the binomial approximation works fine. Otherwise, we can approximate the hypergeometric distribution with a normal distribution using this standard deviation:

$$\sqrt{np(1-p)\left(\frac{N-n}{N-1}\right)}$$

The expression $\sqrt{(N-n)/(N-1)}$ is called the **finite population correction factor**. If the population is much larger than the sample, then it can be ignored. Otherwise, use this formula for the confidence interval for p:

$$\hat{p} \pm 1.96\sqrt{\frac{\hat{p}(1-\hat{p})}{n}}\sqrt{\frac{N-n}{N-1}}$$

We can make a table of these results. (See Table 12-1.)

Table 12-1.   Percent Error for Sample

N	n = 100	n = 500	n = 1,000	n = 5,000	n = 10,000	n = 50,000
10,000	9.8	4.3	2.9	1.0	0	—
50,000	9.8	4.4	3.1	1.3	0.9	0
100,000	9.8	4.4	3.1	1.4	0.9	0.3
500,000	9.8	4.4	3.1	1.4	1.0	0.4
50,000,000	9.8	4.4	3.1	1.4	1.0	0.4
200,000,000	9.8	4.4	3.1	1.4	1.0	0.4

The table assumes that $p = .5$. If the true value of p is different from .5, then the errors will be less than the values listed in the table. The table lists the percentage-point error. For example, for population size 50,000 and sample size 5,000, the table gives the value 1.3. This means that there is a 95 percent chance that the sample proportion will be within 1.3 percentage points of the true value (in other words, between .487 and .513).

• ANALYZING THE PERCENT ERROR IN TERMS OF SAMPLE SIZE

You can see some interesting results if you scan through the table. If you take a sample of 1,000 people, there is a 95 percent chance that the poll result will be within 3.1 percentage points of the true value. If you increase the sample size to 5,000, the error falls to only 1.4 percent. Therefore, the table does tend to reinforce one's faith in polls.

Another interesting result you can see is that a sample of 1,000 does just as well when the population is 200 million as it does when the population is

50,000. You might expect that the sample would become less accurate as the population becomes larger, but this is not the case.

However, the reverse is also true. If you want to get an accurate sample of a population of 50,000, you need just as large a sample as you would if you had a population of 200 million. Making the population smaller does not reduce the number of people you need in the sample in order to get a representative sample. You can get a good cross section of 200 million people by interviewing one person in every 40,000; but if you try to interview one person in 40,000 when the population is 50,000, you will end up with a very unrepresentative sample. What this means is that if you want an accurate view of the opinions of people in every state you will need a much larger sample than you would if you needed to know only the opinions of the entire country.

Another important fact to note is that, although the error goes down as the sample size becomes larger, it reaches a point where large increases in the sample size lead to only small decreases in the error. Therefore, when you decide on what size sample to use, you need to take two factors into consideration. Adding more people will make the sample a bit more accurate, but the cost of taking the sample becomes larger if more people are included.

YOU SHOULD REMEMBER

1. The accuracy of a poll depends on the size of the sample, not the size of the population (provided that the population size is large).

2. If the population is much larger than the sample, then the sample proportion will have an approximately normal distribution.

The confidence interval formula given earlier works when the population size N is much larger than the sample size n [so $(N-n)/(N-1)$ is close to 1] and the sample size is larger than 30. If the sample size is smaller than 30, then the normal approximation is not as accurate, so you need to perform some tricky calculations involving the binomial distribution that we will not discuss in this book.

An opinion poll cannot measure the population proportion if the population proportion is very small. Suppose you would like your poll to be able to tell you the proportion of suburban veterinarians aged 50 to 55 in the nation. The population proportion is very small, and you are quite likely to find that there are zero suburban veterinarians aged 50 to 55 in a national sample of 2,000 people. In that case you don't have the slightest idea what the actual value is (other than the fact that it must be very small, but you knew that even before you took the poll).

In general, if $np > 5$ and $n(1-p) > 5$, then the confidence intervals discussed above based on the normal approximation work fine. For example, suppose $n = 200$ and $p = .025$. Then $np = 5$ and we can calculate the probabilities: (see Figure 12-3)

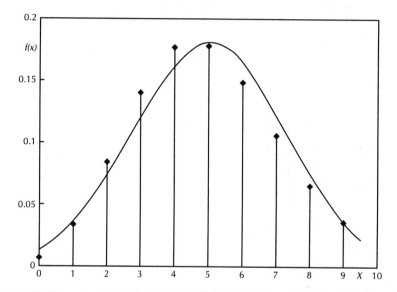

Figure 12-3

	Probability from Binomial Distribution	Normal Approximation
$\Pr(X = 0)$	.006	.014
$\Pr(X = 1)$	.032	.035
$\Pr(X = 2)$	.083	.072
$\Pr(X = 3)$	.140	.120
$\Pr(X = 4)$	.177	.163
$\Pr(X = 5)$	.178	.181
$\Pr(X = 6)$	.148	.163
$\Pr(X = 7)$	.105	.120
$\Pr(X = 8)$	.065	.072
$\Pr(X = 9)$	.036	.035
$\Pr(X = 10)$	.017	.014

The normal approximation still works quite well, and the value of X will probably be close to its expected value of 5. However, now suppose that $n = 200$ and $p = .001$. Then we can find that $E(X) = np = 0.2$. Obviously the actual value of X cannot be equal to 0.2 because it must be a whole number. Here are the probabilities: (see Figure 12-4)

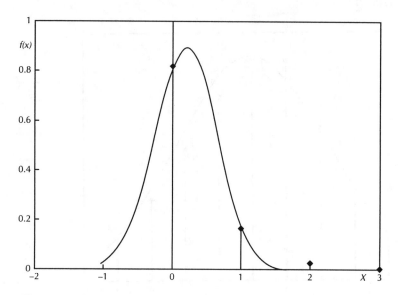

Figure 12-4

	Probability from Binomial Distribution	Normal Approximation
$\Pr(X = -1)$	—	.024
$\Pr(X = 0)$	.819	.807
$\Pr(X = 1)$	.164	.180
$\Pr(X = 2)$	.016	0
$\Pr(X = 3)$	.001	0

The most likely value of X is zero. In this case the normal approximation breaks down because the actual value of X can never be negative even though the normal approximation says it can.

YOU SHOULD REMEMBER

If $\hat{p}$ is the sample proportion and n is the number of items in the sample, then a 95 percent confidence interval for the population proportion is

$$\hat{p} \pm 1.96 \sqrt{\frac{\hat{p}(1 - \hat{p})}{n}}$$

This formula works if the sample was a pure random sample, that is, if all possible samples were equally likely to have been selected.

TYPES OF SAMPLING METHODS

It is very important to remember that all of these methods work only when the sample is a pure random sample. If the sample is not a pure random sample, all bets are off. A good example of a nonrandom sample was the 1936 *Literary Digest* presidential election poll. The *Literary Digest* had 2 million people respond to its poll, which is a much larger number than would have been needed to get an accurate result if the sample had been selected randomly. However, the poll predicted that Alfred Landon would be an easy winner, whereas in fact Franklin D. Roosevelt won by a landslide. The problem was that the *Digest* sample was not a random sample. The magazine mailed out cards to people whose names were obtained from telephone lists and other sources, but the people who had telephones at that time were not representative of the population as a whole. If a sample is not selected randomly there is no way to estimate how far off it might be.

• *CLUSTER SAMPLING*

For statistical purposes it is best to use a pure random sample. However, there are practical reasons why that is often impossible. Modern opinion polls cannot select their samples by putting the name of every person in the country in a hat. For one thing, there is no such thing as a list of names for the whole country. Even if there were, it would be very expensive to interview 3,000 randomly selected people scattered all over the country. Instead, the polls follow a method called **cluster sampling**. In cluster sampling the population is divided into different clusters and the sample is taken only from selected clusters rather than the entire population. Ideally each cluster would be as representative as possible of the entire population. In practice clusters are selected geographically. Some regions are selected at random, and then some sub-regions, and finally some households, are selected. This procedure guarantees that the people in the sample live in clusters, making it possible for one interviewer to interview quite a few people.

• STRATIFIED SAMPLING

Another sampling method is called **stratified sampling**. If the population can be grouped into subgroups that all consist of individuals who are very much alike, a representative sample can be obtained by interviewing a random sample of people in each group. This procedure can produce very accurate samples, but it works only when the population can be divided into homogeneous groups.

• CONVENIENCE SAMPLING

Many other types of sampling methods are also used. Statistical analysis is often incorrectly applied to these samples as if they were pure random samples. We might call these sampling methods **convenience sampling**. For example, colleges often conduct psychological experiments on samples of freshman psychology students. Since there is no reason to expect that freshman psychology students are at all representative of the entire freshman population (let alone the population as a whole), it is inappropriate to make statistical inferences about the population based on these experiments. To take another example, if you set up a table at a shopping mall and ask passersby to stop long enough to fill out a survey form, it is unlikely that you will end up with a random sample of the population. Similarly, magazines often report shocking statistics about behavior based on a survey of their subscribers, but since the subscribers are not a random sample of the population, it is impossible to make valid statistical inferences about the population as a whole from these surveys.

Some television stations conduct pseudopolls by having viewers pay to call a 900 telephone number to vote one way or the other. These polls are worse than misleading because they falsely create the impression of being similar to a real poll. There is no reason why anybody should care about the opinions of the particular group of people who happen to feel like paying to call the phone number that day. Since there has been no random selection, it is not at all appropriate to generalize these results to the larger population.

Many polls never risk the embarrassment of being shown to be wrong because the entire population is seldom checked. However, election prediction polls can be compared with the voting results for the entire voting population.

Opinion polls that try to predict the results of elections have even bigger problems. Not everyone votes, so the poll would rather not count the opinion of anyone who decides not to vote. Therefore, the polls ask some questions that they use to guess whether a person will vote. And there is another obvious problem. If many people change their minds after the poll but before the election, the poll will not be able to predict the election results very well.

Another example of a national sample is the **Current Population Survey**, which is conducted each month by the Census Bureau for the Bureau of Labor Statistics. Among other things, the Survey is used to calculate the unemployment rate every month. The Bureau of Labor Statistics wants to know about the characteristics of the unemployed—how long they have been unemployed, what kind of jobs they used to have, and so on. To get data on that many different categories, the Current Population Survey needs to use a much larger sample than it would need if it just wanted to know the total number of unemployed people in the country. For this reason, the Current Population Survey uses a sample of more than 100,000 people.

YOU SHOULD REMEMBER

1. In practice it is expensive to select pure random samples, so two other sampling methods, cluster sampling and stratified sampling, are used.

2. In *cluster sampling* the population is divided into clusters; some clusters are selected at random, and then some members of the chosen clusters are selected at random to make up the sample.

3. In *stratified sampling* the population is divided into strata that are as much alike as possible.

4. It is not appropriate to apply statistical analysis to samples that consist of the persons or other units who were most conveniently accessible to the interviewer.

KNOW THE CONCEPTS

DO YOU KNOW THE BASICS?

Test your understanding of Chapter 12 by answering the following questions:

1. If a friend refused to believe in the accuracy of polls, could you convince him?

2. What is the problem with applying statistical analysis to convenience samples?

3. If you need to choose a representative sample of 100 people, could you do so by choosing one name at random and then picking the following 99 names from the phone book?

4. When you take a sample, should you set quotas for different groups to make sure that the sample represents the population?

5. Why does a poll have problems if the population proportion is very small?

6. What are some of the different types of errors that can arise with polls, making them unrepresentative of the population?

7. When should you use the finite population correction factor?

8. When you are conducting a poll, should you select your sample with or without replacement?

9. Does the accuracy of the sample depend on the size of the population or on the size of the sample?

10. If you double the sample size, how much more accurate will your sample become?

11. Why is it impossible for national opinion polls to select pure random samples?

TERMS FOR STUDY

cluster sampling
confidence interval for proportion
convenience sampling

Current Population Survey
finite population correction factor
stratified sampling

PRACTICAL APPLICATION

COMPUTATIONAL PROBLEMS

For each of the following exercises you are given the sample size (n) and the number of items in the sample that have the characteristic you are interested in (m). Calculate 95 percent and 99 percent confidence intervals for the unknown value of the proportion of items in the entire population that have this characteristic. Assume that the population size is much larger than the sample size.

	n	m		n	m		n	m
1.	100	36	7.	53	45	13.	824	422
2.	200	72	8.	97	16	14.	560	328
3.	500	180	9.	48	43	15.	705	451
4.	1,000	360	10.	56	18	16.	234	128
5.	5,000	1,800	11.	158	100	17.	657	503
6.	10,000	3,600	12.	652	311			

ANSWERS
KNOW THE CONCEPTS

1. Try this argument: If you take a random sample of 1,000 people from a large population, the chance that this sample will be highly unrepresentative is about as good as the chance that you can flip 1,000 coins with the number of heads being far from 500.

2. Statistical inference techniques are not valid with nonrandom samples, as the 1936 polls demonstrated.

3. This would not be a pure random sample, since all possible samples of 100 people were not equally likely to have been chosen.

4. You should not set quotas. There is no way that you could set a quota for every conceivable population characteristic. The samples most likely to be representative of the population are pure random samples.

5. When the population proportion for a particular characteristic is very small, the expected number of items in the sample with that characteristic will be 0 or 1. Since the number of items in the sample must be a whole number, the sample proportion will not be able to represent the population proportion.

6. Any random sample is subject to some uncertainty. In practice any sample is also subject to difficulties such as the fact that some of the people selected may not want to participate in the sample, or they may not give truthful answers.

7. The finite population correction factor is used when the sample is not significantly smaller than the population. One rule is as follows: You do not need to use this correction factor if the sample is less than one-twentieth the size of the population.

8. The sample should be chosen without replacement. However, if the population size is much larger than the sample size, it does not make much difference whether the sample is chosen with or without replacement.

9. The accuracy of the sample depends primarily on the size of the sample, not the size of the population.

10. Doubling the sample will increase the accuracy of the sample, but it will not cause the width of the confidence interval to be cut in half. The width of the confidence interval is proportional to $1/\sqrt{n}$.

11. A pure random sample would be scattered all over the country, and it would be too expensive for the interviewers to reach all the persons.

PRACTICAL APPLICATION

Use the formulas for the confidence interval for the proportion:

$$\hat{p} \pm 1.96 \sqrt{\frac{\hat{p}(1 - \hat{p})}{n}} \text{ (95 percent)}, \quad \hat{p} \pm 2.6 \sqrt{\frac{\hat{p}(1 - \hat{p})}{n}} \text{ (99 percent)}$$

Below are the calculations for Exercise 1. The other exercises can be done similarly.

1. 95 percent interval:
    $.36 \pm 1.96 \sqrt{.36 \times .64/100}$; .266–.454
    99 percent interval:
    $.36 \pm 2.6 \sqrt{.36 \times .64/100}$; .236–.484

	95% Confidence Interval	99% Confidence Interval
2.	.293–.427	.272– .448
3.	.318–.402	.305– .415
4.	.330–.390	.321– .399
5.	.347–.373	.342– .378
6.	.351–.369	.348– .372
7.	.753–.945	.722– .976
8.	.091–.239	.068– .262
9.	.809–.982	.782–1.000
10.	.199–.444	.160– .482
11.	.558–.708	.534– .732
12.	.439–.515	.427– .527
13.	.478–.546	.467– .557
14.	.545–.627	.532– .639
15.	.604–.675	.593– .686
16.	.483–.611	.463– .631
17.	.733–.798	.723– .808

13
HYPOTHESIS TESTING

KEY TERMS

chi-square test a statistical method to test the hypothesis that two factors are independent

goodness-of-fit test a statistical procedure to test the hypothesis that a particular probability distribution fits an observed set of data

null hypothesis the hypothesis that is being tested in a hypothesis-testing situation; often the null hypothesis is of the form "There is no relation between two quantities"

test statistic a quantity calculated from observed quantities used to test a null hypothesis; the test statistic is constructed so that it will come from a known distribution if the null hypothesis is true; therefore the null hypothesis is rejected if it seems implausible that the observed value of the test statistic could have come from that distribution

In Chapter 3 we considered a specific hypothesis-testing problem: If you toss a coin many times, how can you tell whether or not the coin is fair? Now we'll consider a more general treatment of the methodology that statisticians use when they formulate and test hypotheses.

Remember that the hypothesis that we want to test is called the **null hypothesis** (or H_0), and the hypothesis that says, "The null hypothesis is wrong" is called the **alternative hypothesis**. Examples of null hypotheses include:

- A coin is fair.

- The mean number of raisins in boxes of a particular brand of raisin cereal is 7.

- The difference in effectiveness between four cold medicines occurred entirely by chance.

- The rate of appointments to the U.S. Supreme Court fits the Poisson distribution.

If we decide to reject the null hypothesis, this means that we are almost sure the hypothesis is not true. More specifically, we usually design our test so that there is only a 5 percent chance that we will reject the hypothesis if it is really true. However, if we decide to accept the hypothesis, this does not mean for sure that the hypothesis is true. It just means that we have not yet found statistical evidence to reject it.

One important principle of hypothesis testing: you should state your hypothesis before you collect the data. If you state your hypothesis after you have collected the data, then you can easily mistake a coincidence for a specific result. For example, suppose you collect some data and observe that 20 percent of the people in your sample are named Johnson. Then you state the null hypothesis that 20 percent of the people in the population are named Johnson, and proceed to test it on your data. Obviously you will accept the hypothesis, because the hypothesis was stated after the data was collected.

You will need to modify your hypotheses as you collect more data, but the point is that you can't use data to test a hypothesis if that hypothesis was based on the same data in the first place. The solution is: collect more data. If the hypothesis seems to work even for data that was collected after it was stated, you might be onto something.

Another important principle: you should state your significance level before collecting the data. This way you will be sure you are objective and don't decide on the significance level after seeing the test statistic value, where you can alter its value depending on what you hope the result will be.

TEST STATISTICS

The normal procedure in statistics is to calculate a specific quantity called a **test statistic**. There are several common test statistics. The one that you use depends on the problem you are facing.

The test statistic is designed so that, *if* the null hypothesis is true, the test statistic will be a random variable that comes from a known distribution. After you observe the test statistic you have to ask: "Would you believe that this test statistic value came from this distribution?" If the observed test statistic value is implausible, then you must answer: "I find that hard to believe." Then you must reject the hypothesis.

For example, suppose you toss a coin 100 times, and let X be the number of heads. Then the reasoning works like this (an example from classical logic is also included for comparison):

STATEMENT 1

If the coin is fair, X has a binomial distribution with $n = 100$ and $p = .5$.

STATEMENT 1

If an animal is a frog, then it is green.

STATEMENT 2

The value of $X = 90$ is observed; I find it hard to believe that X came from a binomial distribution with $n = 100$, $p = .5$.

CONCLUSION

I find it hard to believe that the coin is fair.

STATEMENT 2

Fozzie has been observed, and he is not green.

CONCLUSION

Fozzie is not a frog.

The logical structure of each argument is the same. The difference is that the statement "Fozzie is not green" is made with certainty; the statement "X probably did not come from a binomial distribution with $n = 100$, $p = .5$" is not made with certainty. As is usually the case in statistics, some doubt remains.

In this chapter we will see several examples of test statistics in practical problems.

• *TESTING A NULL HYPOTHESIS*

For example, suppose you are testing a null hypothesis using a test statistic Z. Suppose you know that Z will have a standard normal distribution if the null hypothesis is true. Calculate the value of Z. If, for example, the value of Z turns out to be .878, then everything is fine. There is a reasonably good chance of drawing the number .878 from a standard normal distribution. Since the observed value is not particularly implausible, you have no grounds for rejecting the hypothesis.

However, suppose the observed value of the test statistic Z turns out be 3. Then you should begin to get suspicious. You can see from the standard normal table that there is a probability of only .0026 that a standard normal random variable will be outside 3. (We will use the terms *inside* and *outside* in the following fashion. We will say that Z is *inside* a value c if $-c < Z < c$. We will say that Z is *outside* a value c if $Z < -c$ or if $Z > c$. In other words, Z is inside c if $|Z| < c$, and Z is outside c if $|Z| > c$.) You should say to the advocates of the null hypothesis, "You can't pull the wool over my eyes. I know that this test statistic value is very unlikely to have occurred if the null hypothesis were true, so I'm going to reject the hypothesis."

The null hypothesis advocates might respond, "If you reject the null hypothesis, you will be committing a type 1 error, since we think that the null hypothesis is really true. We admit that we had bad luck with our test statistic, and it turned out to have an implausible value. But it is still possible that you might draw the number 3 from a standard normal distribution."

• *AVOIDING TYPE 1 AND TYPE 2 ERRORS*

Of course, there is no way that you can prove them wrong. There still is a slight possibility that the null hypothesis might be true, so you could commit a **type 1** error by erroneously rejecting the hypothesis. But that is the risk that you will have to take. (Remember that a type 1 error occurs if you reject the null hypothesis when it is really true. A **type 2** error occurs if you accept the null hypothesis when it is really false. See Chapter 3.) Normally, we design our test so that the risk of committing a type 1 error is less than 5 percent. The risk of committing a type 1 error is called the **level of significance** of the test. Therefore, we can say that our test is designed to be at the 5 percent significance level.

From the standard normal tables we can see that there is a 95 percent chance that Z will be inside 1.96. For this reason we will design our test so that the null hypothesis is accepted if Z is inside 1.96 and rejected otherwise. Therefore, we will call the region inside 1.96 the **zone of acceptance** and the region outside 1.96 the **rejection region** or **critical region**. (See Figure 13–1.) Sometimes the number that is the boundary between the critical region and the zone of acceptance is called the **critical value** of the test statistic. In this case the critical values are 1.96 and –1.96.

We can see that with this test

$$\text{Pr(rejecting } H_0 \text{ if it is really true)}$$

$$= \text{Pr}[(Z > 1.96) \text{ or } (Z < 1.96)]$$

$$= .025 + .025$$

$$= .05$$

which is the result we want. If the observed value of the test statistic turns out to be outside 1.96, we will say that we can reject the hypothesis at the 5 percent significance level.

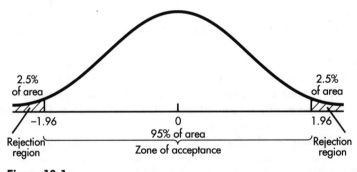

Figure 13-1

However, suppose that we want to be more cautious. Suppose that it is very costly for us to reject the hypothesis erroneously, so we want to make sure that the probability of this event happening is only 1 percent. Then we need to widen the zone of acceptance. (See Figure 13-2.) There is a 99 percent chance that Z will be inside 2.58. Therefore, we can ensure that there is only a 1 percent chance of committing a type 1 error if we design our test so that the zone of acceptance runs from -2.58 to 2.58. If the value of the test statistic turns out to be -2.6, we can reject the hypothesis at the 1 percent significance level. (Unfortunately, this is confusing terminology, since a *more* significant test corresponds to a lower significance level.)

However, suppose the value of the test statistic Z turns out to be 2. In that case we cannot reject the hypothesis at the 1 percent level. If we want a test at that level, we must accept the hypothesis. However, as we saw earlier, with a test statistic of 2 we can reject the hypothesis at the 5 percent level. Test statistics like this one are in a sort of gray area. Is the hypothesis really true? Nobody knows, and this time we are not even sure whether to accept the hypothesis. If we're willing to risk a 5 percent chance of a type 1 error, we can reject the hypothesis. However, if we are more cautious we will have to accept the hypothesis.

The situation is much more clear-cut when the test statistic is 3 or larger. In that case we can reject the hypothesis at every significance level.

It will be helpful to remember these critical values: If Z has a standard normal distribution when the null hypothesis is true, then

- if $-1.96 < Z < 1.96$, accept the hypothesis at the 5 percent level.

- if $-2.58 < Z < 2.58$, accept the hypothesis at the 1 percent level.

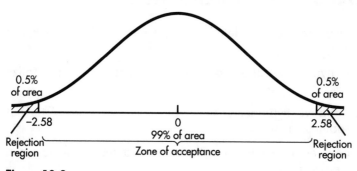

Figure 13-2

The previous example illustrates a **two-tailed test**, because the rejection region includes both tails of the distribution of the test statistic. We will also see some examples of **one-tailed tests**, where the null hypothesis is rejected only if the test statistic takes on an extreme value in one direction. In that case, the rejection region contains only one tail of the distribution (it may be either the right tail or the left tail, depending on the specific circumstances).

When a hypothesis test is performed, the **p-value** is the probability of getting a sample as extreme as the given one or worse, given that the null hypothesis is true. If the p-value is very small (specifically, if it is smaller than the significance level), then the null hypothesis should be rejected.

For example, suppose you are performing a right-tailed hypothesis test based on a standard normal (Z) random variable. If the test statistic value is $Z = 3$, then the p value is .0013 (the area to the right of 3, found from the table on page 472). A p value this small gives you strong evidence to reject the null hypothesis. A smaller p value increases your confidence that rejecting the null hypothesis is the right thing to do.

If you were performing a two-tailed test and found a test statistic of 3, then the p value would be .0026 (the area in both tails that are outside the range −3 to 3).

The **power** of a hypothesis test is the probability that the null hypothesis will be rejected, expressed as a function of the parameter being investigated. (You can't express the probability of a type 2 error as a single number, since its value will vary as the true parameter value changes.) For example, if you were testing for the value of the population mean μ, then ideally the power function would equal 0 at the true value of μ, and 1 everywhere else. This power function would guarantee that the correct decision would always be made, but you can seldom expect such a nice situation in practice. In general, increasing your sample size will improve the power function by making it more like the ideal power function.

YOU SHOULD REMEMBER

1. These are the key definitions you need to know in hypothesis testing:

   null hypothesis: the hypothesis you want to test (symbolized by H_0)

   alternative hypothesis: the hypothesis that says, "The null hypothesis is wrong"

   type 1 error: saying that the null hypothesis is *false* when it is really *true*

   type 2 error: saying that the null hypothesis is *true* when it is really *false*

2. The normal hypothesis-testing procedure is to calculate a quantity, based on your observations, called a *test statistic*.

> **3.** If the null hypothesis is true, then the test statistic will be a random variable with a known distribution.
>
> **4.** If it seems plausible that the calculated value could have come from this distribution, then you accept the null hypothesis.
>
> **5.** If the calculated test-statistic value is not likely to have come from this distribution, then you reject the null hypothesis.

TESTING THE VALUE OF THE MEAN

Now we'll see what test statistics arise in actual practice. Suppose that we can observe a sequence of numbers drawn from a normal distribution. Suppose that we know the variance, but not the mean, of the distribution. We need to test the hypothesis that μ equals a particular value μ^*. Using the H_0 notation we would write:

$$H_0: \mu = \mu^*$$

For example, suppose that we're quality control inspectors investigating the number of raisins in each (small) box of a raisin cereal. If there are too few raisins in the box, customers will complain. If there are too many, the company will lose money on each box of cereal sold. The raisins are put into the boxes by an Automatic Raisin Packer. We know that the machine works in such a way that the number of raisins in each box has a normal distribution with variance 16.16. On the average, each box is supposed to have 7 raisins. Our mission is to test the null hypothesis that the mean μ is equal to 7. We have $n = 13$ observations for the mean:

$$9, 11, 6, 10, 7, 4, 0, 7, 8, 6, 8, 2, 18$$

The sample average $\bar{x}$ is 7.38. Is that close enough to 7 so that we should accept the hypothesis? Or is it too far away? We know that if the hypothesis is true, $\bar{x}$ will have a normal distribution with mean $\mu = 7$ and variance $16.16/n$. Therefore,

$$Z = \frac{\sqrt{n}(\bar{x} - \mu^*)}{\sigma} = \frac{\sqrt{13}(7.38 - 7)}{4.02}$$

will have a standard normal distribution. Therefore, Z will be our test statistic. In our case, the computed value of Z is .341, which is well within the zone of acceptance. Consequently we can accept the hypothesis that $\mu = 7$.

Of course, in general we cannot use the statistic $Z = \sqrt{n}\left(\bar{x} - \mu^*\right)/\sigma$, because we ordinarily won't know the true value of σ. However, if the null hypothesis $\mu = \mu^*$ is true, then the test statistic

$$t = \sqrt{n}\,\frac{\bar{x} - \mu^*}{s_2}$$

will have a t distribution with $n - 1$ degrees of freedom. (See Chapter 11.)

For example, suppose that you have the following data points representing the weights of 27 sample players on a particular football team:

160, 185, 235, 208, 170, 185, 204, 180, 205,
215, 185, 188, 180, 220, 220, 221, 205, 235,
225, 190, 180, 205, 250, 210, 230, 210, 218

You want to test the hypothesis that these weights were selected from a normal distribution with mean 220. You need to calculate the two statistics $\bar{x} = 204.4$ and $s_2 = 22.1$. Then you can calculate the test statistic t:

$$t = \frac{204.4 - 220}{22.1}\sqrt{27} = -3.67$$

If the hypothesis is true, t will have a t distribution with 26 degrees of freedom. If you look up the results in Table A3-5, you can see that the critical value for a 1 percent test is 2.779. In other words, you can reject the null hypothesis at the 1 percent level if the value of the test statistic is outside 2.779. Since –3.67 is in the rejection region, you have good statistical evidence to reject the hypothesis that the sample of football players was selected from a population with mean 220.

General Procedure to Test the Hypothesis That the Mean $\mu = \mu^*$, When You Have Observed n Values Taken from a Normal Distribution

Method 1. Use this method if you *know* the variance (σ^2) of the distribution.

1. Calculate the sample average $\bar{x}$.

2. Calculate the test statistic Z:

$$Z = \frac{\sqrt{n}\,(\bar{x} - \mu^*)}{\sigma}$$

---

3. If you want to test the hypothesis at the 5 percent significance level, then accept the hypothesis that $\mu = \mu^*$ if Z is between -1.96 and 1.96; otherwise reject the hypothesis.

4. If you want to test the hypothesis at another significance level, then look in Table A3-2 to find the critical value for Z.

Method 2. Use this method if you *don't know* the variance of the distribution.

1. Calculate the sample average $\bar{x}$.

2. Calculate the sample variance $s_2{}^2$:

$$s_2{}^2 = \frac{(X_1 - \bar{x})^2 + (X_2 - \bar{x})^2 + \cdots + (X_n - \bar{x})^2}{n - 1}$$

3. Calculate the statistic t:

$$t = \sqrt{n}\,\frac{(\bar{x} - \mu^*)}{s_2}$$

4. The t statistic will have a t distribution with $n - 1$ degrees of freedom. Look in Table A3-5 to find the critical value for the t distribution with the appropriate degrees of freedom.

There is a close connection between hypothesis testing for means and the confidence intervals for means we developed in Chapter 11. We could calculate a 95 percent confidence interval for the unknown mean weight of the population of football players:

$$204.4 - 2.056 \times 22.1/\sqrt{27} \quad \text{to} \quad 204.4 + 2.056 \times 22.1/\sqrt{27}$$

or

$$195.7 \quad \text{to} \quad 213.1$$

Suppose that we wanted to test the hypothesis that $\mu = 210$ at the 5 percent significance level. The value 210 is inside the confidence interval, and if we calculate the t statistic:

$$\frac{\sqrt{27}\,(204.4 - 210)}{22.1} = -1.32$$

you can see that we would accept this hypothesis.

On the other hand, suppose we wanted to test the hypothesis that $\mu = 215$. The value 215 is outside the confidence interval, so you probably can guess that we would reject the hypothesis. If you calculate the t statistic:

$$\frac{\sqrt{27}\,(204.4 - 215)}{22.1} = -2.49$$

you can see that it is in the critical region, so our guess was correct and we would indeed reject the hypothesis.

In general, when we have calculated a 95 percent confidence interval for the mean, we will accept at the 5 percent significance level any null hypothesis where the hypothesized value of the mean is inside the confidence interval. Likewise we will reject any hypothesis where the hypothesized value of the mean is outside the confidence interval. (If we wanted to test the hypothesis at the 1 percent significance level, we would accept the hypothesis if the hypothesized value for the mean was inside the 99 percent confidence interval.)

YOU SHOULD REMEMBER

1. To test the hypothesis that the mean μ of a random variable with a normal distribution is equal to μ^*, calculate this test statistic:

$$\frac{\sqrt{n}\,(\bar{x} - \mu^*)}{s_2}$$

If the null hypothesis is true, this statistic will have a t distribution with $n - 1$ degrees of freedom.

2. If you have calculated a 95 percent interval for μ, then you will accept the hypothesis $\mu = \mu^*$ at the 5 percent level if μ^* is contained in the confidence interval.

THE ONE-TAILED TEST

Suppose that you are a quality control inspector for a semiconductor firm that buys silicon wafers from a particular supplier. Each wafer has a certain number of defects. If there are too many defects, you must reject the wafer. The supplier tells you that, on the average, there are 11 defects per wafer. Your job is to find out whether the supplier is right. You have checked the number of defects for a sample of 17 wafers, with these results:

7, 16, 19, 12, 15, 9, 6, 16, 14, 7, 2, 15, 23, 15, 12, 18, 9

You want to test the hypothesis that the number of defects on each wafer has a normal distribution with mean $\mu = 11$. However, suppose it turns out that you can reject the hypothesis that $\mu = 11$ because the sample average is significantly *less* than 11. In that case you'll be totally happy—you surely won't complain to the supplier if the number of defects is less than is advertised. Therefore, you don't really want to test the null hypothesis that $\mu = 11$.

Instead, you want to test the null hypothesis that $\mu \leq 11$. If you can reject this null hypothesis, then you will complain to the supplier. You can use the same *t* statistic again. The only difference is that, this time, you will reject the null hypothesis only if the value of the *t* statistic is in the region where the top 5 percent of the area is located. Figure 13-3 illustrates the rejection region and the zone of acceptance for this test. For your problem you will have a *t* distribution with 16 degrees of freedom. If you look in Table A3-4, you can see that the rejection region is located for values of *t* above 1.746.

This type of test is called a **one-tailed test**, because the rejection region consists of only one tail of the distribution. In a one-tailed test the null hypothesis is rejected only if the test statistic has a value significantly greater than expected. (Or we could do a one-tailed test using only the left-hand tail, in which case the null hypothesis is rejected only if the test-statistic value is significantly less than expected.)

The type of test that we did before is called a **two-tailed test**. In a two-tailed test, you reject the null hypothesis if the test-statistic value is either very low or very high. Which type of test you should use depends on your situation. Normally, if the null hypothesis involves an inequality, such as $\mu \geq \mu^*$ or $\mu \leq \mu^*$,

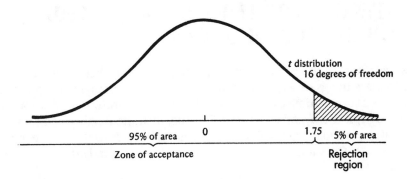

Figure 13-3

you will want to use a one-tailed test. If the null hypothesis involves an equality, such as $\mu = \mu^*$, then you will use a two-tailed test.

In the semiconductor case, the sample average $\bar{x} = 12.647$. The value of s_2 is 5.396, so the test-statistic value is

$$T = \sqrt{n}\,\frac{\bar{x}-11}{s_2} = 1.26$$

Therefore, we can accept the null hypothesis since 1.26 < 1.75.

YOU SHOULD REMEMBER

1. A *two-tailed test* is used to test a hypothesis involving an equality, such as $\mu = \mu^*$. The hypothesis is rejected if the calculated test statistic falls in either the upper tail or the lower tail of the t distribution.

2. A *one-tailed test* is used to test a hypothesis involving an inequality, such as $\mu < \mu^*$. The hypothesis is rejected only if the calculated test statistic falls in the appropriate tail.

TESTING HYPOTHESES ABOUT THE PROBABILITY OF SUCCESS

Now we return to the problem of attempting to tell whether or not a coin is fair. To make things easier, let's assume that we have performed enough tosses so that we can approximate the binomial distribution by the normal distribution. In that case, if the hypothesis $p = p^*$ is true, then the number of heads (X) will have a normal distribution with mean $\mu = p^*N$ and variance $\sigma^2 = Np^*(1 - p^*)$. The variable $Z = (X - \mu)/\sigma$ will have a standard normal distribution.

Now, suppose that we have tossed the coin 10,000 times, and we ended up with 5,056 heads. Should we accept the hypothesis that the coin is fair? The fair-coin hypothesis says that $p = .5$, so X will have a normal distribution with $\mu = 5,000$ and $\sigma^2 = 10,000/4$. We need to calculate the test statistic Z:

$$Z = \frac{X - 5,000}{\sqrt{10,000/4}} = \frac{56}{50} = 1.12$$

This value is within the 95 percent zone of acceptance, so we can accept this hypothesis at the 5 percent significance level.

For another example, suppose that 4,884 heads resulted from the 10,000 tosses. In that case the value of the test statistic is $-116/50 = -2.32$. This value is outside 1.96, so we can reject the fair-coin hypothesis at the 5 percent level (and the person who supplied the coin has some explaining to do). However, the value -2.32 is not outside the zone of acceptance at the 1 percent level, so if we want to be more cautious we cannot reject the fair-coin hypothesis.

Let's return to the first situation, in which 5,056 heads appeared in 10,000 flips. Suppose that we don't actually believe that the coin is fair. Instead, we think that the coin is unbalanced so that heads are slightly more likely to appear. In particular, we'll test the hypothesis that $p = .51$. If this null hypothesis is true, then X will have a normal distribution with mean 5,100 and variance 10,000 $\times .51 \times .49 = 2,499$. In this case the value of our test statistic is $-44/49.98 = -0.880$. This value is well within the zone of acceptance, so we can't reject the hypothesis that $p = .51$. However, we have already found that we also can't reject the hypothesis that $p = .5$. It is not possible that both of these hypotheses can be right, but we have no way to tell the two apart using only the information available to us. You probably could have guessed that it would be very difficult to tell the difference between these two hypotheses.

This fact illustrates that hypothesis-testing methods can do a good job of proving hypotheses wrong, but they often are not much help in proving hypotheses right. Even if you decide to accept the null hypothesis, this does not mean that there is not some other hypothesis that can also adequately account for the data. If you want to make a very convincing case that your hypothesis is true, then you will have to be able to reject all of the likely competing hypotheses. Since we haven't been able to do that in the coin example, we cannot say for sure that the coin is fair.

Some writers will say that you "fail to reject," rather than "accept," the null hypothesis, to emphasize that the mere fact that your test statistic falls in the zone of acceptance does not mean that you can guarantee that this null hypothesis is true, rather than some other possible null hypothesis.

We have only one hope—if we flip the coin many, many times, then finally we will reach the point where we can tell the difference between the hypotheses $p = .5$ and $p = .51$.

For example, suppose you flip the coin 39,000 times, with 19,680 heads. If the null hypothesis is $p = .5$, our test statistic becomes

$$(19,680 - 39,000 \times .5)/\sqrt{39,000 \times .5 \times .5} = 180/\sqrt{9750} = 1.82,$$

which falls in the acceptance region. If the null hypothesis is $p = .51$, our test statistic becomes

$$(19{,}680 - 39{,}000 \times .51)/\sqrt{39{,}000 \times .51 \times .49} = -210/\sqrt{9746.1} = 2.12,$$

which falls in the rejection region. Therefore, with 39,000 tosses we are finally able to distinguish between the hypotheses $p = .5$ and $p = .51$. However, in the real world you often cannot increase your sample size greatly. If you find yourself with two competing hypotheses neither of which can be rejected by the available data, then you're stuck.

YOU SHOULD REMEMBER

1. It is easier to prove that a hypothesis is wrong than to prove that it is right.

2. If neither of two competing hypothesis can be proved wrong, an impasse has been reached.

TESTING FOR THE DIFFERENCE BETWEEN TWO MEANS

Suppose that you have two different populations whose means you want to compare. Assume that the random variables X_a (mean μ_a, variance σ_a^2) and X_b (mean μ_b, variance σ_b^2) have approximately normal distributions. If you take samples of sizes n_a and n_b, respectively, then the sample means $\bar{x}_a$ and $\bar{x}_b$ are normal random variables and their difference is a normal random variable with mean $\mu_a - \mu_b$ and variance $\sigma_a^2/n_a + \sigma_b^2/n_b$.

Often we will want to test hypotheses about the difference between the population means: $\mu_a - \mu_b$. For example, our null hypothesis might be that the population means are equal:

$$\mu_a = \mu_b \quad \text{or} \quad \mu_a - \mu_b = 0$$

More generally, let our null hypothesis be that

$$\mu_a - \mu_b = D$$

If the population variances σ_a^2 and σ_b^2 are known, we can form the test statistic Z:

$$Z = \frac{\bar{x}_a - \bar{x}_b - D}{\sqrt{\sigma_a^2/n_a + \sigma_b^2/n_b}}$$

which will have a standard normal distribution if the null hypothesis is true.

For example, suppose we have two six-sided dice, A and B, which may be biased. We suspect that die A gives, on average, a value 0.7 larger than that of die B. Die A is rolled 20 times and gives the following values:

$$4, 5, 3, 6, 3, 5, 6, 3, 3, 6, 5, 1, 4, 2, 6, 1, 5, 5, 6, 2$$

Die B is rolled 15 times and gives the following values:

$$4, 3, 5, 4, 3, 2, 5, 1, 4, 1, 5, 6, 3, 6, 1$$

The manufacturer tells us that $\sigma_a^2 = 3.0$ and $\sigma_b^2 = 2.8$. We will assume that the manufacturer is statistically honest even though the dice may be shady. We can calculate our test statistic:

$$\bar{x}_a = 4.05, \qquad \bar{x}_b = 3.53, \qquad \sqrt{\frac{\sigma_a^2}{n_a} + \frac{\sigma_b^2}{n_b}} = .580$$

Therefore:

$$Z = \frac{4.05 - 3.53 - 0.7}{0.58} = -0.310$$

Since this result is within the zone of acceptance, we will accept the null hypothesis.

If the population variances are unknown, we must turn once again to the t statistic. If the null hypothesis is true, and if we also know that $\sigma_a^2 = \sigma_b^2$, then

$$T = \frac{(\bar{x}_a - \bar{x}_b) - D}{\sqrt{s_p^2(1/n_a + 1/n_b)}}$$

has a t distribution with $n_a + n_b - 2$ degrees of freedom, where

$$s_p^2 = \frac{(n_a - 1)s_a^2 + (n_b - 1)s_b^2}{n_a + n_b - 2}$$

(See Chapter 11.)

Example: Testing for the Difference Between Two Means

PROBLEM   Upon sober reflection, we decide that the values for σ_a^2 and σ_b^2 given us by the dice manufacturer mentioned above were not reliable. We also decide that it is reasonable to assume that $\sigma_a^2 = \sigma_b^2$. Once again we will test the hypothesis that $\mu_a - \mu_b = 0.7$.

SOLUTION   We can calculate that

$$s_a^2 = 2.892, \quad s_b^2 = 2.981, \quad s_p^2 = 2.930$$

and

$$T = \frac{4.05 - 3.53 - 0.7}{\sqrt{2.930(1/20 + 1/15)}} = -0.308$$

which will have a t distribution with 33 degrees of freedom if our hypothesis is correct. Checking Table A3-5, we see that we will accept the null hypothesis at the 5 percent level since -0.308 is inside the critical value of 2.030.

General Procedure for Hypothesis Testing for the Difference Between Two Means

Method 1 (if the values of σ_a^2 and σ_b^2 are known):

1. Calculate $\bar{x}_a$ and $\bar{x}_b$.
2. Calculate the test statistic Z:

$$Z = \frac{\bar{x}_a - \bar{x}_b - D}{\sqrt{\sigma_a^2/n_a + \sigma_b^2/n_b}}$$

3. If you want to test the hypothesis at the 5 percent significance level, then accept the hypothesis that $\mu_a - \mu_b = D$ if Z is between -1.96 and 1.96; otherwise reject the hypothesis.

4. If you want to test the hypothesis at another significance level, then look in Table A3-2 to find the critical value for Z.

Method 2 (if the values of σ_a^2 and σ_b^2 are unknown but assumed to be equal):

1. Calculate $\bar{x}_a$ and $\bar{x}_b$.

2. Calculate the sample variances (version 2) for both samples: s_a^2 and s_b^2.

3. Calculate the pooled estimator s_p^2:

$$s_p^2 = \frac{(n_a - 1)s_a^2 + (n_b - 1)s_b^2}{n_a + n_b - 2}$$

4. Calculate the test statistic T:

$$T = \frac{\bar{x}_a - \bar{x}_b - D}{\sqrt{s_p^2(1/n_a + 1/n_b)}}$$

5. If the null hypothesis is true, then T will have a t distribution with $n_a + n_b - 2$ degrees of freedom. Look in Table A3-5 to find the critical values for the appropriate t distribution.

STATISTICALLY SIGNIFICANT VERSUS IMPORTANT

The use of the term "significant" can often be misleading. For example, suppose you read this statement: "A study has been conducted of juggling ability, which found the statistically significant result that broccoli lovers have greater juggling ability than broccoli haters."

If this statement was worded precisely, it would say: "Our study indicates that, at the 5 percent significance level, we can reject the null hypothesis that the mean juggling ability is the same for broccoli lovers and broccoli haters." This means we believe the means of the two groups are not exactly the same, but this does not necessarily mean that there is any important difference between the means. The two means may differ by a trivial amount. Suppose the broccoli lovers are group a and the haters are group b. A sample of 1,000 from each group has taken a juggling test, with these results:

$$n_a = 1,000 \qquad n_b = 1,000$$
$$\bar{x}_a = 64 \qquad \bar{x}_b = 62$$
$$\sigma_a = 20.3 \qquad \sigma_b = 20.3$$

Now, test the hypothesis that $\mu_a = \mu_b$:

$$Z = \frac{\bar{x}_a - \bar{x}_b}{\sqrt{\sigma_a^2/n_a + \sigma_b^2/n_b}} = \frac{2}{\sqrt{.824}} = 2.2$$

This value of Z falls in the rejection region, so we reject the null hypothesis that the two means are the same (at the 5 percent significance level). Whenever you have a large enough sample, there is a good chance you will find a statistically significant difference between the means. However, the sample averages are very close to each other, so even if there is a statistically significant difference there is not really any important difference. One way we can determine if the difference is important is to calculate $\Pr(A > B)$, where A is the score of a randomly selected broccoli lover and B is the score of a randomly selected broccoli hater:

$$\Pr(A > B) = \Pr((A - B) > 0))$$

Let $U = A - B$; U has a normal distribution with $E(U) = 2$ and

$$\text{Var}(U) = 20.3^2 + 20.3^2 = 824.18.$$
$$\text{Standard deviation} = \sqrt{824.18} = 28.709$$
$$\Pr(U > 0) = \Pr(Z > (0 - 2)/28.709) = .5279$$

Therefore, there is only about a 53 percent chance that a randomly selected broccoli lover will demonstrate greater juggling ability than a randomly selected broccoli hater.

This issue is important because unfortunate stereotypes often arise from the use of the word *significant* in statistical studies. For example, studies often find statistically significant differences in certain abilities between men and women, but the variation in ability within each sex is much greater than the variation between the means of the two sexes. It would be wrong to use the results of such a study to prejudge the ability of any particular individual.

PAIRED SAMPLES

Suppose you are investigating whether a random sample of students score higher on tests given on Friday or on Monday. The sample of 8 students gives this result:

Initials	Friday test (a)	Monday test (b)	Difference
M.Y.	98	90	8
B.K.	94	84	10
R.T.	91	90	1
G.A.	88	83	5
R.S.	86	80	6
T.J.	82	77	5
L.S.	80	76	4
J.B.	76	72	4

Your first temptation might be to use the test procedure for the difference of two means, using the null hypothesis $\mu_a - \mu_b = 0$, as in the preceding section:

$$s_p^2 = \frac{(n_a - 1)s_a^2 + (n_b - 1)s_b^2}{n_a + n_b - 2} = \frac{7 \times 7.3957^2 + 7 \times 6.6027^2}{8 + 8 - 2} = 48.491$$

$$T = \frac{x_a - x_b - 0}{\sqrt{s_p^2(1/n_a + 1/n_b)}} = 1.54$$

This result is inside the acceptance region for a t distribution with $8 + 8 - 2 = 14$ degrees of freedom, which could lead you to believe there is no statistically significant difference in test scores on the different days of the week.

However, glancing at the table, you can see that each student has a higher score on Friday than on Monday, which suggests there is a problem with the previous analysis. The problem is: there is so much variability in the scores between the different students that it tends to overwhelm the difference between the different days. The preceding analysis is inappropriate, because we are not dealing with a random sample from Friday and an independent random sample on Monday. Instead, we have a paired sample; that is, a sample where we have pairs of values, one each for Friday and Monday, for each student. In this case, we should perform the following t test:

1. Assume that the difference in scores between Monday and Friday for each student comes from a normal distribution with mean μ_D and variance σ_D^2.

2. The null hypothesis is that $\mu_D = 0$.

3. Let X_{Di} be the difference in scores for student i. Calculate the average $\overline{x_D} = (\Sigma X_{Di})/n$; this average will have a normal distribution with mean μ_D and variance σ_D^2/n.

4. Calculate the test statistic:

$$\frac{\overline{x_D}}{s_D/\sqrt{n}} = \frac{\sqrt{n}(\overline{x_D})}{s_D}$$

where s_D is the sample standard deviation (with $n - 1$ in the denominator) for the differences.

If the null hypothesis is true, this test statistic will come from a t distribution with $n - 1$ degrees of freedom.

For our example, we have $\overline{x_D} = 5.375$; $s_D = 2.722$; $n = 8$, and:

$$T = \frac{\sqrt{8} \times 5.375}{2.722} = 5.58$$

This is well within the rejection region, so we reject the null hypothesis that there is no difference between Friday scores and Monday scores.

TESTING FOR THE DIFFERENCE BETWEEN TWO PROPORTIONS

Suppose now that we perform two series of tests. Suppose that test A is performed n_a times, with each test having an unknown probability p_a of success. If X_a is the number of successes, then $\hat{p}_a = X_a/n_a$ is an estimate of p_a. If n_a is large, then the central limit theorem says that $\hat{p}_a$ will have a normal distribution with mean p_a and variance $p_a(1 - p_a)/n_a$. Similarly, if X_b is the number of successes in n_b trials for test B, each with probability p_b of success, then, for large n_b, $\hat{p}_b = X_b/n_b$ has a normal distribution with mean p_b and variance $p_b(1 - p_b)/n_b$. Then $\hat{p}_a - \hat{p}_b$ also has a normal distribution with mean $(p_a - p_b)$ and variance equal to $[p_a(1 - p_a)/n_a + p_b(1 - p_b)/n_b]$.

If $p_a - p_b = D$, then this quantity:

$$\frac{\hat{p}_a - \hat{p}_b - D}{\sqrt{p_a(1 - p_a)/n_a + p_b(1 - p_b)/n_b}}$$

has a standard normal distribution. If we hypothesize that $p_a - p_b = D$, then we want to test this hypothesis using the above statistic. Unfortunately, we need to know the values of p_a and p_b to calculate that statistic. But, if we knew these values, we wouldn't need to test the hypothesis in the first place.

How far off would we be if we substituted $\hat{p}_a$ for p_a and $\hat{p}_b$ for p_b? If n_a is large, then $|\hat{p}_a - p_a|$ will be small (by the law of large numbers), $|(p_a - \hat{p}_a) \times (1 - p_a - \hat{p}_a)|$ will be still smaller, and $|(p_a - \hat{p}_a)(1 - p_a - \hat{p}_a)/n_a|$ will be even smaller. The same reasoning applies to substituting $\hat{p}_b$ for p_b, and thus, if the null hypothesis is true and $p_a - p_b = D$, then the statistic Z:

$$Z = \frac{\hat{p}_a - \hat{p}_b - D}{\sqrt{\hat{p}_a(1 - \hat{p}_a)/n_a + \hat{p}_b(1 - \hat{p}_b)/n_b}}$$

has a standard normal distribution.

Examples: Testing for the Difference Between Two Proportions

PROBLEM   Suppose that two manufacturers, Abercrombie and Bayes, supply electric light bulbs to your multinational corporation, which has many lamps. You suspect that Bayes's bulbs are less reliable than Abercrombie's and, in fact, that the probability of a Bayes bulb being defective is .001 greater than that of a defective Abercrombie bulb. Is your suspicion justified?

SOLUTION   A random sample of 1,000 of Abercrombie's bulbs turns up 15 defective bulbs, while a random sample of 2,000 of Bayes's bulbs turns up 36 defective bulbs. Then

$$\hat{p}_a = \frac{15}{1,000} = .015 \qquad \hat{p}_b = \frac{36}{2,000} = .018$$

$$\frac{\hat{p}_a(1 - \hat{p}_a)}{n_a} = .0000148 \qquad \frac{\hat{p}_b(1 - \hat{p}_b)}{n_b} = .00000883$$

and the value of Z is

$$Z = \frac{.015 - .018 - (-.001)}{\sqrt{.0000148 + .00000883}}$$

$$= -.412$$

Thus you would accept the hypothesis that $p_a - p_b = -.001$ at the 5 percent significance level, since $-1.96 < -.412 < 1.96$.

If $D = 0$, however, we can find a better estimate for p_a and p_b (which are assumed to be equal in this case). We will use the estimator $\hat{p} = (X_a + X_b)/(n_a + n_b)$. Substituting this for $\hat{p}_a$ and $\hat{p}_b$ in the denominator, we get

$$Z = \frac{\hat{p}_a - \hat{p}_b}{\sqrt{\hat{p}(1 - \hat{p})(1/n_a + 1/n_b)}}$$

PROBLEM   Suppose we wish to test the null hypothesis that $p_a = p_b$.
SOLUTION   Then

$$\hat{p} = \frac{15 + 36}{1,000 + 2,000} = .017 \qquad Z = \frac{.015 - .018}{.005007} = -.599$$

$$\sqrt{\hat{p}(1 - \hat{p})\left(\frac{1}{n_a} + \frac{1}{n_b}\right)} = .005007$$

Thus we would accept the null hypothesis at the 5 percent level, since $-1.96 < -.599 < 1.96$.

General Procedure for Testing the Hypothesis That $p_a - p_b = D$, Given That Test A Has X_a Successes in n_a Trials, Test B has X_b Successes in n_b Trials, and n_a and n_b Are Large

Method 1 (if D does not equal zero):

1. Calculate $\hat{p}_a$ and $\hat{p}_b$:

$$\hat{p}_a = \frac{X_a}{n_a} \quad \text{and} \quad \hat{p}_b = \frac{X_b}{n_b}$$

2. Calculate Z:

$$Z = \frac{\hat{p}_a - \hat{p}_b - D}{\sqrt{\hat{p}_a(1 - \hat{p}_a)/n_a + \hat{p}_b(1 - \hat{p}_b)/n_b}}$$

3. If you want to test the hypothesis at the 5 percent significance level, then accept the hypothesis if Z is between -1.96 and 1.96; otherwise reject the hypothesis.

4. If you want to test the hypothesis at another significance level, then look in Table A3-2 to find the critical value of Z.

Method 2 (if you are hypothesizing that $p_a = p_b$, that is, $D = 0$):

1. Calculate $\hat{p}$, $\hat{p}_a$, and $\hat{p}_b$:

$$p = \frac{X_a + X_b}{n_a + n_b}, \quad \hat{p}_a = \frac{X_a}{n_a}, \quad \text{and} \quad \hat{p}_b = \frac{X_b}{n_b}$$

2. Calculate Z:

$$Z = \frac{\hat{p}_a - \hat{p}_b}{\sqrt{\hat{p}(1 - \hat{p})(1/n_a + 1/n_b)}}$$

3. If you want to test the hypothesis at the 5 percent significance level, then accept the hypothesis if Z is between -1.96 and 1.96; otherwise reject the hypothesis.

4. If you want to test the hypothesis at another significance level, then look in Table A3-2 to find the critical value of Z.

THE CHI-SQUARE TEST

• *THE CONTINGENCY TABLE*

Let's suppose that we are trying to test whether there is any difference between four competing cold-prevention medicines. None of the medicines is guaranteed to work—instead, each one just promises to reduce the chances of getting a cold. Therefore, the number of people who try each kind of medicine and then get colds can be regarded as a random variable. The null hypothesis is: there is no difference between the medicines. Another way of saying it: the condition of a person (whether they get sick or not) is independent of the medicine they took. Suppose we have checked with a sample of 495 people. We asked them what kind of medicine they used, and then whether or not they got colds. The results were as follows:

	Medicine 1	Medicine 2	Medicine 3	Medicine 4	Total
How many got colds	15	26	9	14	64
How many did not	111	107	96	117	431
total	126	133	105	131	495

(This type of table is a **contingency table**—in this case with two rows and four columns. Each location is a **cell**. This table has eight cells.)

We can see from the table that medicine 3 seemed to be the most effective; only 8.5 percent of the people who tried medicine 3 got colds. However, there are many other factors that could have determined whether those people got colds. Maybe the people who used medicine 3 just happened to be exposed to fewer cold germs; in that case, the fact that they got fewer colds is a chance occurrence that has nothing to do with the fact that they used medicine 3.

Therefore, our null hypothesis will be as follows: There is basically no difference between the four medicines. In that case, the observed differences between the medicines arose solely by chance.

• *DEVELOPING A TEST STATISTIC*

Now we need to develop a test statistic to check this hypothesis. We can observe that in the total sample the fraction of people who got colds was .129 and the fraction who did not was .871. If there really was no difference between the medicines, then the fraction of people who got colds or did not get colds in each group should be close to these fractions. We can make a table comparing the actual and predicted values for the number of people in each group.

	Medicine 1	Medicine 2	Medicine 3	Medicine 4
Number with colds				
actual	15	26	9	14
predicted	16.254	17.157	13.545	16.899
Number without colds				
actual	111	107	96	117
predicted	109.746	115.843	91.455	114.101

We want to base our test statistic on the difference between the observed frequencies and the frequencies that are predicted if there is in fact no difference between the medicines. If this difference is small, we can reasonably accept the no-difference hypothesis. If this difference is large, then we should reject the hypothesis.

Let f_i represent the observed frequency in cell i, and let f_i^* represent the predicted frequency for cell i. Then, if there are N cells, we will use this test statistic:

$$S = \frac{\left(f_1 - f_1^*\right)^2}{f_1^*} + \frac{\left(f_2 - f_2^*\right)^2}{f_2^*} + \cdots + \frac{\left(f_N - f_N^*\right)^2}{f_N^*}$$

$$= \sum_{i=1}^{N} \frac{\left(f_i - f_i^*\right)^2}{f_i^*}$$

In our case we have eight cells, and the value of the test statistic is

$$\frac{(15 - 16.254)^2}{16.254} + \frac{(26 - 17.157)^2}{17.157} + \frac{(9 - 13.545)^2}{13.545} + \frac{(14 - 16.899)^2}{16.899}$$

$$+ \frac{(111 - 109.746)^2}{109.746} + \frac{(107 - 115.843)^2}{115.843} + \frac{(96 - 91.455)^2}{91.455}$$

$$+ \frac{(117 - 114.101)^2}{114.101} = 7.666$$

As it turns out, if the null hypothesis is true this test statistic has approximately a chi-square distribution, so it is called the chi-square statistic. The number of degrees of freedom is given by

Degrees of freedom = (number of rows − 1) × (number of columns − 1)

In our case, we have two rows and four columns, so our chi-square statistic has $(2 - 1) \times (4 - 1) = 3$ degrees of freedom.

Now that we have the test statistic, we need to look up the critical values in the chi-square tables. A χ_3^2 random variable has a 5 percent chance of being greater than 7.8. Since our test statistic is less than this value, we cannot reject the hypothesis at the 5 percent level. Using these data we cannot establish that there is any difference between the four medicines. However, the observed test statistic value of 7.666 is almost as large as 7.8, so these data indicate that there is almost more variation between the medicines than would be expected to happen by pure chance. Therefore, these data do tend to suggest that we should investigate this question further. Note that the **chi-square test** is a one-tailed test, because we reject the null hypothesis only if the calculated test statistic is too big.

General Procedure for Chi-square Test

Suppose you are given a contingency table with *m* rows (categories) and *n* columns (groups):

Category	Group 1	Group 2	Group 3	$\cdots$	Group n
Category a	a_1	a_2	a_3	$\cdots$	a_n
Category b	b_1	b_2	b_3	$\cdots$	b_n
Category c (and so on)	c_1	c_2	c_3	$\cdots$	c_n

The chi-square test is used to test the hypothesis that there is *no* significant difference between the groups. In other words, any observed difference in the proportion of each group belonging to a particular category arose solely by chance.

1. Calculate the total number of observations in each category:

$$a_{total} = a_1 + a_2 + \cdots + a_n$$

$$b_{total} = b_1 + b_2 + \cdots + b_n$$

and so on.

2. Calculate the total number of observations in each group:

$$t_1 = a_1 + b_1 + c_1 + \cdots$$

$$t_2 = a_2 + b_2 + c_2 + \cdots$$

$$\cdots\cdots\cdots\cdots\cdots\cdots\cdots\cdots$$

$$t_n = a_n + b_n + c_n + \cdots$$

3. Calculate the grand-total number of observations:

$$T = t_1 + t_2 + t_3 + \cdots + t_n$$

4. Calculate the proportion in each category:

$$p_a = \frac{a_{total}}{T}$$

$$p_b = \frac{b_{total}}{T}$$

and so on.

5. Calculate the predicted frequency of occurrence for each cell:

$$f_{a1} = p_a t_1 \quad f_{a2} = p_a t_2 \cdots f_{an} = p_a t_n$$

$$f_{b1} = p_b t_1 \quad f_{b2} = p_b t_2 \cdots f_{bn} = p_b t_n$$

$$\cdots$$

6. Calculate the value of the chi-square statistic S:

$$S = \frac{(a_1 - f_{a1})^2}{f_{a1}} + \frac{(a_2 - f_{a2})^2}{f_{a2}} + \cdots + \frac{(a_n - f_{an})^2}{f_{an}}$$

$$+ \frac{(b_1 - f_{b1})^2}{f_{b1}} + \frac{(b_2 - f_{b2})^2}{f_{b2}} + \cdots + \frac{(b_n - f_{bn})^2}{f_{bn}} + \cdots$$

7. If the null hypothesis is true, the statistic S will have a chi-square distribution with $(m - 1) \times (n - 1)$ degrees of freedom. Look up the critical value in Table A3-3. If the observed value is greater than the critical value, then you should reject the hypothesis.

We can summarize the formula for the chi-square statistic with this expression:

$$\text{Chi-square statistic} = \sum \frac{(\text{observed} - \text{expected})^2}{\text{expected}}$$

where we calculate the observed entry for each cell in the contingency table and the expected frequency of occurrence for that cell, assuming the null hypothesis is true, and then take the summation over all of the cells.

• APPLICATION OF THE CHI-SQUARE TEST

In general, the chi-square test can be used to test whether two factors are independent. As we found in Chapter 5, two random events are independent if a knowledge that one of the events has occurred does not provide any information about whether the other event occurred. In our previous example we wanted to test whether knowledge of one factor (which medicine was taken) provided any information about the other factor (whether the person got a cold). In other words, we were trying to test whether or not the two factors were independent.

Here are two extreme examples. Suppose we run a factory, and we would like to test whether parts made on Mondays and Fridays are more likely to be defective than parts made in the middle of the week. Here is our table of observations:

Day	Defective Parts	Acceptable Parts
Monday	16	132
Tuesday	4	140
Wednesday	5	138
Thursday	2	149
Friday	13	126

It clearly looks as if there is a significantly higher proportion of defective parts on Mondays and Fridays. We will use a chi-square test to test the null hypothesis that the chance of a part being defective is independent of the day of the week when it was made. Here is a table comparing the observed frequencies to the expected frequencies:

	Defective Parts		Acceptable Parts	
Day	Actual	Predicted	Actual	Predicted
Monday	16	8.166	132	139.835
Tuesday	4	7.945	140	136.055
Wednesday	5	7.890	138	135.110
Thursday	2	8.331	149	142.669
Friday	13	7.669	126	131.331

The calculated value of the chi-square statistic is 20.16. There are $(5 - 1) \times (2 - 1) = 4$ degrees of freedom. The 95 percent critical value from the table is 9.48, so we can clearly reject the null hypothesis.

For another example, suppose that we are evaluating the effect of three worker training programs. Our null hypothesis is that the evaluation of a worker is not affected by which training program he or she took. Here are the data:

	Evaluation		
Program	Above Average	Average	Below Average
1	36	78	29
2	24	53	21
3	33	67	28

A casual glance at these numbers does not indicate much difference between the training programs. The calculated value of the chi-square statistic is 0.19, so this confirms our suspicion that we should accept the null hypothesis.

YOU SHOULD REMEMBER

1. The chi-square test is used to test whether there is any significant difference between several groups or whether the observed differences could have happened by chance.
2. The chi-square test is based on the difference between the observed frequencies in a contingency table and the expected frequencies that would occur if the null hypothesis were true.
3. The null hypothesis is rejected if the calculated value for the chi-square statistic is greater than the critical value.

GOODNESS-OF-FIT TESTS

The chi-square test can also be used to test whether a particular probability distribution fits the observed data very well. This type of test is called a **goodness-of-fit test**. Once again, we want to compare the observed frequencies f of a particular occurrence with the frequencies f^* that are predicted to occur if the alleged distribution really does fit the data well. Once again, we compute the statistic

$$\sum_{i=1}^{n} \frac{(f_i - f_i^*)^2}{f_i^*}$$

If the null hypothesis is true, this statistic will have approximately a chi-square distribution. If the value of the test statistic turns out to be too large, there is too much of a discrepancy between the actual results and the predicted results, so we can reject the hypothesis that the predicted distribution fits the data. The number of degrees of freedom for the chi-square statistic is

$$n - 1 - \text{(number of parameters that you have to estimate using the sample)}$$

For example, if you use the sample to estimate the mean of the distribution you are using, then the χ^2 statistic will have $n - 2$ degrees of freedom.

Let's perform a goodness-of-fit test to see if the Poisson distribution is appropriate for predicting the number of United States Supreme Court justices that will be appointed in a five-year period. Table 18-1 shows the number of court appointments that have been made during each five-year period in U.S. history.

TABLE 18-1: Appointments to United States Supreme Court

Period	Number of appointments	Period	Number of appointments
1790–94	3	1835–39	5
1795–99	4	1840–44	1
1800–04	2	1845–49	3
1805–09	2	1950–54	2
1810–14	2	1855–59	1
1815–19	0	1860–64	5
1820–24	1	1865–69	0
1825–29	2	1870–74	4
1830–34	1	1875–79	1

(continued)

Period	Number of appointments	Period	Number of appointments
1880–84	4	1935–39	4
1885–89	3	1940–44	5
1890–94	4	1945–49	4
1895–99	2	1950–54	1
1900–04	2	1955–59	4
1905–09	2	1960–64	2
1910–14	6	1965–69	3
1915–19	2	1970–74	3
1920–24	4	1975–79	1
1925–29	1	1980–84	1
1930–34	3	1985–89	2
		1990–94	4

The mean is 2.585, so that on average 2.585 Supreme Court appointments are made in a five-year period. Here is the frequency distribution of these data. The upper figure is the number of appointments, the lower figure the number of periods in which there were that many appointments:

$$\frac{0}{2} \quad \frac{1}{9} \quad \frac{2}{11} \quad \frac{3}{6} \quad \frac{4}{9} \quad \frac{5}{3} \quad \frac{6}{1}$$

If the number of Supreme Court appointments is really given by a Poisson distribution with mean $\lambda = 2.585$, the predicted frequency is given by the formula $41 \times e^{-\lambda}\lambda^i/i!$ (since there are 41 five-year periods). Here are the values of the predicted frequencies:

0	1	2	3	4	5	6
3.090	7.989	10.327	8.900	5.752	2.974	1.282

We can now calculate the chi-square test statistic:

$$\frac{(2 - 3.090)^2}{3.090} + \frac{(9 - 7.989)^2}{7.98} + \frac{(11 - 10.327)^2}{10.327} + \frac{(6 - 8.900)^2}{8.900} +$$

$$\frac{(9 - 5.752)^2}{5.752} + \frac{(3 - 2.974)^2}{2.974} + \frac{(1 - 1.282)^2}{1.282}$$

$$= 0.384561 + 0.127925 + 0.034815 + 0.944927 +$$

$$1.833451 + 0.00022 + 0.061899 = 3.396798$$

It looks as though the observed frequencies match the predicted frequencies quite well. We have $n = 7$ categories, and we had to use the sample data to estimate the mean, so that leaves us with $7 - 1 - 1 = 5$ degrees of freedom. We can see from a chi-square table that a χ_5^2 random variable has a 95 percent chance of being less than 11.07, so the critical region occurs for values of the test statistic above 11.07. The observed value is well within this limit, so we will accept the hypothesis that the rate of Supreme Court appointments can be described by the Poisson distribution. (Note that this was a one-tailed test, since we only wanted to reject the hypothesis if the test statistic was larger than expected. If the test statistic is very small, that means that the predicted frequencies are very close to the observed frequencies.)

YOU SHOULD REMEMBER

1. The goodness-of-fit test using the chi-square statistic is used to test whether a particular distribution is appropriate for a given set of observations.

2. The chi-square statistic is based on the differences between the actual frequencies and the frequencies that would be expected to occur if the null hypothesis were true.

KNOW THE CONCEPTS

DO YOU KNOW THE BASICS?

Test your understanding of Chapter 13 by answering the following questions:

1. If you decide to accept the null hypothesis, can you be sure that it is really true?

2. If you decide to reject the null hypothesis, can you be sure that it is really false?

3. Why is it necessary to choose a test statistic whose distribution is known if the null hypothesis is true?

4. When is a one-tailed test appropriate? When is a two-tailed test appropriate?

5. Is the chi-square test a one-tailed or a two-tailed test?

6. Suppose you have found statistical evidence to reject the hypothesis that the population mean μ is equal to 7. Is it possible that the true value of the mean may still be close to 7?

TERMS FOR STUDY

alternative hypothesis	null hypothesis
cell	one-tailed test
chi-square test	test statistic
contingency table	two-tailed test
critical region	type 1 error
critical value	type 2 error
goodness-of-fit test	zone of acceptance
level of significance	

PRACTICAL APPLICATION

COMPUTATIONAL PROBLEMS

Given the following samples, test the hypothesis that the mean is as given:

1. 1, 5, 17, 9, 23, 17, 4, 3, 8, 8, 7, 8, 6, 0, −1: mean 7
2. 4, 30, −17, −29, 8, 7, −5, 4, 3, −6: mean −2
3. 15, 22, −19, 0, 1, 2, 4, 3, −3, 7: mean 14
4. 17, −9, −8, −10, 8, 5, 4, −7, 3, 4, −5, −7, −3, 2, 3: mean 0
5. Suppose that four new pesticides are being tested in a laboratory, with the following results:

Result	Type 1	Type 2	Type 3	Type 4	Total
Insects killed	139	100	73	98	410
Insects surviving	15	50	80	47	192
total tested	154	150	153	145	602

   Is pesticide 1 significantly better than the rest?

6. Suppose five different meteorological theories are tested to see whether they predict the weather correctly. The results are as follows:

Result	Theory 1	Theory 2	Theory 3	Theory 4	Theory 5	Total
Reports correct	50	48	53	47	46	244
Reports incorrect	76	74	75	76	77	378
total reports	126	122	128	123	123	622

   Is theory 3 significantly better than the rest?

7. Estimate the mean of the scores of your favorite football team last year. Test the hypothesis that the mean score is greater than 14.

8. Test the hypothesis that the mean number of pages in your favorite daily newspaper is greater than 50.

9. Estimate the mean price of a gallon of milk in your city over the past month. Test the hypothesis that the price this month is significantly greater than the price last month.

10. Consider a random variable with a binomial distribution with p unknown. How big does n have to be for you to be able to tell the difference (at the 5 percent significance level) between the hypothesis $p = .5$ and the hypothesis $p = .51$?

11. Perform a goodness-of-fit test to see whether the normal distribution is appropriate for the heights of a sample of people you know.

12. Perform a goodness-of-fit test to see whether the Poisson distribution is appropriate for the number of phone calls that you receive at your house in a week.

13. Perform a goodness-of-fit test to see whether the uniform distribution is appropriate for the numbers that you roll on a die.

14. Test the hypothesis that there is no significant difference between groups a, b, c, and d:

	Group a	Group b	Group c	Group d
Number of successes	16	12	7	13
Number of failures	54	94	66	49

15. Make a list of the first letters of the last names of 20 people you know and assign each letter a number ($A = 1, B = 2$, etc.) Test the hypothesis that the resulting numbers come from a distribution with mean 13.

16. Repeat the same procedure as in Exercise 15, only this time use the first names, and assume that the variance of the distribution is 25.

17. Divide your friends into categories by hair color and eye color. Perform a chi-square test to see whether there is a significant difference in eye color for people with different hair colors.

18. Perform a goodness-of-fit test to see whether the weights of a group of your friends fit the normal distribution.

In each exercise below you are given a table listing the results of a survey of some people in different occupational groups. Each exercise indicates the preferences of the people in the survey with regard to a particular item. For each exercise perform a chi-square test to see whether the preferences are independent of the groups.

19.

Favorite Color	Group 1	Group 2	Group 3
Red	16	8	20
Blue	48	21	54
Yellow	12	40	16

20.

Favorite Candidate	Group 1	Group 2	Group 3	Group 4
Smith	36	17	44	12
Jones	132	69	178	49

21.

Favorite Music	Group 1	Group 2	Group 3
Classical	26	16	13
Contemporary	58	91	30

22.

Favorite Fast Food	Group 1	Group 2	Group 3
Hamburger	30	16	17
Chicken	10	48	17

23.

Favorite Ice Cream	Group 1	Group 2	Group 3	Group 4
Vanilla	25	62	48	12
Strawberry	8	18	12	6
Chocolate	16	42	28	11

ANSWERS

KNOW THE CONCEPTS

1. No. There is always a possibility of committing a type 2 error.
2. No. There is always a possibility of committing a type 1 error.

3. You choose a test statistic of known distribution so that you can put an upper limit on the probability of a type 1 error.

4. A one-tailed test is appropriate if the null hypothesis is to be rejected only if the test statistic has an extreme value in one direction, but not the other direction. A two-tailed test is appropriate if the null hypothesis is to be rejected if the test statistic is either too big or too small. For example, if you are testing a hypothesis about a population mean that uses an equality sign, you will generally use a two-tailed test.

5. It is a one-tailed test. The null hypothesis is rejected only if the test-statistic value is large.

6. Yes. You should always check the value of the sample average in addition to the value of the test statistic. For example, if the sample average is 7.01, you can be sure that the true value of the average is close to 7 even if your statistical evidence led you to reject the hypothesis that the true mean is exactly equal to 7.

PRACTICAL APPLICATION

Calculations are shown for Exercises 1 and 5. The same methods can be applied to similar exercises.

1. Test statistic $= \sqrt{15} \, (7.667 - 7)/6.7365 = 0.383$; accept.
2. Test statistic $= 0.380$; accept.
3. Test statistic $= -3.15$; reject.
4. Test statistic $= -0.101$; accept.

5.

Type	Actual	Predicted
1		
Killed	139	104.8837
Survived	15	49.1163
2		
Killed	100	102.1595
Survived	50	47.8405
3		
Killed	73	104.2027
Survived	80	48.7973
4		
Killed	98	98.7542
Survived	47	46.2458

$$\text{Chi-square statistic} = \frac{(139 - 104.8837)^2}{104.8837} + \frac{(15 - 49.1163)^2}{49.1163}$$

$$+ \frac{(100 - 102.1595)^2}{102.1595} + \frac{(50 - 47.8405)^2}{47.8405}$$

$$+ \frac{(73 - 104.2027)^2}{104.2027} + \frac{(80 - 48.7973)^2}{48.7973}$$

$$+ \frac{(98 - 98.7542)^2}{98.7542} + \frac{(47 - 46.2458)^2}{46.2458}$$

$$= 64.251$$

Degrees of freedom $= 3$
Reject null hypothesis that there is no difference.

6.	Theory	Actual	Predicted
	1		
	Correct	50	51.9324
	Incorrect	76	74.0676
	2		
	Correct	48	50.2838
	Incorrect	74	71.7162
	3		
	Correct	53	52.7568
	Incorrect	75	75.2432
	4		
	Correct	47	38.3311
	Incorrect	46	54.6689
	5		
	Correct	46	50.6959
	Incorrect	77	72.3041

Chi-square statistic $= 4.3759$ Degrees of freedom $= 4$
Accept null hypothesis that there is no difference.

14.	Group	Actual	Predicted
	A		
	Success	16	10.8039
	Failure	54	59.1961
	B		
	Success	12	16.3601
	Failure	94	89.6399

Group	Actual	Predicted
C		
Success	7	11.2669
Failure	66	61.7331
D		
Success	13	9.5691
Failure	49	52.4309

Chi-square statistic = 7.6947   Degrees of freedom = 3
The 95 percent critical value for a chi-square distribution with 3 degrees of freedom is 7.8, so at the 5 percent level we will accept the null hypothesis that there is no difference. However, note that the computed chi-square statistic is very close to the critical value.

19. Group	Actual	Predicted
1		
Red	16	14.2298
Blue	48	39.7787
Yellow	12	21.9915
2		
Red	8	12.9191
Blue	21	36.1149
Yellow	40	19.9660
3		
Red	20	16.8511
Blue	54	47.1064
Yellow	16	26.0426

Chi-square statistic = 40.2300   Degrees of freedom = 4
95 percent critical value = 9.48
Reject hypothesis that groups have same preferences.

20. Chi-square statistic: 0.1938   Degrees of freedom = 3
95 percent critical value = 7.8
Accept hypothesis that groups have some preferences.

21. Chi-square statistic = 8.0258   Degrees of freedom = 2
95 percent critical value = 5.99
Reject hypothesis that groups have same preferences.

22. Chi-square statistic = 25.1467   Degrees of freedom = 2
95 percent critical value = 5.99
Reject hypothesis that groups have same preferences.

23. Chi-square statistic = 1.7933   Degrees of freedom = 6
95 percent critical value = 12.59
Accept hypothesis that groups have same preferences.

14
ANALYSIS OF VARIANCE

KEY TERMS

analysis of variance a method for testing the hypothesis that several different groups all have the same mean

ANOVA table a table that summarizes the results of an analysis of variance calculation

two-way analysis of variance a test procedure that can be applied to a table of numbers in order to test two hypotheses: (1) there is no significant difference between the rows; and (2) there is no significant difference between the columns

---

TESTING FOR THE EQUALITY OF SEVERAL MEANS

Suppose that we have observed scores on a particular aptitude test for three different groups of 10 people each. The results were as follows:

Group *a*: 88, 92, 91, 89, 89, 86, 92, 86, 89, 89
Group *b*: 91, 92, 85, 94, 93, 87, 87, 92, 91, 89
Group *c*: 87, 88, 95, 88, 92, 87, 89, 88, 87, 88

The average scores for the three groups are close together. It seems reasonable to suppose that there is in fact no difference in aptitude between the groups, and that the observed difference in the average score has arisen solely by chance.

Suppose that we check the scores for three different groups and find these results:

Group *d*: 87, 94, 91, 89, 89, 84, 92, 86, 89, 89
Group *e*: 82, 76, 84, 79, 77, 84, 81, 69, 79, 74
Group *f*: 69, 79, 67, 64, 65, 69, 69, 64, 72, 66

In this case it seems clear that there is a real difference in aptitude between the three groups. In other words, we can reject the hypothesis that the observed differences between the groups arose solely by chance.

In both of these cases it was obvious whether or not there was a significant difference between the average scores for the three groups. However, in general it will be more difficult to tell if the observed difference in scores is significant or random. We need to develop a new method, called **analysis of variance**. For now, let's assume that we have $m = 3$ independent groups (call them group a, group b, and group c), and that there are n people in each group. Assume that we know that the aptitude scores for the people in each group are selected from a normal distribution, and assume that the variance of the distribution is the same for all three groups.

Let's say that μ_a is the unknown mean aptitude test score for group a, μ_b is the mean for group b, and μ_c is the mean for group c. Our mission is to test the null hypothesis:

$$\mu_a = \mu_b = \mu_c = \mu$$

In other words, the null hypothesis states that the mean score for each group is the same. The alternative hypothesis simply states that the means are not all the same.

First, one obvious thing to do is to calculate $\bar{a}$, $\bar{b}$, and $\bar{c}$ (the sample average for each sample). If $\bar{a}$, $\bar{b}$, and $\bar{c}$ are close to each other, we will be more willing to accept the hypothesis that μ_a, μ_b and μ_c are all equal. We can calculate $\bar{x}$, the average for all the numbers:

$$\bar{x} = \frac{\bar{a} + \bar{b} + \bar{c}}{m} = \frac{\bar{a} + \bar{b} + \bar{c}}{3}$$

We can also calculate the sample variance (version 2) for these three averages (we'll call that variance S^{*2}):

$$S^{*2} = \frac{(\bar{a} - \bar{x})^2 + (\bar{b} - \bar{x})^2 + (\bar{c} - \bar{x})^2}{m - 1}$$

The larger S^{*2} is, the *less* likely we will be to accept the no-difference null hypothesis.

We should also look at the sample variance for each individual sample:

$$S_a^2 = \sum_{i=1}^{n} \frac{(a_i - \bar{a})^2}{n - 1}, \qquad S_b^2 = \sum_{i=1}^{n} \frac{(b_i - \bar{b})^2}{n - 1}, \qquad S_c^2 = \sum_{i=1}^{n} \frac{(c_i - \bar{c})^2}{n - 1}$$

It will turn out to be useful to calculate the average of the three sample variances (call it S^2):

$$S^2 = \frac{S_a^2 + S_b^2 + S_c^2}{3}$$

The larger these three variances are, the more likely we are to see $\bar{a}$, $\bar{b}$, and $\bar{c}$ spread out, even if they really do come from distributions with the same mean. For example, suppose that the observed values of $\bar{a}$, $\bar{b}$, and $\bar{c}$ are 500, 400, and 450. If $s_a^2 = s_b^2 = s_c^2 = 1$, we know right away that it is extremely unlikely that $\bar{a}$, $\bar{b}$, and $\bar{c}$ could have the observed values if they really did come from distributions with the same mean. On the other hand, if $s_a^2 = s_b^2 = s_c^2 = 10{,}000$, we would be quite likely to see $\bar{a}$, $\bar{b}$, and $\bar{c}$ spread out by this much even if the null hypothesis is true. Therefore, the larger S^2 is, the *more* likely we will be to accept the null hypothesis.

We will calculate the following statistic (call it F):

$$F = \frac{n\,S^{*2}}{S^2}$$

Recall from the central limit theorem that the variance of $\bar{x}$ is σ^2/n, where σ^2 is the variance of the population. Thus the numerator of this fraction is an approximation for $n\,\sigma^2/n = \sigma^2$, and the denominator is also an approximation for σ^2. If the means are truly equal, then F should be close to 1. If it is much larger than 1 (it can be shown that it won't get much smaller), then we reject our null hypothesis.

This F statistic will have $m - 1$ degrees of freedom in the numerator (since we calculate a sample variance from m means for it) and $m(n - 1)$ degrees of freedom in the denominator (since we calculate a sample variance of n items in each of the m groups). The F distribution is described in Chapter 8 and Table A3-6 lists some values for the cumulative distribution function.

In the examples we discussed earlier, we had $m = 3$ and $n = 10$. Therefore, the F statistic will have 2 and 27 degrees of freedom. Table A3-6 shows that this type of F statistic has a 95 percent chance of being less than about 3.3. Therefore, if the observed F-statistic value is greater than 3.3, we will reject the null hypothesis; otherwise we will accept the null hypothesis.

In the first example, the F statistic is $4.133/6.693 = .6175$. Just as we suspected all along, we should accept the hypothesis. In the second example, the F value is $1061/16.996 = 62.426$, so we should reject the hypothesis.

General Procedure for an Analysis-of-Variance Test

(Assume that you have m groups, each with n members.)

1. Calculate the sample average for each group:

$$\bar{a} = \frac{a_1 + a_2 + \cdots + a_n}{n}$$

$$\bar{b} = \frac{b_1 + b_2 + \cdots + b_n}{n}$$

$$\bar{c} = \frac{c_1 + c_2 + \cdots + c_n}{n}$$

and so on.

2. Calculate the average of all the averages:

$$\bar{x} = \frac{\bar{a} + \bar{b} + \bar{c} + \cdots}{m}$$

3. Calculate the sample variance of the averages:

$$S^{*2} = \frac{(\bar{a} - \bar{x})^2 + (\bar{b} - \bar{x})^2 + (\bar{c} - \bar{x})^2 + \cdots}{m - 1}$$

4. Calculate the sample variance for each group:

$$s_a^2 = \frac{(a_1 - \bar{a})^2 + (a_2 - \bar{a})^2 + \cdots + (a_n - \bar{a})^2}{n - 1}$$

$$s_b^2 = \frac{(b_1 - \bar{b})^2 + (b_2 - \bar{b})^2 + \cdots + (b_n - \bar{b})^2}{n - 1}$$

$$s_c^2 = \frac{(c_1 - \bar{c})^2 + (c_2 - \bar{c})^2 + \cdots + (c_n - \bar{c})^2}{n - 1}$$

and so on.

5. Calculate the average of all of the sample variances:

$$S^2 = \frac{s_a^2 + s_b^2 + s_c^2 + \cdots}{m}$$

6. Calculate the value of the F statistic:

$$F = \frac{nS^{*2}}{S^2}$$

7. Look in Table A3-6 to find the critical value for an F distribution with $m - 1$ and $m(n - 1)$ degrees of freedom.

8. If the observed value of the F statistic is greater than the critical value, reject the null hypothesis. Otherwise accept it.

YOU SHOULD REMEMBER

1. The analysis of variance procedure is used to test whether several groups of observations all come from distributions with the same mean.

2. In analysis of variance an F statistic is calculated.

3. If the null hypothesis is true and the means for all the groups really are the same, then the F statistic will have an F distribution.

4. If the calculated F-statistic value is larger than the critical value found in Table A3-6, then the null hypothesis (that the mean of each group is the same) is rejected.

SUM OF SQUARES

The analysis of variance approach can be further understood by looking at the sum of squared deviations (or the **sum of squares**). Again assume that we have m groups, each with n members. Each group has been treated differently in some manner, and our goal is to see if the treatments really made any difference. For example, we may have groups of people who were given different medicines. We will use x_{ij} to represent the ith item in the jth group. Note that we need two subscripts to identify each element uniquely. For example, if we have $m = 3$ groups with $n = 5$ members:

Group 1	Group 2	Group 3
16	38	19
13	21	14
36	36	17
29	39	15
18	26	12

we can represent each element like this:

$$
\begin{array}{lll}
x_{11} = 16 & x_{12} = 38 & x_{13} = 19 \\
x_{21} = 13 & x_{22} = 21 & x_{23} = 14 \\
x_{31} = 36 & x_{32} = 36 & x_{33} = 17 \\
x_{41} = 29 & x_{42} = 39 & x_{43} = 15 \\
x_{51} = 18 & x_{52} = 26 & x_{53} = 12
\end{array}
$$

To represent the sum of all of the items, we need to use two sigmas:

$$T = \sum_{i=1}^{n} \sum_{j=1}^{m} x_{ij}$$

This expression can be broken down like this:

$$T = \sum_{i=1}^{n} (x_{i1} + x_{i2} + \cdots + x_{im})$$

$$
\begin{aligned}
= \quad & x_{11} + x_{12} + \cdots + x_{1m} + \\
& x_{21} + x_{22} + \cdots + x_{2m} + \\
& \qquad \cdots \qquad\qquad\qquad + \\
& x_{n1} + x_{n2} + \cdots + x_{nm}
\end{aligned}
$$

For our example:

$$
\begin{aligned}
T = \ & 16 + 38 + 19 + \\
& 13 + 21 + 14 + \\
& 36 + 36 + 17 + \\
& 29 + 39 + 15 + \\
& 18 + 26 + 12 \\
= \ & 349
\end{aligned}
$$

We will use $\bar{\bar{x}}$ (x double bar) to represent the overall mean of all the elements (we could call it the **grand mean**):

$$\bar{\bar{x}} = \frac{T}{mn}$$

$$= \frac{\sum_{i=1}^{n} \sum_{j=1}^{m} x_{ij}}{mn}$$

$$= \frac{349}{15} = 23.267 \quad \text{for our example}$$

• *TOTAL SUM OF SQUARES*

For each element in the list we can calculate how far away it is from the grand mean $\bar{\bar{x}}$:

$$(\text{distance from } x_{ij} \text{ to } \bar{\bar{x}}) = |x_{ij} - \bar{\bar{x}}|$$

As you may have begun to suspect, we would like to square this distance:

$$(\text{squared distance from } x_{ij} \text{ to } \bar{\bar{x}}) = (x_{ij} - \bar{\bar{x}})^2$$

and then add all of the squared distances:

$$\text{TSS} = \sum_{i=1}^{n} \sum_{j=1}^{m} (x_{ij} - \bar{\bar{x}})^2$$

We will call this quantity the **total sum of squares**, or **TSS**. For our example we can calculate:

$$
\begin{aligned}
\text{TSS} = &(16 - 23.267)^2 + (38 - 23.267)^2 + (19 - 23.267)^2 + \\
&(13 - 23.267)^2 + (21 - 23.267)^2 + (14 - 23.267)^2 + \\
&(36 - 23.267)^2 + (36 - 23.267)^2 + (17 - 23.267)^2 + \\
&(29 - 23.267)^2 + (39 - 23.267)^2 + (15 - 23.267)^2 + \\
&(18 - 23.267)^2 + (26 - 23.267)^2 + (12 - 23.267)^2 \\
= &\ 1358.93
\end{aligned}
$$

We will look at several different sum of squares statistics. Note that a sum of squares is similar to a variance except that there is no division by n. The variance is calculated from the sum of squares about the mean, but we will see that there are several other types of sums of squares.

To analyze the total sum of squares, we need to break it into two parts. We know that, if the null hypothesis is true, the population mean is the same for each group and the deviation of any one element from the grand mean arises only by chance. On the other hand, if the population means are different, there are two reasons why an individual element may deviate from the grand mean: (1) because the mean of its own group is different from the overall population mean, and (2) because there is chance variation within its own group. Therefore, we will split the total sum of squares into two parts: the part arising from deviations of individual elements from their group mean, and the part arising from deviations of the group means from the grand mean. We can write the TSS formula like this:

$$\text{TSS} = \sum_{i=1}^{n} \sum_{j=1}^{m} (x_{ij} - \bar{\bar{x}})^2$$

$$= \sum_{i=1}^{n} \sum_{j=1}^{m} [(x_{ij} - \bar{x}_j) + (\bar{x}_j - \bar{\bar{x}})]^2$$

We will use $\bar{x}_j$ to stand for the sample average of group j. We have not done anything to our original expression but add and subtract $\bar{x}_j$. Now we have

$$\text{TSS} = \sum_{i=1}^{n} \sum_{j=1}^{m} [(x_{ij} - \bar{x}_j)^2 + (\bar{x}_j - \bar{\bar{x}})^2 + 2(x_{ij} - \bar{x}_j)(\bar{x}_j - \bar{\bar{x}})]$$

It turns out quite conveniently that, when the double summation is taken over the last term, the result is always zero. Hence we are left with:

$$\text{TSS} = \sum_{i=1}^{n} \sum_{j=1}^{m} [(x_{ij} - \bar{x}_j)^2 + (\bar{x}_j - \bar{\bar{x}})^2]$$

We can break that up into two double summations:

$$\text{TSS} = \sum_{i=1}^{n} \sum_{j=1}^{m} (x_{ij} - \bar{x}_j)^2 + \sum_{i=1}^{n} \sum_{j=1}^{m} (\bar{x}_j - \bar{\bar{x}})^2$$

Since the last term does not contain anything that depends on i, we can replace the summation $\sum_{i=1}^{n}$ by simply multiplying by n:

$$\text{TSS} = \sum_{i=1}^{n} \sum_{j=1}^{m} (x_{ij} - \bar{x}_j)^2 + n \sum_{j=1}^{m} (\bar{x}_j - \bar{\bar{x}})^2$$

• ERROR SUM OF SQUARES

We have now broken the total sum of squares into two components. The first part represents the deviation of each element from its group mean. Since we can think of these differences as arising because of unknown random factors (called statistical error), we will call this the **error sum of squares**, or **ERSS**. Then

$$\text{ERSS} = \sum_{i=1}^{n} \sum_{j=1}^{m} (x_{ij} - \bar{x}_j)^2$$

• TREATMENT SUM OF SQUARES

The second part of the total sum of squares represents the deviation of each group mean from the grand mean. We can think of these deviations as arising because the groups were given different treatments, so we will call this the **treatment sum of squares**, or **TRSS**. Then

$$\text{TRSS} = n \sum_{j=1}^{m} (\bar{x}_j - \bar{\bar{x}})^2$$

We can see that TSS = ERSS + TRSS. The larger the treatment sum of squares becomes, the less likely we will be to accept the null hypothesis that the treatments don't matter.

• MEAN SQUARE VARIANCE

Each sum-of-squares statistic has an associated quantity called its degrees of freedom. The degrees of freedom of TRSS is $m - 1$. The degrees of freedom of ERSS is $m(n - 1)$. (We are adding together m different sum of squares for each sample. Each individual sum of squares has $n - 1$ degrees of freedom.) When a sum of squares is divided by its degrees of freedom, the result is called a **mean square variance**:

$$(\text{treatment mean square variance}) = \frac{\text{TRSS}}{m - 1}$$

$$(\text{error mean square variance}) = \frac{\text{ERSS}}{m(n - 1)}$$

When we take the treatment mean square variance divided by the error mean square variance, then the result is exactly the same F statistic we used before:

$$F = \frac{\dfrac{\text{TRSS}}{m - 1}}{\dfrac{\text{ERSS}}{m(n - 1)}}$$

For our example, we have:

$$\text{TRSS} = 694.533, \quad m - 1 = 2$$

$$\text{ERSS} = 664.400, \quad m(n - 1) = 12$$

$$F = \frac{\dfrac{694.533}{2}}{\dfrac{664.400}{12}} = 6.272$$

If the null hypothesis is true and the treatments do not matter, then the F statistic will have an F distribution with 2 and 12 degrees of freedom. The critical value for a 5 percent test is 3.9, so in this case we can reject the null hypothesis.

• ANOVA TABLE

The information from the test can be summarized in a table known as an **ANOVA table**. (ANOVA is short for "analysis of variance.") The ANOVA table for our example looks like this:

Source of Variation	Sum of Squares	Degrees of Freedom	Mean Square Variance	F Ratio
Between means (treatment)	694.533	2	347.267	6.272
Within samples (error)	664.400	12	55.367	
total	1,358.933	14		

The table arranges the information about the sums of squares, their degrees of freedom, and the F statistic. Note that the total sum of squares has $mn - 1 = 14$ degrees of freedom, which is the sum of the degrees of freedom for the TRSS and ERSS.

The general formula for the ANOVA table looks like this:

ANOVA Table for m Different Groups Each with n Members

Source of Variation	Sum of Squares	Degrees of Freedom	Mean Square Variance	F Ratio
Between means (treatment)	$TRSS = n \sum_{j=1}^{m} (\bar{x}_j - \bar{\bar{x}})^2$	$m - 1$	$\dfrac{TRSS}{m - 1}$	$\dfrac{TRSS/(m - 1)}{ERSS/[m(n - 1)]}$
Within samples (error)	$ERSS = \sum_{i=1}^{n} \sum_{j=1}^{m} (x_{ij} - \bar{x}_j)^2$	$m(n - 1)$	$\dfrac{ERSS}{m(n - 1)}$	
total	$TSS = TRSS + ERSS$	$mn - 1$		

• *TWO CONSIDERATIONS IN USING ANALYSIS OF VARIANCE TESTS*

Here are two things to keep in mind when using analysis of variance tests. First, remember that the analysis of variance model assumes that the variance is the same for each group. If you observe that the sample variances for the different groups are markedly different, then the analysis of variance approach is not reliable.

Second, if you decide to reject the null hypothesis, then you are sure that the means of the groups are not all the same. However, you don't know whether all of the means are unequal, or whether they are all equal except for one particular mean that is different from the others, or whether there is perhaps some other pattern for the means. You will need to perform further investigations to determine exactly how the means differ.

• *ANALYSIS OF VARIANCE WITH UNEQUAL SAMPLE SIZES*

There may be times when the samples we have are not all the same size. In that case the analysis of variance formulas become more complicated, but we can still use our analysis of sums of squares to extend the applicability of the method. Suppose we have m different samples, with n_1 elements in sample 1, n_2 elements in sample 2, and so on, up to n_m elements in sample m. We will let x_{ij} represent the ith element in the jth sample. The total number of elements in all of the samples will be N:

$$N = n_1 + n_2 + \cdots + n_m = \sum_{j=1}^{m} n_j$$

We can calculate the mean for each group:

$$\bar{x}_j = \frac{\sum_{i=1}^{n_j} x_{ij}}{n_j}$$

and the grand mean:

$$\bar{\bar{x}} = \frac{\sum_{j=1}^{m} \sum_{i=1}^{n_j} x_{ij}}{N}$$

Note that the double summation is a bit more complicated now. We can calculate the total sum of squares:

$$\text{TSS} = \sum_{j=1}^{m} \sum_{i=1}^{n_j} (x_{ij} - \bar{\bar{x}})^2$$

and again we can partition the total sum of squares:

$$\text{TSS} = \sum_{j=1}^{m} \sum_{i=1}^{n_j} (x_{ij} - \bar{x}_j)^2 + \sum_{j=1}^{m} n_j (\bar{x}_j - \bar{\bar{x}})^2$$

The first part is the error sum of squares:

$$\text{ERSS} = \sum_{j=1}^{m} \sum_{i=1}^{n_j} (x_{ij} - \bar{x}_j)^2$$

which has $N - m$ degrees of freedom. The second part is the treatment sum of squares:

$$\text{TRSS} = \sum_{j=1}^{m} n_j (\bar{x}_j - \bar{\bar{x}})^2$$

which has $m - 1$ degrees of freedom. Now we can calculate the F statistic:

$$F = \frac{\dfrac{\text{TRSS}}{m - 1}}{\dfrac{\text{ERSS}}{N - m}}$$

which has $m - 1$ degrees of freedom in the numerator and $N - m$ degrees of freedom in the denominator.

YOU SHOULD REMEMBER

1. The total sum of squares (TSS) for a group of numbers consists of the sum of the squares of the deviations of all the elements from the grand mean.

2. In analysis of variance the total sum of squares is divided into two parts: the treatment sum of squares (TRSS) and the error sum of squares (ERSS).

3. The treatment sum of squares indicates the effects of the differences between the groups.

4. The null hypothesis that there is no difference between the groups is more likely to be rejected when the treatment sum of squares is large relative to the error sum of squares.

5. The analysis of variance results can be summarized in an ANOVA table (see page 278).

TWO-WAY ANALYSIS OF VARIANCE

Consider a situation where we need to test the effectiveness of three different types of factory-air-pollution-control devices. To conduct a test we must install one of the devices on a factory smokestack for a test period of 1 week. Six different factories are available to conduct tests. The factories are all very similar, but we cannot be sure that they are exactly the same. Therefore, it would not be a good idea to design a test like this:

> Put device a on factories 1 and 2 during the test period.
> Put device b on factories 3 and 4 during the test period.
> Put device c on factories 5 and 6 during the test period.

Suppose that factories 1 and 2 happen to produce less pollution than the other factories. In that case this test procedure would tend to make device a look better than the other devices even if they were all the same. To be complete, therefore, we must install each device for a test period on each factory. Our table of measurements for the pollution looks like this:

Factory	Device a	Device b	Device c
1	50.8	53.0	49.7
2	49.5	49.2	49.1
3	51.5	51.7	49.6
4	48.3	49.8	49.3
5	48.8	48.1	47.1
6	48.4	52.2	47.9

We could start by assuming that all six factories are identical and that the output of pollution from each device follows a normal distribution as follows:

$$\text{device } a\text{: mean} = \mu_a, \quad \text{variance} = \sigma^2$$
$$\text{device } b\text{: mean} = \mu_b, \quad \text{variance} = \sigma^2$$
$$\text{device } c\text{: mean} = \mu_c, \quad \text{variance} = \sigma^2$$

Now we can employ the standard analysis of variance procedure to test the null hypothesis that there is no difference between the devices:

$$H_0: \mu_a = \mu_b = \mu_c$$

The ANOVA table is as follows:

Source of Variation	Sum of Squares	Degrees of Freedom	Mean Square Variance	F Ratio
Between means (treatment)	10.763	2	5.382	2.472
Within samples (error)	32.657	15	2.177	
total	43.420	17		

The 5 percent critical value for an F distribution with 2 and 15 degrees of freedom is 3.68. Since the observed value is less than the critical value, we can accept the null hypothesis that there is no difference between the devices.

However, we still should have some nagging doubts. What if the factories really are not identical? What if some of the observed variation in the pollution data came about because of differences in the factories? In our ANOVA table we have assumed that all of the variation not explained by differences in the control devices arises because of statistical error—that is, random, mysterious forces that we know nothing about. It would be very helpful if we could determine how much of the variation might be explainable by differences in the factories. That line of thought leads us to a new method: **two-way analysis of variance.** This new method provides an added bonus: it also allows us to test the hypothesis that the factories are identical.

In general, suppose we have m different levels of the first type of treatment and n different levels of the second type of treatment. In our case, the first type of treatment represents the difference in air-pollution-control devices (so $m = 3$) and the second type of treatment represents the different factories (so $n = 6$). We have a total of mn different possible levels of the combined treatments, and we will assume that we have one observation for each possibility. In our case, then, we have $3 \times 6 = 18$ observations. (In some cases you will have or will need more than one observation per treatment possibility, and in other cases you may not have observations for all of the possibilities. More advanced books can tell you what to do in these cases.) We will arrange the data in an array with m columns and n rows, using x_{ij} to represent the element in row i and column j:

m levels of first treatment

	1	2	$\cdots$	m
1	x_{11}	x_{12}		x_{1m}
2	x_{21}	x_{22}		x_{2m}
$\vdots$				
n	x_{n1}	x_{n2}		x_{nm}

n levels of second treatment

In theory we will assume that each observation consists of the sum of four different effects:

$$x_{ij} = \mu + \mu_{ri} + \mu_{cj} + e_{ij}$$

where μ represents the overall mean, and μ_{ri} represents the specific effect that row i has on all observations in that row. We put the r in the expression μ_{ri} to remind us that it represents row i. If the different levels of the second treatment have no effect, then all of these row effects μ_{r1} to μ_{rn} are zero. Suppose, however, that in our example factory 1 pollutes a little more than the average and factory 2 pollutes a little less than the average. The μ_{r1} will be a positive number and μ_{r2} will be a negative number.

Likewise, μ_{cj} represents the specific effect of column j. If the three pollution-control devices are the same, then $\mu_{c1} = \mu_{c2} = \mu_{c3} = 0$. On the other hand, if device 1 lets a higher than average level of pollution escape, then μ_{c1} will be positive.

There still will be mysterious factors (other than the device or the factory) that affect the level of pollution. Therefore we need to include the random error term e_{ij} to take all of these other factors into account.

We can calculate the grand mean of all the observations:

$$\bar{\bar{x}} = \frac{\sum_{i=1}^{n} \sum_{j=1}^{m} x_{ij}}{mn}$$

and once again we can calculate the total sum of squares:

$$\text{TSS} = \sum_{i=1}^{n} \sum_{j=1}^{m} (x_{ij} - \bar{\bar{x}})^2$$

• *ROW, COLUMN, AND ERROR SUM OF SQUARES*

Now we need to divide TSS into three parts: the part that can be explained by the effects of the columns, the part that can be explained by the effects of the rows, and the remaining part that can be explained only by error. We'll use ROWSS to represent the row sum of squares, COLSS to represent the column sum of squares, and ERSS to represent the error sum of squares. Then

$$\text{TSS} = \text{ROWSS} + \text{COLSS} + \text{ERSS}$$

The formula for the column sum of squares is basically the same as the formula for the treatment sum of squares with one-way analysis of variance:

$$\text{COLSS} = n \sum_{j=1}^{m} (\bar{x}_{cj} - \bar{\bar{x}})^2$$

In this formula $\bar{x}_{cj}$ represents the average of all the elements in column j. The formula for ROWSS looks very similar:

$$\text{ROWSS} = m \sum_{i=1}^{n} (\bar{x}_{ri} - \bar{\bar{x}})^2$$

where $\bar{x}_{ri}$ is the average of all the elements in row i. If the factories really are identical, then there is no reason for the row averages to be significantly different from $\bar{\bar{x}}$, so ROWSS will be small. COLSS has $m - 1$ degrees of freedom, and ROWSS has $n - 1$ degrees of freedom.

To get the error sum of squares, we need to subtract three parts from x_{ij}. We will use $\bar{\bar{x}}$ to estimate the overall population mean, $\bar{x}_{cj} - \bar{\bar{x}}$ to estimate the specific effect of column j, and $\bar{x}_{ri} - \bar{\bar{x}}$ to estimate the specific effect of row i. Then

$$\text{ERSS} = \sum_{i=1}^{n} \sum_{j=1}^{m} [x_{ij} - \bar{\bar{x}} - (\bar{x}_{cj} - \bar{\bar{x}}) - (\bar{x}_{ri} - \bar{\bar{x}})]^2$$

ERSS has $(m - 1)(n - 1)$ degrees of freedom.

Now we can perform the calculations for our example. We need to calculate the average of each row and column:

Factory	Device a	Device b	Device c	Average
1	50.8	53.0	49.7	51.167
2	49.5	49.2	49.1	49.267
3	51.5	51.7	49.6	50.933
4	48.3	49.8	49.3	49.133
5	48.8	48.1	47.1	48.000
6	48.4	52.2	47.9	49.500
average	49.55	50.667	48.783	49.667

The grand mean is $\bar{\bar{x}} = 49.667$. Now we can calculate:

$$\begin{aligned}
\text{TSS} = &(50.8 - 49.667)^2 + (53.0 - 49.667)^2 + (49.7 - 49.667)^2 \\
&+ (49.5 - 49.667)^2 + (49.2 - 49.667)^2 + (49.1 - 49.667)^2 \\
&+ (51.5 - 49.667)^2 + (51.7 - 49.667)^2 + (49.6 - 49.667)^2 \\
&+ (48.3 - 49.667)^2 + (49.8 - 49.667)^2 + (49.3 - 49.667)^2 \\
&+ (48.8 - 49.667)^2 + (48.1 - 49.667)^2 + (47.1 - 49.667)^2 \\
&+ (48.4 - 49.667)^2 + (52.2 - 49.667)^2 + (47.9 - 49.667)^2
\end{aligned}$$

$$= 43.4$$

$$\text{COLSS} = 6 \times [(49.55 - 49.667)^2 + (50.667 - 49.667)^2 + (48.783 - 49.667)^2]$$

$$= 10.8$$

$$\begin{aligned}
\text{ROWSS} = 3 \times [&(51.167 - 49.667)^2 + (49.267 - 49.667)^2 + (50.933 - 49.667)^2 \\
&+ (49.133 - 49.667)^2 + (48 - 49.667)^2 + (49.5 - 49.667)^2]
\end{aligned}$$

$$= 21.3$$

$$\text{ERSS} = 43.4 - 10.8 - 21.3 = 11.3$$

• *TWO-FACTOR ANOVA TABLE*

Now we can construct an ANOVA table. The two-factor ANOVA table is very similar to the one-factor table. The only difference is that we now have another source of variation. We will include the sum of squares for the row along with the degrees of freedom and the mean square variance.

Two-factor ANOVA Table

Source of Variation	Sum of Squares	Degrees of Freedom	Mean Square Variance	F Ratio
Treatment 1 effect: columns (control devices)	10.763	2	5.382	4.472
Treatment 2 effect: rows (factories)	21.313	5	4.263	3.758
Error	11.343	10	1.134	
total	43.420	17		

Note that the total sum of squares and the sum of squares resulting from the different control devices are the same here, as they were in the one-way ANOVA table presented earlier. The difference is that now the error sum of squares is much smaller because a large part of the sum of squares previously attributed to error is now attributed to the effects of the different factories. We now have two *F*-statistic values. The first *F* statistic tests the null hypothesis that there is no difference between the columns. It is equal to

(F statistic for columns)

$$= \frac{(\text{column sum of squares})/(\text{column SS degrees of freedom})}{(\text{error sum of squares})/(\text{error SS degrees of freedom})}$$

$$= \frac{\dfrac{\text{COLSS}}{m-1}}{\dfrac{\text{ERSS}}{(m-1)(n-1)}}$$

The second *F* statistic tests the hypothesis that there is no difference between the rows:

(F statistic for rows)

$$= \frac{(\text{row sum of squares})/(\text{row SS degrees of freedom})}{(\text{error sum of squares})/(\text{error SS degrees of freedom})}$$

$$= \frac{\dfrac{\text{ROWSS}}{n-1}}{\dfrac{\text{ERSS}}{(m-1)(n-1)}}$$

Remember that our one-factor ANOVA table indicated that there was no difference between the three devices. Now the two-factor table gives us a clearer story. The F statistic for the columns has 2 and 10 degrees of freedom, and the 5 percent critical value from the F table is 4.1. The calculated value is greater than the critical value, so we can reject the null hypothesis that all columns are the same. Likewise, the F statistic for the rows has 5 and 10 degrees of freedom, for which the critical value is 3.3. Hence we can also reject the second null hypothesis, and we can conclude that there is a difference between the factories.

The problem with the one-factor test arose because the factories themselves added much variability. At that time we viewed all that variability as being random. There was so much random variability that it obscured the fact that the devices really were different. With the two-factor table we can separate the effects of the different factories from the effects of the different devices.

We can find alternative formulas for the sums of squares that are sometimes more convenient for calculation purposes.

Two-way Analysis of Variance

1. Let $N = mn$ be the number of values for x, T_x be the sum of all values of x, T_{x2} be the sum of the squares of all values of x, T_{ri} be the sum of row i, and T_{cj} be the sum of column j:

$$T_x = \sum_{i=1}^{n} \sum_{j=1}^{m} x_{ij}, \qquad T_{x2} = \sum_{i=1}^{n} \sum_{j=1}^{m} x_{ij}^2,$$

$$T_{ri} = \sum_{j=1}^{m} x_{ij}, \qquad T_{cj} = \sum_{i=1}^{n} x_{ij}$$

2. Then calculate:

$$T_r^2 = \sum_{i=1}^{n} T_{ri}^2 \qquad \text{(sum of squares of row totals)}$$

$$T_c^2 = \sum_{j=1}^{m} T_{cj}^2 \qquad \text{(sum of squares of column totals)}$$

3. Now use these formulas:

$$TSS = T_{x^2} - \frac{(T_x)^2}{N}$$

$$COLSS = \frac{T_c^2}{n} - \frac{(T_x)^2}{N}$$

$$ROWSS = \frac{T_r^2}{m} - \frac{(T_x)^2}{N}$$

$$ERSS = T_{x^2} - \frac{T_c^2}{n} - \frac{T_r^2}{m} + \frac{(T_x)^2}{N}$$

Here is an example of the summary formulas. Our data matrix is as follows:

First Treatment (3 levels)

		1	2	3
	1	18	14	16
Second	2	56	59	53
treatment	3	46	48	50
(5 levels)	4	100	90	95
	5	28	31	25

We can clearly see from this table that there is much more variability between the rows than there is between the columns, so we can guess in advance how the ANOVA result will come out. We need to calculate these totals:

	1	2	3	Total
1	18	14	16	48
2	56	59	53	168
3	46	48	50	144
4	100	90	95	285
5	28	31	25	84
Total	248	242	239	729

Therefore, $T_x = 729$.
Now we can calculate:

$$T_c^2 = 248^2 + 242^2 + 239^2 = 177{,}189$$

$$T_r^2 = 48^2 + 168^2 + 144^2 + 285^2 + 84^2 = 139{,}545$$

We also need to calculate T_{x^2}:

$$T_{x^2} = 18^2 + 14^2 + 16^2 + 56^2 + 59^2 + 53^2$$
$$+ 46^2 + 48^2 + 50^2 + 100^2 + 90^2 + 95^2$$
$$+ 28^2 + 31^2 + 25^2$$

$$= 46{,}617$$

Now we can use the formulas:

$$\text{TSS} = 46{,}617 - \frac{729^2}{15} = 11{,}187.6$$

$$\text{COLSS} = \frac{177{,}189}{5} - \frac{729^2}{15} = 8.4$$

$$\text{ROWSS} = \frac{139{,}545}{3} - \frac{729^2}{15} = 11{,}085.6$$

$$\text{ERSS} = 46{,}617 - \frac{177{,}189}{5} - \frac{139{,}545}{3} + \frac{729^2}{15} = 93.6$$

Here is the ANOVA table:

Source of Variation	Sum of Squares	Degrees of Freedom	Mean Square Variance	F Ratio
Columns	8.4	2	4.2	0.359
Rows	11,085.6	4	2,771.4	236.872
Error	93.6	8	11.7	
total	11,187.6	14		

We can clearly see that the variation between the rows is significant while the variation between the columns is not.

YOU SHOULD REMEMBER

The two-way analysis of variance procedure is used to test these two hypotheses about an array of numbers:

null hypothesis 1: there is no significant difference between the columns in the array

null hypothesis 2: there is no significant difference between the rows in the array

KNOW THE CONCEPTS

DO YOU KNOW THE BASICS?

Test your understanding of Chapter 14 by answering the following questions:
For questions 1–5, consider normal (one-way) analysis of variance.

1. Explain intuitively why your likelihood of accepting the null hypothesis depends on the relation between the sample averages for all the groups involved.

2. Explain intuitively why your likelihood of accepting the null hypothesis depends on the degree of dispersion within each group.

3. If you decide to reject the null hypothesis, what conclusion can you reach about the population means?

4. What assumption does the analysis of variance procedure make about the variance of each population?

5. How would the analysis of variance results be affected if you doubled the size of each sample?

6. When do you apply the two-way analysis of variance procedure?

7. Could you apply two-way analysis of variance to the test score example presented in the chapter?

8. Suppose a situation really calls for two-way analysis of variance, but you use regular one-way analysis of variance instead. How will your results be affected?

TERMS FOR STUDY

analysis of variance	sum of squares
ANOVA table	total sum of squares (TSS)
error sum of squares (ERSS)	treatment sum of squares (TRSS)
grand mean	two-way analysis of variance
mean square variance	

PRACTICAL APPLICATION

COMPUTATIONAL PROBLEMS

In Exercise 1–5 you are given lists of numbers. In each case perform an analysis of variance test to determine whether the numbers came from distributions with the same mean.

1.

	B_1	B_2	B_3
1:	37	56	86
2:	36	57	93
3:	36	62	90
4:	32	59	91
5:	39	57	90
6:	43	54	94
7:	34	51	90
8:	32	64	98
9:	46	53	86
10:	37	52	96
11:	35	66	96
12:	31	55	94
13:	41	50	89
14:	46	64	99
15:	37	62	94
16:	50	52	90

2.

	C_1	C_2	C_3	C_4
1:	116	128	141	122
2:	107	120	141	109
3:	115	123	133	116
4:	101	117	138	123
5:	108	119	145	117
6:	105	110	126	118
7:	108	129	126	115
8:	105	124	131	115
9:	106	121	144	115
10:	103	120	132	113
11:	109	120	142	119
12:	119	116	126	127

3.

	D_1	D_2	D_3
1:	152	156	160
2:	130	158	153
3:	155	161	161
4:	150	164	158
5:	142	162	158
6:	144	155	148
7:	147	141	158
8:	136	154	169

	D_1	D_2	D_3
9:	150	152	160
10:	149	156	164
11:	152	142	162
12:	149	151	170

4.

	E_1	E_2	E_3	E_4
1:	301	219	224	225
2:	320	211	227	223
3:	296	229	221	213
4:	318	222	207	221
5:	303	212	227	228
6:	313	219	207	224
7:	298	215	208	218
8:	298	207	211	230
9:	308	218	216	215
10:	292	216	203	201
11:	296	224	208	217
12:	308	211	218	217
13:	311	204	222	217
14:	294	211	226	227
15:	304	211	206	201
16:	311	229	222	209
17:	313	203	219	211
18:	296	213	226	224

5.

	F_1	F_2	F_3
1:	142	131	145
2:	128	143	138
3:	126	149	141
4:	132	150	147
5:	137	155	137
6:	139	133	158
7:	135	152	160
8:	131	157	161
9:	135	136	145
10:	141	135	139
11:	146	158	140
12:	138	151	161
13:	141	144	158
14:	145	138	138
15:	149	134	146

Perform a two-way analysis of variance test for Exercises 6 and 7:

6.

A_1	A_2	A_3
12	17	16
18	19	16
35	29	27
2	8	5
95	121	130
61	69	75
73	70	68
125	108	105
17	17	19
62	63	60

7.

B_1	B_2	B_3
126	130	147
118	132	149
136	147	152
101	129	145
98	128	140
129	135	149
117	137	138
128	130	136
112	135	139
131	139	148
100	134	143
97	129	129

ANSWERS

KNOW THE CONCEPTS

1. If the sample averages are close together, you will be more likely to accept the null hypothesis that the means are all the same.

2. If there is greater spread within each group, you will be more likely to accept the null hypothesis, because the observed difference between the sample averages is more likely to have occurred randomly even if the means are all the same.

3. You can conclude that the means of the different groups are not all the same, but you need to investigate further to determine the exact relation between the different means.

4. It assumes the variances are all the same.

5. If you have larger samples, you can say with more confidence that the observed differences (if any) among the sample averages arose because the population means are different, not because of random factors.

6. The two-way analysis of variance procedure is applied when there are two separate factors that affect each observation and you generally have one observation for each possible combination of the two factors.

7. No. The two-way analysis of variance procedure would be appropriate only if there was some treatment common to the first student in each sample, some other treatment common to the second student in each sample, and so on.

8. The error sum of squares will be larger because you are not recognizing the effect of the second factor. There is a greater chance that you will accept the null hypothesis even if there really is a difference between the populations.

PRACTICAL APPLICATION

Calculations are shown below for Exercise 1. The same methods can be used for similar exercises.

1. $\bar{\bar{x}} = 62.452$

$$
\begin{aligned}
\text{TSS} = &\ (37 - 62.452)^2 + (56 - 62.452)^2 + (86 - 62.452)^2 \\
&+ (36 - 62.452)^2 + (57 - 62.452)^2 + (93 - 62.452)^2 \\
&+ (36 - 62.452)^2 + (62 - 62.452)^2 + (90 - 62.452)^2 \\
&+ (32 - 62.452)^2 + (59 - 62.452)^2 + (91 - 62.452)^2 \\
&+ (39 - 62.452)^2 + (57 - 62.452)^2 + (90 - 62.452)^2 \\
&+ (43 - 62.452)^2 + (54 - 62.452)^2 + (94 - 62.452)^2 \\
&+ (34 - 62.452)^2 + (51 - 62.452)^2 + (90 - 62.452)^2 \\
&+ (32 - 62.452)^2 + (64 - 62.452)^2 + (98 - 62.452)^2 \\
&+ (46 - 62.452)^2 + (53 - 62.452)^2 + (86 - 62.452)^2 \\
&+ (37 - 62.452)^2 + (52 - 62.452)^2 + (96 - 62.452)^2 \\
&+ (35 - 62.452)^2 + (66 - 62.452)^2 + (96 - 62.452)^2 \\
&+ (31 - 62.452)^2 + (55 - 62.452)^2 + (94 - 62.452)^2 \\
&+ (41 - 62.452)^2 + (50 - 62.452)^2 + (89 - 62.452)^2 \\
&+ (46 - 62.452)^2 + (64 - 62.452)^2 + (99 - 62.452)^2 \\
&+ (37 - 62.452)^2 + (62 - 62.452)^2 + (94 - 62.452)^2 \\
&+ (50 - 62.452)^2 + (52 - 62.452)^2 + (90 - 62.452)^2
\end{aligned}
$$

$$= 25{,}120$$

$\bar{x}_1 = 38.25 \quad \bar{x}_2 = 57.125, \quad \bar{x}_3 = 92.25$

$\text{TRSS} = 16[(38.25 - 62.452)^2 + (57.125 - 62.452)^2 + (92.25 - 62.452)^2]$

$\quad = 24{,}032$

ANOVA table:

Source of Variation	Sum of Squares	Degrees of Freedom	Mean Square Variance	F Ratio
Treatment	24,032.167	2	12,016.083	497.103
Error	1,087.750	45	24.172	
total	25,119.917	47		

The 95 percent critical value for an F distribution with 2 and 45 degrees of freedom is about 3.2, so we can clearly reject the null hypothesis that all numbers came from distributions with the same mean.

2. ANOVA table:

Source of Variation	Sum of Squares	Degrees of Freedom	Mean Square Variance	F Ratio
Treatment	4,512.229	3	1,504.076	45.336
Error	1,459.750	44	33.176	
total	5,971.979	47		

Reject null hypothesis.

3. ANOVA table:

Source of Variation	Sum of Squares	Degrees of Freedom	Mean Square Variance	F Ratio
Treatment	1,144.500	2	572.250	12.197
Error	1,548.250	33	46.917	
total	2,692.750	35		

Reject null hypothesis.

4. ANOVA table:

Source of Variation	Sum of Squares	Degrees of Freedom	Mean Square Variance	F Ratio
Treatment	104,385.486	3	34,795.162	509.216
Error	4,646.500	68	68.331	
total	109,031.986	71		

Reject null hypothesis.

5. ANOVA table:

Source of Variation	Sum of Squares	Degrees of Freedom	Mean Square Variance	F Ratio
Treatment	771.244	2	385.622	5.288
Error	3,062.533	42	72.917	
total	3,833.778	44		

Reject null hypothesis.

6.

	A_1	A_2	A_3	Average
1:	12	17	16	15.00
2:	18	19	16	17.67
3:	35	29	27	30.33
4:	2	8	5	5.00
5:	95	121	130	115.33
6:	61	69	75	68.33
7:	73	70	68	70.33
8:	125	108	105	112.67
9:	17	17	19	17.67
10:	62	63	60	61.67
average	50.00	52.10	52.10	51.40

Two-factor ANOVA table:

Source of Variation	Sum of Squares	Degrees of Freedom	Mean Square Variance	F Ratio
Columns	29.400	2	14.700	0.251
Rows	44,367.867	9	4,929.763	84.195
Error	1,053.933	18	58.552	
total	45,451.200	29		

95% Critical value for F distribution with 2 and 18 d.f. = 3.5
   There is no significant difference between columns.
95% Critical value for F distribution with 9 and 18 d.f. = 2.4
   There is a significant difference between rows.

7.

	B_1	B_2	B_3	Average
1:	126	130	147	134.33
2:	118	132	149	133.00
3:	136	147	152	145.00
4:	101	129	145	125.00
5:	98	128	140	122.00
6:	129	135	149	137.67
7:	117	137	138	130.67
8:	128	130	136	131.33
9:	112	135	139	128.67
10:	131	139	148	139.33
11:	100	134	143	125.67
12:	97	129	129	118.33
average	116.08	133.75	142.92	130.92

Two-factor ANOVA table:

Source of Variation	Sum of Squares	Degrees of Freedom	Mean Square Variance	F Ratio
Columns	4,464.667	2	2,232.333	43.283
Rows	1,909.417	11	173.583	3.366
Error	1,134.667	22	51.576	
total	7,508.750	35		

95% Critical value for F distribution with 2 and 22 d.f. = 3.4
  There is a significant difference between columns.
95% Critical value for F distribution with 11 and 22 d.f. = 2.3
  There is a significant difference between rows.

15
SIMPLE LINEAR REGRESSION

KEY TERMS

r^2 a measure of how well the independent variable in a simple linear regression can explain changes in the dependent variable; its value is between 0 (meaning poor fit) and 1 (meaning perfect fit)

regression line a line calculated in regression analysis that is used to estimate the relation between two quantities (the independent variable and the dependent variable)

simple linear regression a method for analyzing the relation between one independent variable and one dependent variable

Often in statistics we want to investigate questions such as these: Is there a relation between two quantities? Do changes in one quantity cause changes in the other quantity? For example, do changes in interest rates affect the demand for houses? Other times we will want to know if we can use one variable to predict the value of another variable. For example, suppose you want to guess next year's value of consumption spending. Consumption spending may be difficult to predict directly; it may be easier to predict next year's value of disposable income. If you can find a relation between income and consumption, you're in business. You can use that relation plus the knowledge of income to predict the value of consumption.

THE REGRESSION LINE

In economics it is assumed that the level of income affects the quantity that will be demanded of a particular good. For most goods, higher income leads to higher demand. This is not always the case, though. There are some goods (called **inferior goods**) that people buy less often when their incomes increase. (Presumably they buy something better when they can afford it. For example, hamburgers might be inferior goods if people would rather buy steaks when

311

their incomes rise.) The only way to tell whether a particular good is an inferior good is to collect some data.

Suppose we have observations of average income and total pizza sales for a 1-month period for eight different towns:

Town	Income (in thousands of dollars)	Pizza Sales (in thousands)
1	5	27
2	10	46
3	20	73
4	8	40
5	4	30
6	6	28
7	12	46
8	15	59

When faced with a real problem it would be better to have more observations, but this sample of eight observations will work well to illustrate the calculations.

One good way to determine whether there is a relation between income and pizza sales is to draw a picture. We will draw a diagram with income measured along the horizontal axis, pizza sales along the vertical axis, and a dot marking each observation. This type of graph is called a **scatter diagram** or **scatterplot**. (See Figure 15-1.)

From the diagram we can see clearly that there is a relationship between income and pizza sales, and we can see also that an increase in income leads to an increase in pizza sales. Therefore, pizzas are not an inferior good.

As the saying goes, "One picture is worth a thousand words," and we should not underestimate the value of a diagram for illustrating and clarifying the nature of a relationship. However, it would help to develop a more precise way of characterizing this type of relationship. The tool we will use is called *regression analysis*. In this chapter we will investigate **simple linear regression**, which applies when there is only one independent variable that affects the value of the dependent variable. In the next chapter we will look at situations with more than one independent variable, for which we use *multiple linear regression*.

The goal of simple regression analysis is to approximate a pattern of dots by a straight line. In the pizza case we cannot draw a line that passes through all the points, but we can find a line that comes close to most of them (See Figure 15-2.) This type of line is called a **regression line**. It is possible to determine a regression line by visual inspection, but we will develop an easier method.

In some cases it is possible to find a line that fits a pattern of dots very well.

Income (x)			
x	y	x	y
5	27	4	30
10	46	6	28
20	73	12	46
8	40	15	59

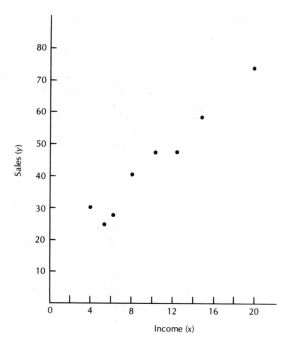

Figure 15-1

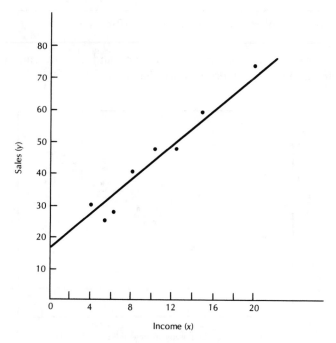

Figure 15-2

For example, economic models sometimes assume that personal consumption is determined by disposable income. Here is a list of observations for total national consumption spending and disposable income for several recent years:

Year	Disposable Income (billion $)	Consumption Spending (billion $)
1981	2,200.2	1,941.3
1982	2,347.3	2,076.8
1983	2,522.4	2,283.4
1984	2,810.0	2,492.3
1985	3,002.0	2,704.8
1986	3,187.6	2,892.7
1987	3,363.1	3,094.5
1988	3,640.8	3,349.7
1989	3,894.5	3,594.8
1990	4,166.8	3,839.3
1991	4,343.7	3,975.1
1992	4,613.7	4,219.8
1993	4,789.3	4,454.1
1994	5,018.8	4,698.7
1995	5,307.4	4,924.3

Source: Economic Report of the President, 1996

We can draw a scatter diagram (Figure 15-3), and on it we can draw a line that closely matches this pattern of dots. (Figure 15-4).

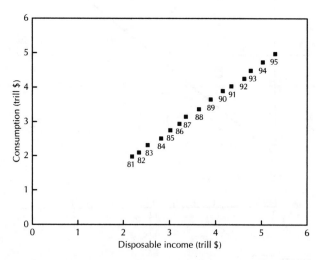

Figure 15-3

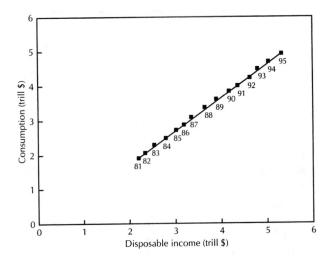

Figure 15-4

In other cases the pattern of dots will not look at all like a line. For example, suppose we wished to investigate whether more people live in states that have higher elevations. Here is a table comparing population with highest elevation for the 50 states:

State	1990 Population (thousands)	Highest Elevation
Alabama	4,041	2,407
Alaska	550	20,320
Arizona	3,665	12,633
Arkansas	2,351	2,753
California	29,760	14,494
Colorado	3,294	14,433
Connecticut	3,287	2,380
Delaware	666	442
Florida	12,938	345
Georgia	6,478	4,784
Hawaii	1,108	13,796
Idaho	1,007	12,662
Illinois	11,431	1,235
Indiana	5,544	1,257
Iowa	2,777	1,670
Kansas	2,478	4,039
Kentucky	3,685	4,145

(continued)

State	1990 Population (thousands)	Highest Elevation
Louisiana	4,220	535
Maine	1,228	5,268
Maryland	4,781	3,360
Massachusetts	6,016	3,491
Michigan	9,295	1,980
Minnesota	4,375	2,301
Mississippi	2,573	806
Missouri	5,117	1,772
Montana	799	12,799
Nebraska	1,578	5,426
Nevada	1,202	13,143
New Hampshire	1,109	6,288
New Jersey	7,730	1,803
New Mexico	1,515	13,161
New York	17,990	5,344
North Carolina	6,629	6,684
North Dakota	639	3,506
Ohio	10,847	1,550
Oklahoma	3,146	4,973
Oregon	2,842	11,239
Pennsylvania	11,882	3,213
Rhode Island	1,003	812
South Carolina	3,487	3,560
South Dakota	696	7,242
Tennessee	4,877	6,643
Texas	16,987	8,749
Utah	1,723	13,528
Vermont	563	4,393
Virginia	6,187	5,729
Washington	4,867	14,410
West Virginia	1,793	4,863
Wisconsin	4,892	1,951
Wyoming	454	13,804

The scatter diagram is shown in Figure 15-5. We can try to draw a line that best fits these dots, but we can see that the line does not follow the pattern of dots very closely. (Figure 15-6).

It is not good enough just to find the best line. We must also find a way to measure how well the line fits the points.

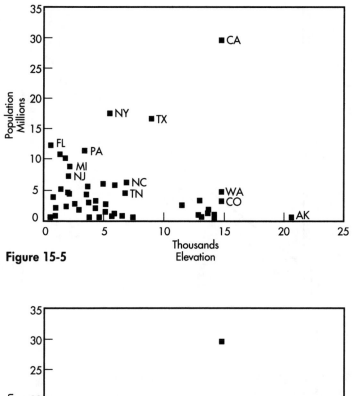

Figure 15-5

Figure 15-6

CALCULATING A REGRESSION LINE

Figure 15-7 illustrates the general situation. We will use x to represent the independent variable, which will be measured along the horizontal axis. We will use y to represent the dependent variable—that is, the variable that depends on x. In the pizza example, income is the independent variable and pizza sales

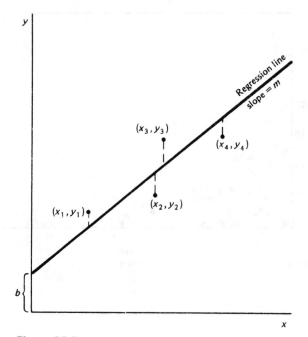

Figure 15-7

are the dependent variable. We will measure the dependent variable along the vertical axis. Suppose we have four observations. Then the scatter diagram contains four points, which we will call (x_1, y_1), (x_2, y_2), (x_3, y_3), and (x_4, y_4).

Any line can be described by specifying two numbers: the **slope** and the **vertical intercept**. We will use $\hat{m}$ to represent the slope and $\hat{b}$ to represent the intercept. (We will explain later why we put the hats on the letters.) The equation of the line can be written as

$$y = \hat{m}x + \hat{b}$$

Suppose we guess that the best regression line is the line shown in Figure 15-8. This line looks like a good choice, but it doesn't fit the data points perfectly. For each point there is a certain amount of vertical distance between the point and the line. We'll call that distance the **error** or **residual** of the line relative to that point. A larger value for the error means that the line does a worse job of representing the points. Each point has its own error. (We'll call these $error_1$, $error_2$, $error_3$, and $error_4$.) We'd like to choose the line so that the total error is as small as possible. The normal procedure in statistics is to minimize the sum of the *squares* of all the errors. The square of the error for the point (x_i, y_i) is

$$(error_i)^2 = [y_i - (\hat{m}x_i + b)]^2$$

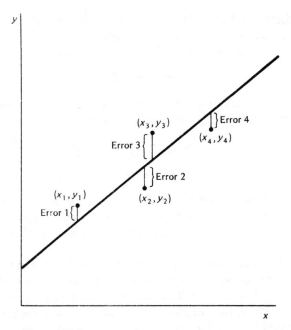

Figure 15-8

We will call the sum of the squares of all of the errors SE_{line} (short for "squared error about the line"):

$$SE_{line} = [y_1 - (\hat{m}x_1 + \hat{b})]^2 + [y_2 - (\hat{m}x_2 + \hat{b})]^2$$
$$+ [y_3 - (\hat{m}x_3 + \hat{b})]^2 + [y_4 - (\hat{m}x_4 + \hat{b})]^2$$

We originally looked at the question, Is there a relation between x and y? Here is another way of looking at this question: Does knowing the value of x help you predict the value of y? Suppose you have discovered that there is a very clear relation between x and y that can be represented by this line:

$$y = 2x + 15$$

Then, if you know that next year's value of x will be 10, you can predict that next year's value of y will be $2 \times 10 + 15 = 35$. If next year's value of x will be 30, you can predict that next year's value of y will be 75.

In general, if the variable x has the value x_i, we will use the sumbol $\hat{y}_{x_i}$ to represent the **predicted value** of y for that given value of x. In general:

$$\hat{y}_{x_i} = \hat{m}x_i + \hat{b}$$

Note that there will be a different value of $\hat{y}_{x_i}$ for each different value of x_i. (The predicted value of y is also called the **fitted value**.)

We can write our expression for the first error like this:

$$\text{Error}_1 = y_1 - \hat{y}_{x1}$$

In general, we will use n to represent the number of data points, so we can write the expression for SE_{line} with summation notation:

$$SE_{line} = \Sigma[y_i - (\hat{m}x_i + \hat{b})]^2$$
$$= \Sigma(y_i - \hat{y}_{x_i})^2$$

All through this chapter we will be taking summations from $i = 1$ to n. For convenience we will leave out the little numbers above and below the sigma. Whenever you see a capital sigma (Σ) in this chapter remember that it means $\Sigma_{i=1}^{n}$.

We have n observations for x and y, so the only unknowns in the expression for SE_{line} are $\hat{m}$ and $\hat{b}$. Somehow we need to find the values of $\hat{m}$ and $\hat{b}$ that will result in SE_{line} being as small as possible. Remember that we can choose the values of $\hat{m}$ and $\hat{b}$, but we can't change the values of the x's and the y's because we found these values when we conducted our observations. It requires calculus to find the optimum values of $\hat{m}$ and $\hat{b}$. The result is that $\hat{m}$ and $\hat{b}$ must satisfy these two equations:

$$\hat{m} = \frac{\overline{xy} - \overline{x}\,\overline{y}}{\overline{x^2} - \overline{x}^2}$$

$$\hat{b} = \overline{y} - \hat{m}\overline{x}$$

Now we can calculate the slope and the intercept for the pizza example. We will assume that pizza sales depend on income, so we will call income the independent variable (x) and pizza sales the dependent variable (y). We need to calculate $\overline{x}, \overline{y}, \overline{x^2},$ and $\overline{xy}$.

	x	y	x^2	xy	y^2
	5	27	25	135	729
	10	46	100	460	2,116
	20	73	400	1,460	5,329
	8	40	64	320	1,600
	4	30	16	120	900
	6	28	36	168	784
	12	46	144	552	2,116
	15	59	225	885	3,481
total	80	349	1,010	4,100	17,055
average	10.000	43.625	126.250	512.500	2,131.875

Therefore, $\bar{x} = 10, \bar{y} = 43.625, \overline{x^2} = 126.25$, and $\overline{xy} = 512.5$. Note that we also calculated $\overline{y^2} = 2131.88$ because it will be useful later.

We can use the formulas for $\hat{m}$ and $\hat{b}$:

$$\hat{m} = \frac{512.5 - (10 \times 43.625)}{126.25 - 10^2} = 2.905$$

$$\hat{b} = \bar{y} - \hat{m}\bar{x} = 43.625 - 2.905 \times 10 = 14.577$$

Therefore, the equation of the regression line can be written as

$$y = 2.905x + 14.577$$

There are many other ways to construct a regression line. One way is to pick a line that minimizes the sum of the absolute values of the errors instead of the sum of the squares of the errors. This method is not as seriously affected by points that are far from the main body of the data. Another method is the median-median line: the sample data is divided into three parts (according to the x values), and a **median point** found for each part (a median point having the median of that part's x values for its x coordinate and the median of that part's y values for its y coordinate). A line is then constructed, parallel to the two outside median points, moved a third of the way toward the middle point. For small samples, the median-median line is easier to construct by hand than the least-squares line.

Both of these methods become very difficult with large sample sizes and are difficult to analyze mathematically. The least-squares line is much more commonly used.

YOU SHOULD REMEMBER

1. If you have a scatter diagram with values of x along the horizontal axis and values of y along the vertical axis, then the slope and vertical intercept of the line that best fits these points can be found from these formulas:

$$\text{slope} = \hat{m} = \frac{\overline{xy} - \bar{x}\,\bar{y}}{\overline{x^2} - \bar{x}^2}$$

$$\text{intercept} = \hat{b} = \bar{y} - \hat{m}\bar{x}$$

The bars over the letters represent average values.

2. This line minimizes the sum of the squares of the errors (the vertical distance from each point to the line).

ACCURACY OF THE REGRESSION LINE

As we pointed out earlier, knowing the slope and the intercept of the regression line doesn't tell us anything about how well the line fits the data. Therefore, we need to develop another measure to tell how well the line fits. Our first inclination is just to use SE_{line}, since that formula measures how much discrepancy there is between the points on the line and the actual data points:

$$SE_{line} = \Sigma(y_i - \hat{y}_{xi})^2$$

If SE_{line} is zero, then the line fits the data points perfectly. However, if the numerical value of SE_{line} is greater than zero, then we need something to compare this number with so that we can tell whether the fit of the line is any good.

We can compare the predictions of our regression line with those of a very simple-minded prediction plan: We could always predict that the value of y will be $\bar{y}$. For example, suppose you need to predict the total rainfall in your city next year. If you know nothing about what the weather conditions will be next year, but you do know the average rainfall in your city over the last several years, then your best prediction will be to guess that the rainfall next year will be the same as the average.

Let's have a contest between ourselves (using the regression line) and the simple-minded person who always will predict that the value of y will be equal to $\bar{y}$. We can calculate the total squared error of the simple-minded method (call it SE_{av}, since it is the total squared error of y about its average:

$$SE_{av} = \Sigma(y_i - \bar{y})^2$$

[We could also write: $SE_{av} = n \, Var(y)$.]

If y really does depend on x, and our regression line describes that relation accurately, then the contest will not even be close. We will do a much better job predicting the value of y using our regression line than our simple-minded opponent will be able to do without using the line. In that case our squared error (SE_{line}) will be much less than the simple-minded squared error (SE_{av}). However, suppose that y really does not depend on x. In that case knowing the regression line will not help us. The simple-minded prediction plan will work almost as well, and SE_{line} will be almost as big as SE_{av}. Therefore, we'll define our measure of the accuracy of the regression line as follows:

$$r^2 = 1 - \frac{SE_{line}}{SE_{av}}$$

The quantity r^2 is called the **coefficient of determination**. This measure has two features that our fitness measure should have:

1. If $SE_{line} = 0$, then $r^2 = 1$, and the line fits perfectly.

2. If $SE_{line} = SE_{av}$, then $r^2 = 0$, and the line fits very poorly.

The value of r^2 will always be between 0 and 1. The higher the value of r^2, the better the line fits. Here is another interpretation: the value of r^2 is the fraction of the variation in y that can be explained by variations in x. For example, an r^2 value of .75 means that 75 percent of the variations in y can be explained by variations in x. (The symbol r^2 is used because it is the square of the sample correlation coefficient between these two variables. The correlation coefficient will be discussed in the next section.)

We can calculate the r^2 value for the pizza example by finding the sum of the squares of all of the residuals. For each value of x we can calculate the predicted value of y from the formula

$$\hat{y}_{x_i} = \hat{m}x_i + \hat{b} = 2.905x_i + 14.577$$

Then we can calculate the residual by subtracting the actual value of y from the predicted value. Here is a table of the results:

x	Actual Value y	Predicted Value $\hat{y}_{x_i} = \hat{m}x_i + \hat{b}$	Residual $y_i - \hat{y}_{x_i}$	Squared Residual $(y_i - \hat{y}_{x_i})^2$
5	27	29.102	−2.102	4.418
10	46	43.627	2.373	5.631
20	73	72.677	0.323	0.104
8	40	37.817	2.183	4.765
4	30	26.197	3.803	14.463
6	28	32.007	−4.007	16.056
12	46	49.437	−3.437	11.813
15	59	58.152	0.848	0.719
total 80	349	349	0	57.970

This table illustrates two interesting properties of a regression line: (1) the sum of the residuals is always zero, and (2) the sum of the predicted values of y is always the same as the sum of the actual values of y.

We can calculate the variance of the y values from the formula

$$\text{Var}(y) = \overline{y^2} - \bar{y}^2 = 2131.875 - 43.625^2 = 228.734$$

Since $\text{Var}(y) = \text{SE}_{av}/n$, we can calculate:

$$\text{SE}_{av} = n \, \text{Var}(y) = 8 \times 228.734 = 1829.875$$

Therefore:

$$r^2 = 1 - \frac{57.970}{1829.875} = .968$$

Just as we suspected, the regression line fits the data points very well; 96.8 percent of the variation in pizza sales can be explained by variations in income. The value of r^2 can also be calculated from either of these formulas:

$$r^2 = \frac{(\overline{xy} - \bar{x}\,\bar{y})^2}{(\overline{x^2} - \bar{x}^2)(\overline{y^2} - \bar{y}^2)}$$

or

$$r^2 = \frac{(nT_{xy} - T_x T_y)^2}{(nT_{x^2} - T_x^2)(nT_{y^2} - T_y^2)}$$

where $T_x = \Sigma x$, $T_y = \Sigma y$, $T_{xy} = \Sigma xy$, $T_{x^2} = \Sigma x^2$, and $T_{y^2} = \Sigma y^2$.

Note that we cannot calculate r^2 if all of the y values are the same, because then $\text{SE}_{av} = 0$. If y is always constant, we don't really need the regression line to predict its value anyway.

YOU SHOULD REMEMBER

1. The r^2 value is a number between 0 and 1 that indicates whether or not knowledge of the values of x helps you to estimate the values of y.

2. An r^2 value close to 1 indicates that the regression line fits the dots well.

3. The r^2 value can be calculated from one of these formulas:

$$r^2 = \frac{(\overline{xy} - \overline{x}\,\overline{y})^2}{(\overline{x^2} - \overline{x}^2)(\overline{y^2} - \overline{y}^2)}$$

or

$$r^2 = 1 - \frac{SE_{line}}{SE_{av}}$$

where $SE_{line} = \Sigma(y_i - \hat{y}_{xi})^2$ and $SE_{av} = \Sigma(y_i - \overline{y})^2$.

CORRELATION

The **correlation** (or correlation coefficient) also measures the extent of the linear relation between two variables. The correlation (represented by r) is always between -1 and 1. The r^2 value for a regression is equal to the square of the correlation between the two regression variables. If the slope of the regression line is positive, then r is positive; if the slope is negative, then r is negative. The correlation is zero if the regression line has zero slope. A correlation near zero means that there is not much of a linear relation between the two variables. (There could perhaps be a nonlinear relation between the two variables even if the correlation is near zero. We will discuss that possibility later.) A correlation value near 1 or -1 indicates that there is a very strong linear relation between the two variables.

Here is an example of the correlation coefficient. We have observations of the height, weight, experience, and age for the 49 players on an NFL football team. In Figures 15-9 to 15-11 we have drawn scatter diagrams comparing height to weight, experience to age, and weight to experience. As we would have predicted, there is an almost perfect linear relation between experience and age; the correlation coefficient between these two variables is .970. There is also a strong linear relation between height and weight (correlation .715). On the other hand, there is practically no relation between weight and experience (correlation .043) since there is no reason why more experienced players would be heavier than less experienced players.

The correlation coefficient can be calculated from this formula:

$$r = \frac{\overline{xy} - \overline{x}\,\overline{y}}{\sqrt{\mathrm{Var}(x)\,\mathrm{Var}(y)}} = \frac{\overline{xy} - \overline{x}\,\overline{y}}{\sqrt{(\overline{x^2} - \overline{x}^2)(\overline{y^2} - \overline{y}^2)}}$$

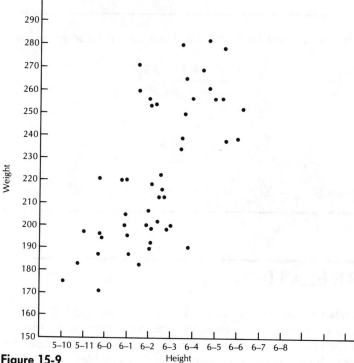

Figure 15-9

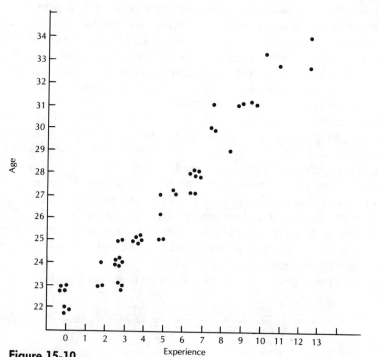

Figure 15-10

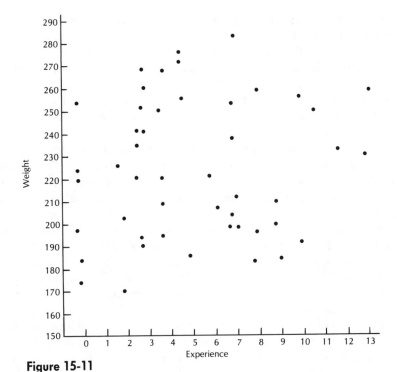

Figure 15-11

We can calculate the correlation for the pizza example:

$$r = \frac{\overline{xy} - \bar{x}\,\bar{y}}{\sqrt{(\overline{x^2} - \bar{x}^2)(\overline{y^2} - \bar{y})}}$$

$$= \frac{512.5 - (10 \times 43.625)}{\sqrt{(126.25 - 10^2)(2131.88 - 43.625^2)}}$$

$$= .984$$

STATISTICAL ANALYSIS OF REGRESSIONS

Now we would like to perform statistical tests on our regression results. To do this we need to make some assumptions about the process relating x and y. In the standard regression model we assume that the true nature of the relation between x and y can be described by this equation:

$$y_i = mx_i + b + e_i$$

where x_i is the ith observation for the variable x, y_i is the ith observation for the variable y, and e_i is known as the random error term. We are assuming that x is the dominant influence on y, and the relationship can be represented by a straight line with slope m and y-intercept b. (Thus, each e_i should be independent of each x_i. This can always be checked by plotting e_i against x_i to see if there is any pattern.) Unfortunately the true values of m and b are unknown, but as you have probably guessed, we will use statistical procedures to estimate their values.

If the true equation was $y = mx + b$, then x would be the only factor that affected the value of y. Every single increase or decrease in y could be explained by an increase or a decrease in x. However, there are almost always some other factors that affect the value of the dependent variable. If the regression line represents the relationship well, these other factors will not be very important. Because all these other factors are mysterious and unknown, we will call them random error. In the equation given above e is a random variable that represents all of these other factors. (When we discuss multiple regression, we will see how it is possible to include some of these other factors in the regression model; but even so there will always be some remaining unexplained factors that make up the random error term.)

If we have n observations, then there will be n different random errors. We will let e_i represent the random error effect on the first observation. In general, e_i is a random variable representing the random error effect on the ith observation. We know that the expected value of each random error is zero $[E(e_i) = 0]$ because of the fact that we have included the intercept term b in the equation. We will let $\text{Var}(e_i)$ be equal to σ^2, but unfortunately we don't know the true value of σ^2. If x is a very good predictor of y, then σ^2 will tend to be small. If σ^2 is large, then there are other important factors that affect y and we should try to include them in our model. We will also assume that each random error term has a normal distribution, that the variance σ^2 is the same for each random error, and that each random error is independent of all of the other random errors.

To summarize the situation:

$$y = mx + b + e$$

Here x is known and not random. We have a list of n observations for x. Also, y is known and random. We also have a list of n observations of y, with each observation being paired with one of the observations of x. The presence of the e term means that y is a random variable with a normal distribution.

$$E(y_i) = mx_i + b \quad \text{and} \quad Var(y_i) = \sigma^2$$

Here m and b are not random, but their true values are unknown; e is both random and unknown. We know that $E(e) = 0$, but $Var(e) = \sigma^2$ is unknown. We assume that e has a normal distribution.

Even though the true values of m and b are unknown, we can calculate the slope and the intercept of the regression line:

$$\hat{m} = \frac{\overline{xy} - \overline{x}\,\overline{y}}{\overline{x^2} - \overline{x}^2}$$

$$\hat{b} = \overline{y} - \hat{m}\overline{x}$$

Now you see why we included the hats. In statistical inference we often put a hat over a calculated statistic that is being used to estimate the value of an unknown parameter. The regression line is called the *least-squares* line, so $\hat{m}$ and $\hat{b}$ can be called the **least-squares estimators** of the parameters m and b.

Fortunately, it turns out that $\hat{m}$ and $\hat{b}$ have some very desirable statistical properties. For one thing, $\hat{m}$ and $\hat{b}$ are the maximum likelihood estimators of m and b. It can be shown that $\hat{m}$ has a normal distribution with $E(\hat{m}) = m$ and

$$Var(\hat{m}) = \frac{\sigma^2}{\Sigma(x_i - \overline{x})^2}$$

We can write the expression for $\hat{m}$ like this:

$$\hat{m} = \frac{\Sigma(x_i - \overline{x})(y_i - \overline{y})}{\Sigma(x_i - \overline{x})^2}$$

which can also be written as

$$\hat{m} = \frac{1}{\Sigma(x_i - \overline{x})^2} \Sigma(x_i - \overline{x})y_i$$

Remember that the x's are constants and the y's are random variables with a normal distribution. From the expression given above we can see that $\hat{m}$ is found by adding together a bunch of normal random variables multiplied by different constants, so we know from the properties of normal distributions that $\hat{m}$ will also have a normal distribution.

The fact that $E(\hat{m}) = m$ is important. It means that $\hat{m}$ is an unbiased estimator of the slope. It can also be shown that $\hat{m}$ is the best estimator (because it has the smallest variance) among a certain general class of all unbiased estimators.

The expression for the variance of $\hat{m}$ tells us that the variance of $\hat{m}$ is larger if σ^2 is larger. This seems reasonable, since a larger value of σ^2 means that we can expect more scatter about the true regression line, and the increased scatter will make it more difficult to pin down the true value of m. The expression $\Sigma(x_i - \bar{x})^2$ represents the squared error about the average of the x's, or we could write

$$\Sigma(x_i - \bar{x})^2 = n \, \text{Var}(x)$$

We can see that, when there is a greater spread among the x values, then the variance of $\hat{m}$ will be less and it will be easier to pin down the true value of m. Figure 15-12 illustrates two different situations. In each case the number of observations is the same. However, in example (a) there is not very much spread among the x values. In that case there will be much more uncertainty about the true value of slope than there will be in example (b).

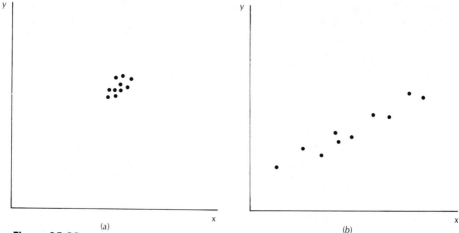

(a)

(b)

Figure 15-12

Since $\hat{m}$ has a normal distribution with mean m and variance $\sigma^2/\Sigma(x_i - \bar{x})^2$, we know that this random variable:

$$\frac{\hat{m} - m}{\sqrt{\dfrac{\sigma^2}{\Sigma(x_i - \bar{x})^2}}}$$

will have a standard normal distribution. Therefore, we can determine a 95 percent confidence interval for m:

$$\hat{m} \pm \frac{1.96\sigma}{\sqrt{\Sigma(x_i - \bar{x})^2}}$$

(See Chapter 11.) However, there is one obvious problem with calculating the confidence interval in this way: we don't know the value of σ. We need a way to estimate it. If we knew the true values of m and b, then we would have n observations of the random variable e:

$$e_i = y_i - (mx_i + b)$$

and so on. Since $E(e) = 0$, we could write

$$\sigma^2 = \text{Var}(e) = E(e^2) - [E(e)]^2 = E(e^2)$$

and we could estimate σ^2 by the average of the squares of the e's. Since we don't know m and b, we can use the residuals from our calculated regression line:

$$[y_1 - (\hat{m}x_1 + \hat{b})], \qquad [y_2 - (\hat{m}x_2 + \hat{b})], \qquad \text{and so on}$$

The sum of the squares of all these residuals is what we called SE_{line}. Here we run into a conflict between the different criteria for estimators. The maximum likelihood estimator of σ^2 is equal to $\text{SE}_{\text{line}}/n$. However, in order to have an unbiased estimator of σ^2, we must use $\text{SE}_{\text{line}}/(n - 2)$. (Here $n - 2$ is the degrees of freedom of the squared error about the line, since we started with n points but lost two degrees of freedom when we had to use our observations to estimate the values of the slope and the intercept.) The quantity $\text{SE}_{\text{line}}/(n - 2)$ is called the **mean square error** (**MSE** for short):

$$\text{MSE} = \frac{\text{SE}_{\text{line}}}{n - 2} = \frac{\Sigma(y_i - \hat{y}_{x_i})^2}{n - 2} = \frac{\Sigma[y_i - (\hat{m}x_i + \hat{b})]^2}{n - 2}$$

The expected value of MSE is σ^2, which makes it an unbiased estimator of σ^2.

As you recall, we had established that this random variable:

$$\frac{\hat{m} - m}{\sqrt{\dfrac{\sigma^2}{\Sigma(x_1 - \bar{x})^2}}}$$

has a standard normal distribution. In similar situations in Chapter 11, we substituted an estimator for σ^2 in place of the unknown value of σ^2, and we ended up with something that had a t distribution. In this case, the random variable

$$\frac{\hat{m} - m}{\sqrt{\dfrac{\text{MSE}}{\Sigma(x_i - \bar{x})^2}}}$$

has a t distribution with $n - 2$ degrees of freedom. Then the confidence interval is

$$\hat{m} \pm a \sqrt{\frac{\text{MSE}}{\Sigma(x_i - \bar{x})^2}}$$

Here a is a number found from Table A3-5 such that

$$\Pr(-a < t < a) = \text{CL}$$

where t is a random variable having a t distribution with $n - 2$ degrees of freedom and CL is the confidence level.

For the pizza sales example we have MSE $= 57.97/6 = 9.662$ and $\Sigma(x_i - \bar{x})^2 = 210$. Since $n = 8$, we will have a t distribution with $8 - 2 = 6$ degrees of freedom. If we choose a 95 percent confidence interval, we can see from Table A3-5 that the value of a is 2.447. We already found that $m = 2.905$, so the confidence interval is

$$2.905 \pm 2.447 \sqrt{\frac{9.662}{210}}$$

which is from 2.38 to 3.43.

We can also perform hypothesis tests on our model. Here is one important null hypothesis: "The values of x do not have any relationship to the values of y." Obviously, if we think our regression is any good, we would like to collect enough statistical evidence to be able to prove this hypothesis wrong. From the equation $y = mx + b + e$, we can see that there will be no relation between x and y if the true value of the slope is zero. Since

$$\frac{(\hat{m} - m)}{\sqrt{\dfrac{\text{MSE}}{\Sigma(x_i - \bar{x})^2}}}$$

has a t distribution with $n - 2$ degrees of freedom, it follows that *if $m = 0$* then the statistic

$$\frac{\hat{m}}{\sqrt{\dfrac{\text{MSE}}{\Sigma(x_i - \bar{x})^2}}}$$

will have a t distribution with $n - 2$ degrees of freedom. We can calculate the value of that statistic. If it seems plausible that the calculated value could have come from the t distribution, then we will accept the null hypothesis; otherwise we will reject it. For the pizza sales the calculated t statistic is

$$2.905 \sqrt{\frac{210}{9.662}} = 13.542$$

For a two-tailed test at the 5 percent significance level the critical value for a t distribution with 6 degrees of freedom is 2.447. Since 13.542 is in the rejection region, well above the critical value, we can safely reject the null hypothesis that the true value of the slope is zero.

YOU SHOULD REMEMBER

1. Assume that the true relation between x and y is given by this formula:

$$y = mx + b + e$$

where e is a normal random variable with mean 0 and unknown variance σ^2.

2. Then the least-squares estimators $\hat{m}$ and $\hat{b}$ are the maximum likelihood estimators of m and b, and they are also unbiased estimators.

3. The mean square error:

$$\text{MSE} = \frac{\Sigma[y_i - (\hat{m}x_i + \hat{b})]^2}{n - 2}$$

is an unbiased estimator of the unknown value of σ^2.

4. To test the hypothesis that the true value of the slope is zero, calculate this statistic:

$$\frac{\hat{m}}{\sqrt{\dfrac{MSE}{\Sigma(x_i - \bar{x})^2}}}$$

If the true value of the slope is zero, then this statistic will have a t distribution with $n - 2$ degrees of freedom.

PREDICTING VALUES OF y

• *FOUR WARNINGS ABOUT PREDICTING VALUES*

Now we will discuss how to use our regression model to predict values of the dependent variable. Before we get carried away making predictions with a regression model, however, we must heed several important warnings.

- Any prediction based on a regression model is a conditional prediction, since the prediction for the dependent variable depends on the value of the independent variable. Suppose you have found a regression relationship that describes perfectly the relation between y and x. In that case you can predict future values of y *if* (but only if) you know the future value of x. If pizza sales do depend on income as predicted by our regression relation, then we can predict next year's pizza sales if we know next year's income. If we cannot predict next year's income, then we cannot predict the actual value of next year's pizza sales (although it still may be very useful to have a regression model that tells us how much income affects demand for pizza).

- The regression line has been estimated using past data. This line will not be able to predict the future if the relationship between x and y changes. A sudden change in people's pizza preferences will ruin the ability of the regression line to predict future values of pizza sales.

- Many regression predictions try to predict values of y in situations where the value of x falls outside the range of values previously observed for x. These predictions, known as **extrapolations**, are much less reliable than predictions based on values of the independent variable that fall within the range of previously observed values.

  For example, part (a) of Figure 15-13 shows a number of observations for pizza sales as a function of income in a new set of towns. It seems

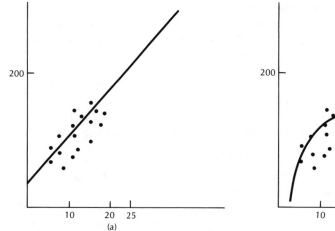 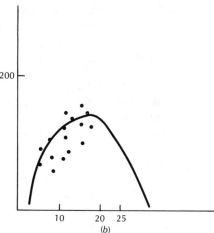

Figure 15-13

reasonable to represent the data with a straight line. Suppose that next year we expect income will be 25. Based on the regression line, it seems logical to predict that pizza sales next year will be 200. However, this prediction is an extrapolation. The regression analysis gives us good evidence that the relation between income and sales can be well represented by a straight line when income is in the range from 10 to 20, but we have no way of knowing for sure whether this linear relationship holds for other levels of income. It is quite possible that, unbeknownst to us, the true relation between income and pizza sales is given by the curve shown in part (b) of Figure 15-13. This curve suggests that pizza sales do not continue to increase as income increases. Instead, it appears that people start going to fancier restaurants and consume less pizza once income rises past a certain point. This curve fits the original data points as well as the regression line does, but it predicts a far different value for pizza sales if income is 25. As long as our observations cover only values of income within the narrow range shown, we have no way of distinguishing between these two situations, and extrapolations based on the regression line may be quite erroneous.

- The mere fact that there is a strong association between two variables does not indicate there is a cause and effect relationship between them. If you find a regression line that fits the relationship between y and x very well, then there are four possibilities:

1. The values of y may really depend on the values of x, as we have assumed so far.

2. The observed relation may have occurred completely by chance. If we have many observations, this is extremely unlikely, but we have seen that in statistics we cannot rule out the possibility that an event that looks significant may have occurred randomly.

3. There may be a third variable that affects both x and y. This is the most likely explanation for situations where two variables are closely correlated but there does not seem to be a causal relationship between them. For example, many unrelated variables tend to increase with time.

4. There may indeed be a causal link between x and y, but it may be that y is causing x. We may have incorrectly determined which was the dependent variable and which was the independent variable. For example, we assumed that higher incomes cause people to buy more pizza. However, it may work this way: when people buy more pizzas, they create more income for all the people who work in pizza places, and this leads to a multiplier effect that increases the income of the whole community. Therefore it could be that higher pizza sales cause higher incomes. Here is another example. You may have found a regression result that seems to indicate that higher advertising levels cause higher sales. However, suppose that firms allocate their advertising budgets on the basis of sales. Then it is quite likely that higher sales cause higher advertising budgets.

• PREDICTING VALUES OF THE DEPENDENT VARIABLE

Now that we have heeded all of these warnings, let's see how we can use our regression model to predict values of y. Assume that we have determined that x really does cause y, that this relationship will still apply in the future, and that it can be accurately described by the regression line $y = 2.905x + 14.577$. If we know that the value of income next year will be 16, then naturally we will predict that next year's value of y will be $2.905 \times 16 + 14.577 = 61.06$.

The next question is: How precise is this prediction? We would like to construct an interval such that there is a 95 percent chance that the value of y will be contained in that interval (given that $x = 16$). This type of interval is called a **prediction interval**. Note that it is similar to the confidence intervals we developed for unknown parameters. Suppose for the moment that we know the true values of m, b, and σ. Then, if $x = x_{new}$, we know that y will have a normal distribution with mean $\hat{y}_{x_{new}} = mx_{new} + b$ and variance σ^2. Therefore, there is a 95 percent chance that the value of y will be between $[(mx_{new} + b) - 1.96\sigma]$ and $[(mx_{new} + b) + 1.96\sigma]$.

Unfortunately, matters become much worse because we don't know the true values of m, b, and σ. Now there are two sources of uncertainty involved with predicting values of y: we don't know the true regression line, and the predicted value of y will deviate randomly about the line. The formula for the estimated variance of y for a given value of x is

$$\text{Var}(y) = \text{MSE}\left[1 + \frac{1}{n} + \frac{(x_{new} - \bar{x})^2}{\Sigma(x_i - \bar{x})^2}\right]$$

Note that the variance becomes larger when the value of x_{new} becomes farther from $\bar{x}$. When x_{new} is closer to $\bar{x}$, we have greater confidence that our estimated regression line is close to the true regression line. However, if our estimate for the slope of the line is slightly different from the true value, then this difference will cause our estimated regression line to depart farther and farther from the true line when we move farther from $\bar{x}$.

When x has the value x_{new} and we have calculated $\hat{y}_{x_{new}} = \hat{m}x_{new} + \hat{b}$ and $\text{Var}(y)$ using the formula above, then the prediction interval for y is

$$\hat{y}_{new} \pm a\sqrt{\text{Var}(y)}$$

where

$$\text{Pr}(-a < t < a) = \text{CL}$$

t is a random variable having a t distribution with
    $n - 2$ degrees of freedom

CL is the confidence level (such as .95)

Here are some sample calculations:

Value of x	Predicted Value of $y = \hat{m}x + \hat{b}$	95 Percent Prediction Interval for y
2	20.39	11.29–29.48
6	32.01	23.67–40.34
10	43.63	35.56–51.69
14	55.25	46.91–63.58
18	66.87	57.77–75.96
20	72.68	63.05–82.30

Figure 15-14 illustrates the prediction intervals compared to the regression line. You can see how the intervals become wider as the value of x is farther from $\bar{x}$.

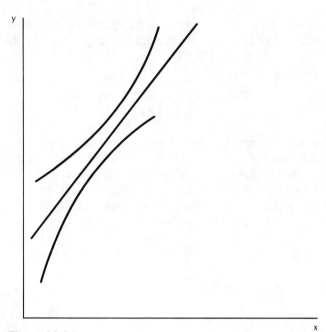

Figure 15-14

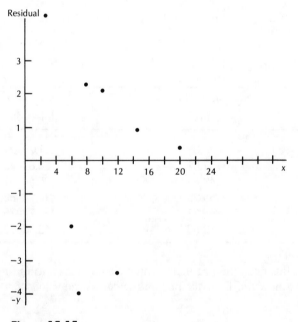

Figure 15-15

ANALYSIS OF RESIDUALS

Another way to gain valuable information about a regression model is to make a graph of the residuals. For each data point (x_i, y_i) we can calculate the residual:

$$(i\text{th residual}) = y_i - \hat{y}_{x_i} = y_i - (\hat{m}x_i + \hat{b})$$

Let's make a scatter diagram that measures values of x along the horizontal axis and the residual along the vertical axis. Many computer regression programs will automatically prepare such a diagram if you wish. In the pizza example we have these values:

x	Residual
5	-2.102
10	2.373
20	0.323
8	2.183
4	3.803
6	-4.007
12	-3.437
15	0.848

Figure 15-15 shows the scatter diagram.

If the assumptions of the regression model are met, the plot of the residuals should look like a random array of dots. There should be no apparent pattern. Because the error term in the model is assumed to have a normal distribution, there should be more values near zero than far from zero. In particular, here are some things to look for in a residual plot:

- Outliers. An **outlier** is a residual much larger (or much more negative) than the others. On the original scatter diagram an outlier will show up as a point that is far from the estimated regression line. When you have identified which observation corresponds to the outlier point, then you should check to make sure that the observation is correct. An outlier may have occurred because you made a typographical error in entering the data for that observation, in which case you can correct the error and run the regression again. If you are convinced that the outlier observation is indeed correct, then you should investigate to see whether there is some special circumstance that caused that observation to depart so far from the others. Perhaps you can identify a rare natural disaster that corresponds with the observation involved. If you are convinced that the circumstance that caused the outlier will not recur, you may drop that observation and then perform the regression again with the remaining points. If, on the other hand, you can identify the cause of the outlier with a variable that should be included in the model, you should set up a multiple regression model. (See the next chapter.) If you cannot identify any cause for the outlier, then you must leave the outlier in the regression and treat it as a random error.

- Nonnormal errors. The original regression model assumed that the error terms had normal distributions, but the plot of the residuals may indicate that this is not the case. The least-squares estimators $\hat{m}$ and $\hat{b}$ are unbiased estimators of the true values of m and b whether or not the errors are normally distributed, but the statistical tests we performed were all based on the assumption of normal errors.

- Nonconstant variances. We assumed that the variances of the error terms associated with the observations were all the same. If the residual diagram shows that the residuals are systematically bigger in one area of the diagram, however, this condition may not be met. The pizza regression residual plot indicates that the residuals tend to be greater for smaller values of x, so perhaps there is a situation of nonconstant variance here. (More observations would be needed to make this conclusion definite.) The impressive technical term for the situation with nonconstant variances among the error terms is **heteroscedasticity**. In situations of nonconstant variance it is sometimes possible to transform the model into an equivalent model with constant variances.

- Omitted variables. It may be worthwhile to make a plot of the residuals compared with another independent variable that seems significant but has not been included in the model. If there is an association between the residuals and the new variable, you should set up a multiple regression model including that variable.

- Nonlinearity. If the true relationship is not a straight line, the residual plot will usually reveal the situation immediately. For example, here are observations for two variables x and y:

x	y	Residual
1	10.000	0.786
2	10.800	0.386
3	11.664	0.050
4	12.597	−0.217
5	13.605	−0.409
6	14.693	−0.521

(continued)

x	y	Residual
7	15.869	−0.546
8	17.138	−0.476
9	18.509	−0.305
10	19.990	−0.024
11	21.589	0.375
12	23.316	0.902

The simple linear regression calculation for these observations gives a slope of 1.2, an intercept of 8.014, and an r^2 value of .986. The scatter diagram is shown in Figure 15-16.

However, the residual plot shown in Figure 15-17 clearly is not random. When the residuals follow a definite curve like this one, there is strong evidence either that the underlying model is not a linear model, or one of the other problems mentioned above has arisen. In this case we can see that for small or large values of x the residual is always positive, but for medium values of x the residual is always negative. This pattern indicates that the relationship between x and y can be better represented by a curve than by a line.

We will now turn our attention to transformations that can turn a model with a curve into an equivalent model with a straight line.

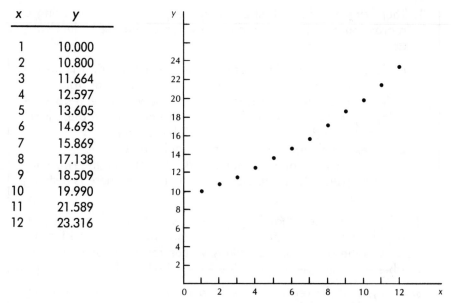

x	y
1	10.000
2	10.800
3	11.664
4	12.597
5	13.605
6	14.693
7	15.869
8	17.138
9	18.509
10	19.990
11	21.589
12	23.316

Figure 15-16

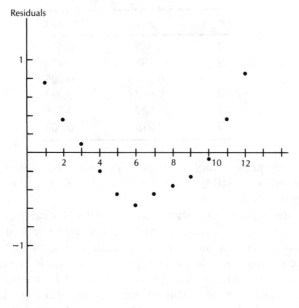

Figure 15-17

YOU SHOULD REMEMBER

1. There are four important points to keep in mind when using a regression equation to predict the value of the dependent variable:

   a. Before the dependent variable can be predicted, it is necessary to make a prediction for the independent variable.

   b. The regression line based on past observations will not reliably predict the future if the relation between the dependent variable and the independent variable changes.

   c. An extrapolation prediction may not be reliable.

   d. There may not be a causal connection between the two variables even if the regression equation indicates there is a strong relation.

2. If it appears likely that the regression equation will reliably predict the future, a *prediction interval* may be constructed that indicates the range of most likely values for the dependent variable.

3. Analysis of the regression residuals is helpful in spotting problems such as outliers, nonnormal errors, nonconstant variance, omitted variables, and nonlinearity.

TRANSFORMATIONS WITH LOGARITHMS

Suppose that the true relationship between x and y can be given by this equation:

$$y = ca^x$$

where c and a are two unknown constants. We cannot use simple linear regression to find the values for c and a. To handle this situation we must dig back into memory to find the concept called **logarithm**. You should have studied logarithms sometime previously, but we will review the basic ideas here.

Consider this question: What number will we get if we raise 2 to the 7th power? (Remember that raising a number to the power n means writing down that number n different times and then multiplying all of those n numbers together.) Here is a table of powers of 2:

$$2^0 = 1, \quad 2^1 = 2, \quad 2^2 = 4, \quad 2^3 = 8, \quad 2^4 = 16,$$
$$2^5 = 32, \quad 2^6 = 64, \quad 2^7 = 128, \quad 2^8 = 256, \quad 2^9 = 512$$

We can see that 128 is the result if we raise 2 to the 7th power. Now let's ask the opposite question: To what power should we raise 2 in order to get 128 as the result? The result is called the logarithm to the base 2 of 128. In this case we happen to know already that the answer is 7. In logarithm notation this is written as $\log_2 128 = 7$. We could also write

$$\log_2 1 = 0, \quad \log_2 2 = 1, \quad \log_2 4 = 2, \quad \log_2 8 = 3,$$
$$\log_2 16 = 4, \quad \log_2 32 = 5, \quad \text{and so on}$$

There are many different logarithm functions. Any positive number except 1 can act as the base for a logarithm function. In general, if a is the base of the logarithm function, then

$$\log_a x = n \quad \text{means} \quad a^n = x$$

The two most common bases for logarithms are 10 and a special number called e, which is about 2.71828. The expression $\log x$, where no base is specified, usually refers to logarithms to the base 10, which are called **common logarithms**. For example, $\log 10 = 1$, $\log 100 = 2$, $\log 1000 = 3$, and so on. The expression $\ln x$ usually refers to logarithms to the base e, which are called **natural logarithms**. Natural logarithms are especially important in calculus. Calculating logarithms is easy if the number you are interested in happens to be a whole-number power of the base. Otherwise the best thing to do is use a calculator or a computer with a built-in logarithm function to calculate the values. If you cannot do that, the next best thing is to look in a book that contains a table of logarithms.

Logarithms are very useful for our purposes because they satisfy these properties:

$$\log(ab) = \log a + \log b$$

$$\log\left(\frac{a}{b}\right) = \log a - \log b$$

$$\log a^n = n \log a$$

These properties are true for logarithms of any base. For example, if the true relation between y and x is given by this formula:

$$y = ca^x$$

we can take the logarithm of both sides:

$$\log y = \log(ca^x)$$

Using the properties of logarithms, we can rewrite as follows:

$$\log y = \log c + \log a^x$$
$$= \log c + x \log a$$

Let's make these definitions:

$$\text{let } b = \log c, \quad \text{let } m = \log a$$

Then

$$\log y = b + mx$$

That looks very familiar: we have transformed our model into a situation where the linear regression model is appropriate. All we have to is use the values of $\log y$ as the dependent variable instead of the original values of y. (See Figure 15-18.)

y	$\log y$	y	$\log y$
10.000	1.000	15.869	1.201
10.800	1.033	17.138	1.234
11.664	1.067	18.509	1.267
12.597	1.100	19.990	1.301
13.605	1.134	21.589	1.334
14.693	1.167	23.316	1.368

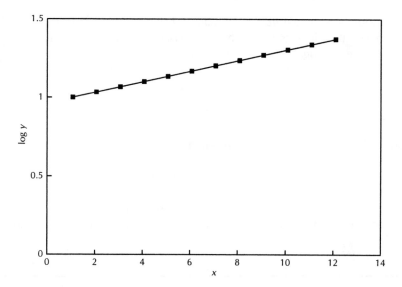

Figure 15-18

Now we can perform the simple linear regression calculations, using x as the independent variable and $\log y$ as the dependent variable. The result is

$$\text{slope} = m = 0.0334, \qquad \text{intercept} = b = 0.966$$

Since the original model is $y = ca^x$ with $b = \log c$ and $m = \log a$, we can calculate that

$$c = 10^{0.966} = 9.3 \qquad \text{and} \qquad a = 10^{0.0334} = 1.08$$

Therefore, our estimate for the true model is

$$y = 9.3 \times (1.08)^x$$

Note that we used common logarithms for our calculations, but you may use natural logarithms if you wish. This function is an example of a situation called *exponential growth* because the independent variable x appears as an exponent.

Here is another situation where a logarithmic transformation is useful. Suppose we have determined that the quantity demanded of a particular good is given by a formula like this:

$$Q = Q_0 P^{-a}$$

where P is the price of the good and Q_0 and a are unknown parameters. (In this formula a is the *elasticity* of demand for this product. When the formula is written like this, it is assumed that the elasticity is constant.) Once again we need to take the logarithm of both sides:

$$\log Q = \log(Q_0 P^{-a})$$

$$= \log Q_0 - a \log P$$

Now let $b = \log Q_0$ and $m = -a$. Then

$$\log Q = b + m \log P$$

Now we can estimate b and m by performing a simple linear regression calculation using $\log P$ as the independent variable and $\log Q$ as the dependent variable.

YOU SHOULD REMEMBER

1. If $a^n = x$, then $\log_a x = n$ (n is the logarithm to the base a of x).

2. Logarithms are particularly useful with regression models because they can be used in transformations that convert curved relations into straight-line relations.

3. If the true relation between x and y is $y = ca^x$, then the transformed equation is

$$\log y = \log c + x \log a$$

4. If the true relation between x and y is $y = cx^n$, then the transformed equation is

$$\log y = \log c + n \log x$$

KNOW THE CONCEPTS

DO YOU KNOW THE BASICS?

Test your understanding of Chapter 15 by answering the following questions:

1. If the r^2 value for a simple regression is .80, does this mean that 80 percent of the changes in the dependent variable are caused by changes in the independent variable?

2. List three ways to make regression estimates more precise.

3. What do you learn from a regression calculation if the estimated slope is zero?

4. What are two sources of uncertainty when you are using a regression equation to predict a value of the dependent variable?

5. What do you use the *t* statistic for?

6. Why are extrapolation predictions especially difficult?

7. Suppose you perform a regression calculation based on time series data, and it indicates that the demand for a particular good goes up when the price goes up. Can you think of an explanation why this might happen?

8. Does a correlation of zero indicate that there is no relation between two quantities?

9. Suppose someone reported a correlation between the diameters of the nine planets in the solar system and the crime rates in nine large cities. What does this analysis show?

10. If you have only two observations, will you be able to get a high r^2 value? Will you be able to obtain precise estimates of the slope and intercept?

11. In what situations will you take the logarithm of the dependent variable and then perform the regression calculation? In what situations will you take the logarithms of both the independent and the dependent variable?

TERMS FOR STUDY

coefficient of determination	natural logarithm
common logarithm	outlier
correlation	predicted value
error	prediction interval
extrapolation	r^2
fitted value	regression line
heteroscedasticity	residual
inferior good	scatter diagram
least-squares estimators	simple linear regression
logarithm	slope
mean square error (MSE)	vertical intercept

PRACTICAL APPLICATION

COMPUTATIONAL PROBLEMS

In Exercises 1–5 you are given lists of observations for two variables: X, an independent variable, and Y, a dependent variable. In each case, perform a simple linear regression calculation. Calculate the slope, intercept r^2 value, *correlation, mean square error, estimated standard deviation, and t statistic used to test the hypothesis that the true value of the slope is zero. Also calculate a 95 percent confidence interval for the true value of the slope.*

1.

	X_1	Y_1			X_1	Y_1
1:	34	6		9:	45	90
2:	26	57		10:	10	25
3:	9	89		11:	47	45
4:	30	60		12:	37	23
5:	47	95		13:	47	52
6:	10	42		14:	8	95
7:	34	31		15:	45	48
8:	34	28				

2.

	X_2	Y_2			X_2	Y_2
1:	37	6		10:	11	29
2:	21	19		11:	39	3
3:	21	28		12:	18	4
4:	28	27		13:	29	7
5:	37	24		14:	4	33
6:	32	14		15:	20	27
7:	4	9		16:	7	19
8:	20	16		17:	13	6
9:	37	26		18:	19	2

3.

	X_3	Y_3			X_3	Y_3
1:	51	59		6:	3	62
2:	34	5		7:	7	70
3:	47	12		8:	11	70
4:	42	42		9:	61	69
5:	59	57		10:	16	59

4.

	X_4	Y_4			X_4	Y_4
1:	75	9		9:	90	5
2:	34	1		10:	87	9
3:	80	5		11:	52	7
4:	33	6		12:	72	3
5:	13	8		13:	81	3
6:	93	4		14:	19	9
7:	60	10		15:	89	3
8:	66	9		16:	10	11

5.

	X_5	Y_5		X_5	Y_5
1:	25	1	11:	32	11
2:	84	2	12:	93	12
3:	66	3	13:	33	13
4:	76	4	14:	1	14
5:	78	5	15:	70	15
6:	56	6	16:	94	16
7:	68	7	17:	62	17
8:	1	8	18:	57	18
9:	28	9	19:	10	19
10:	67	10	20:	58	20

6. Perform a regression calculation for the income/consumption data given at the beginning of the chapter.

7. Perform a regression calculation for the elevation/population data given at the beginning of the chapter.

In each exercise below you are given a set of observations for two variables. Perform a simple regression calculation, using the first variable as the independent variable.

8. Money supply broadly defined (M2); average price level as measured by the consumer price index (CPI):

Year	M2	CPI
1981	1,793.3	90.9
1982	1,953.2	96.5
1983	2,187.7	99.6
1984	2,378.4	103.9
1985	2,576.0	107.6
1986	2,820.3	109.6
1987	2,922.3	113.6
1988	3,083.5	118.3
1989	3,243.0	124.0
1990	3,356.0	130.7
1991	3,457.9	136.2
1992	3,515.3	140.3
1993	3,583.6	144.5
1994	3,617.0	148.2
1995	3,780.7	152.4

9: Unemployment rate (UNEMP); percent change in consumer price index (INF):

Year	UNEMP	INF
1981	7.6	10.3
1982	9.7	6.2
1983	9.6	3.2
1984	7.5	4.3
1985	7.2	3.6
1986	7.0	1.9
1987	6.2	3.6
1988	5.5	4.1
1989	5.3	4.8
1990	5.5	5.4
1991	6.7	4.2
1992	7.4	3.0
1993	6.8	3.0
1994	6.1	2.6
1995	5.6	2.8

10. Percent change in output per hour of all persons, business sector (PROD); percent change in compensation per hour (COMP):

Year	PROD	COMP
1981	2.0	–.7
1982	–.7	1.2
1983	3.3	.8
1984	2.5	.2
1985	1.9	1.3
1986	2.6	3.3
1987	–.1	.2
1988	.5	.4
1989	.8	–1.9
1990	.8	.4
1991	.5	.5
1992	3.4	2.1
1993	.2	–.4
1994	.7	–.4

11. Unemployment rate (UNEMP); capacity utilization rate, total industry (CAP):

Year	UNEMP	CAP
1981	7.6	80.9
1982	9.7	75.0
1983	9.6	75.8
1984	7.5	81.1
1985	7.2	80.3
1986	7.0	79.2
1987	6.2	81.5
1988	5.5	83.7
1989	5.3	83.7
1990	5.5	82.1
1991	6.7	79.2
1992	7.4	80.3
1993	6.8	81.4
1994	6.1	83.9
1995	5.6	83.7

For Exercises 12–14 you are given values for an independent variable X and a dependent variable Y. In each case take the logarithm of the Y value, and then perform a simple linear regression calculation.

12.

	X_{12}	Y_{12}		X_{12}	Y_{12}
1:	1.659	15.882	9:	0.373	1.862
2:	0.207	1.412	10:	1.703	17.098
3:	0.094	1.169	11:	0.515	2.358
4:	1.696	16.879	12:	0.171	1.330
5:	1.318	8.996	13:	0.867	4.243
6:	1.209	7.504	14:	0.639	2.900
7:	1.572	13.737	15:	1.777	19.337
8:	1.577	9.200			

13.

	X_{13}	Y_{13}		X_{13}	Y_{13}
1:	1.000	10.000	9:	9.000	15.938
2:	2.000	10.600	10:	10.000	16.895
3:	3.000	11.236	11:	11.000	17.908
4:	4.000	12.400	12:	12.000	18.983
5:	5.000	12.625	13:	13.000	20.122
6:	6.000	13.382	14:	14.000	21.329
7:	7.000	14.185	15:	15.000	22.609
8:	8.000	15.036			

14.

X_{14}	Y_{14}		X_{14}	Y_{14}
1: 1.000	1,000.000	9:	9.000	272.491
2: 2.000	850.000	10:	10.000	231.617
3: 3.000	722.500	11:	11.000	196.874
4: 4.000	614.125	12:	12.000	158.324
5: 5.000	522.006	13:	13.000	142.242
6: 6.000	443.705	14:	14.000	120.905
7: 7.000	377.150	15:	15.000	102.770
8: 8.000	320.577			

For Exercises 15 and 16 you are given values for an independent variable X and a dependent variable Y. In each case take the logarithms of the X and Y values, and then perform a simple linear regression calculation.

15.

X_{15}	Y_{15}		X_{15}	Y_{15}
1: 6.868	323.960	7:	18.174	6,002.768
2: 16.244	4,286.264	8:	24.758	15,175.628
3: 0.736	0.399	9:	24.034	13,882.835
4: 0.173	0.005	10:	5.711	186.267
5: 2.541	16.406	11:	17.027	4,936.446
6: 8.046	592.883	12:	1.816	5.989

16.

X_{16}	Y_{16}		X_{16}	Y_{16}
1: 17	225,140	5:	25	1,206,000
2: 11	33,888	6:	27	1,684,350
3: 5	1,090	7:	19	365,230
4: 25	1,205,000	8:	28	1,973,000

ANSWERS

KNOW THE CONCEPTS

1. No. You can say that 80 percent of the changes in the dependent variable are associated with changes in the independent variable, but regression calculations do not indicate causation.

2. Obtain more observations; obtain observations with a greater spread; investigate a situation with a smaller value of σ.

3. There is no linear relation between the quantities.

4. One source of uncertainty arises because you do not know the true values of the slope and the intercept, and another source because any particular value of the dependent variable will deviate randomly from the line.

5. The t statistic is used to test the null hypothesis that the true value of the slope is 0.

6. You cannot be sure that the observed past relation between the variables applies in a new situation that you have never observed before.

7. Economic theory predicts that the quantity demanded will go down when the price goes up, provided that all else remains the same. It is possible, however, that other factors have been changing which have been causing the demand for the good to go up despite the price rise. In Chapter 16 we will discuss multiple regression, which makes it possible to take some of these other factors into account.

8. There could be a nonlinear relation.

9. This is a nonsensical calculation. Regression analysis can be applied only to pairs of observations. There must be some connection between the dependent variable and the independent variable in each pair; for example, both may be for the same week or the same town.

10. You will be able to obtain an r^2 value of 1, because you can always draw a line that fits any two points perfectly. However, your estimates of the slope and intercept will not be very reliable. It is important to remember that the r^2 value alone does not tell you whether the regression results are reliable.

11. Let x be the independent variable, and let y be the dependent variable. You will take the logarithms of both x and y if you suspect that the true relation is of the form $y = ax^b$. You will take only the logarithm of y if you suspect that the true relation is of the form $y = ab^x$.

PRACTICAL APPLICATION

Calculations are shown for Exercise 1. The same methods can be applied to similar exercises.

1.

x_1	y_1	x_1^2	y_1^2	x_1y_1
34	6	1,156	36	204
26	57	676	3,249	1,482
9	89	81	7,921	801
30	60	900	3,600	1,800
47	95	2,209	9,025	4,465
10	42	100	1,764	420
34	31	1,156	961	1,054
34	28	1,156	784	952
45	90	2,025	8,100	4,050
10	25	100	625	250
47	45	2,209	2,025	2,115
37	23	1,369	529	851
47	52	2,209	2,704	2,444
8	95	64	9,025	760
45	48	2,025	2,304	2,160
Average: 30.9	52.4	1,162.3	3,510.1	1,587.2

Slope = $(1,587.2 - 30.9 \times 52.4)/(1,162.3 - 30.9^2) = -0.15$
Intercept = $52.4 - (-0.15 \times 30.9) = 57$
$r^2 = (1,587.2 - 30.9 \times 52.4)^2/[(1,162.3 - 30.9^2)(3,510.1 - 52.4^2)]$
$= .006$
Correlation = $-\sqrt{r^2} = -.075$
Squared error = 11,400
Mean square error = $11,400/13 = 876.9$
Estimated standard deviation = $\sqrt{876.9} = 29.6$
t statistic = $-0.15/\sqrt{876.9/3,112.3} = -0.28$

There are 13 degrees of freedom. The critical value for the t distribution is 2.16, so we will accept the null hypothesis that the true value of the slope is 0.
0.950 confidence interval for slope = –1.285–0.997

2. Slope = -0.152 Intercept = 19.966 $r^2 = .028$
Correlation = $-.168$ Squared error = 1,810.011   Mean square error = 113.126
Estimated standard deviation = 10.636
t statistic = -0.680 Degrees of freedom = 16
  t distribution critical value = 2.120
0.950 confidence interval for slope = -0.627–0.322

3. Slope = -0.283 Intercept = 59.854 $r^2 = .070$
Correlation = $-.264$ Squared error = 4,694.245   Mean square error = 586.781   Estimated standard deviation = 24.224
t statistic = -0.775 Degrees of freedom = 8

t distribution critical value = 2.306
0.950 confidence interval for slope = −1.124–0.558

4. Slope = −0.040   Intercept = 8.733 r^2 = .144
   Correlation = −.379   Squared error = 117.946   Mean square error = 8.425
   Estimated standard deviation = 2.903
   t statistic = −1.533   Degrees of freedom = 14
     t distribution critical value = 2.145
   0.950 confidence interval for slope = −0.095–0.016

5. Slope = −0.025   Intercept = 11.823 r^2 = .015
   Correlation = −.122   Squared error = 655.071   Mean square error = 36.393
   Estimated standard deviation = 6.033
   t statistic = −0.522   Degrees of freedom = 18
     t distribution critical value = 2.101
   0.950 confidence interval for slope = −0.125–0.075

6. Disposable income, consumption:
   Slope: 0.963   Intercept: −176.254
   R^2: 0.999
   Correlation: 1.000   Covariance: 907725.375
   Squared error: 9721.347
   Mean square error: 747.796
   Estimated stand deviation: 27.346
   t statistic: 132.443   Degrees of freedom = 13
   F statistic: 17541.080   1, 13 Degrees of freedom

7. Elevation and population:
   Slope: −0.073   Intercept: 5411.558 r^2: 0.005
   Correlation: −0.068   Mean square error: 30289745.218
   Estimated standard deviation: 5503.612
   t statistic: −0.472   Degrees of freedom = 48
     t distribution critical value: 2.011
   0.950 confidence interval for slope = −0.384–0.238

8. *M2* and CPI
   Slope: 0.030   Intercept: 32.119
   R^2: 0.936
   Correlation: 0.967   Covariance: 11614.836
   Squared error: 360.625
   Mean square error: 27.740
   Estimated stand deviation: 5.267
   t statistic: 13.760   Degrees of freedom = 13
   F statistic: 189.331   1, 13 Degrees of freedom

9. UNEMP and INFLATION
   Slope: 0.287   Intercept: 2.218
   R^2: 0.037
   Correlation: 0.192   Covariance: 0.491
   Squared error: 55.330
   Mean square error: 4.256
   Estimated stand deviation: 2.063
   t statistic: 0.704   Degrees of freedom = 13
   F statistic: 0.496   1, 13 Degrees of freedom
10. PROD and COMP
   Slope: 0.404   Intercept: −0.031
   R^2: 0.171
   Correlation: 0.414   Covariance: 0.626
   Squared error: 17.103
   Mean square error: 1.425
   Estimated stand deviation: 1.194
   t statistic: 1.575   Degrees of freedom = 12
   F statistic: 2.481   1, 12 Degrees of freedom
11. UNEMP and CAP
   Slope: −1.832   Intercept: 93.454
   R^2: 0.847
   Correlation: −0.920   Covariance: −3.136
   Squared error: 15.535
   Mean square error: 1.195
   Estimated stand deviation: 1.093
   t statistic: −8.493   Degrees of freedom = 13
   F statistic: 72.137   1, 13 Degrees of freedom

12.

	X_{12}	$\log Y_{12}$		X_{12}	$\log Y_{12}$
1:	1.659	1.201	9:	0.373	0.270
2:	0.207	0.150	10:	1.703	1.233
3:	0.094	0.068	11:	0.515	0.373
4:	1.696	1.227	12:	0.171	0.124
5:	1.318	0.954	13:	0.867	0.628
6:	1.209	0.875	14:	0.639	0.462
7:	1.572	1.138	15:	1.777	1.286
8:	1.577	0.964			

Slope = 0.707   Intercept = 0.006 $r^2 = .990$

13.

	X_{13}	$\log Y_{13}$			X_{13}	$\log Y_{13}$
1:	1.000	1.000		9:	9.000	1.202
2:	2.000	1.025		10:	10.000	1.228
3:	3.000	1.051		11:	11.000	1.253
4:	4.000	1.093		12:	12.000	1.278
5:	5.000	1.101		13:	13.000	1.304
6:	6.000	1.127		14:	14.000	1.329
7:	7.000	1.152		15:	15.000	1.354
8:	8.000	1.177				

Slope = 0.025   Intercept = 0.978 $r^2 = 0.998$

14.

	X_{14}	$\log Y_{14}$			X_{14}	$\log Y_{14}$
1:	1.000	3.000		9:	9.000	2.435
2:	2.000	2.929		10:	10.000	2.365
3:	3.000	2.859		11:	11.000	2.294
4:	4.000	2.788		12:	12.000	2.200
5:	5.000	2.718		13:	13.000	2.153
6:	6.000	2.647		14:	14.000	2.082
7:	7.000	2.577		15:	15.000	2.012
8:	8.000	2.506				

Slope = -0.071 Intercept = 3.072 $r^2 = 1.000$

15.

	$\log X_{15}$	$\log Y_{15}$			$\log X_{15}$	$\log Y_{15}$
1:	0.837	2.510		7:	1.259	3.778
2:	1.211	3.632		8:	1.394	4.181
3:	$-$ 0.133	-0.399		9:	1.381	4.142
4:	$-$ 0.762	-2.301		10:	0.757	2.270
5:	0.405	1.215		11:	1.231	3.693
6:	0.906	2.773		12:	0.259	0.777

Slope = 3.006   Intercept = -0.001 $r^2 = 1.000$

16.

	$\log X_{16}$	$\log Y_{16}$			$\log X_{16}$	$\log Y_{16}$
1:	1.230	5.352		5:	1.398	6.081
2:	1.041	4.530		6:	1.431	6.226
3:	0.699	3.037		7:	1.279	5.563
4:	1.398	6.081		8:	1.447	6.295

Slope = 4.354   Intercept = -0.005 $r^2 = 1.000$

16
MULTIPLE LINEAR REGRESSION

KEY TERMS

F statistic a statistic used to test the hypothesis that the true value of each coefficient in a regression equation is 0

multiple regression a statistical method for analyzing the relation between several independent variables and one dependent variable

t statistic a statistic used to test the hypothesis that the true value of one specific coefficient is 0

SEVERAL INDEPENDENT VARIABLES

In many real situations there will be more than one independent variable that affects the dependent variable you are interested in. In such cases we need to use a technique called **multiple regression**. In Chapter 15 we discussed a situation where income was the only variable that affected the demand for pizza. However, that situation seems very unrealistic. In economic theory many different variables might affect the demand. In addition to income, one of the variables that can be expected to be very important is the price of the good. We will investigate the effect of income and price on the quantity demanded of statistics books.

• AN EXAMPLE OF THE USE OF MULTIPLE REGRESSION

Suppose we have observations on the number of statistics books sold, the price of statistics books, and the per capita income in 15 different towns for a specified period. We will let y represent the independent variable, which is the quantity of statistics books sold. We have two independent variables: x_1 will represent price and x_2 will represent income.

Town	y (Books Demanded)	x_1 (Price)	x_2 (Income)
1	166	10	20
2	180	9	21
3	73	10	12
4	81	14	16
5	229	8	24
6	182	15	24
7	233	6	23
8	102	10	15
9	190	7	20
10	150	10	19
11	221	11	25
12	137	15	21
13	173	8	19
14	150	12	20
15	92	10	14

We will assume that the relation between y, x_1, and x_2 is given by this equation:

$$y_i = B_1 x_{i1} + B_2 x_{i2} + B_3 + e_i$$

where y_i represents the ith value of the dependent variable, and x_{ij} represents the ith observation of the jth independent variable. We need two subscripts now because we must use one subscript to keep track of the observation number and another subscript to keep track of the variable number. For the observations listed above, x_{11} is 10, x_{21} is 9, x_{12} is 20, x_{22} is 21, and so on.

The true values of B_1, B_2, and B_3 are unknown, but we will try to estimate them. B_1 represents the effect that x_1 has on y, assuming that x_2 remains constant. Likewise, B_2 represents the effect that x_2 has on y when x_1 remains constant. If x_1 increases by 1 and everything else remains constant, then y will increase by B_1. When we design the model this way, we are assuming that the effects of x_1 and x_2 on y are additive. This means that the amount that x_1 affects y does not depend on the level of x_2, and vice versa. We expect that B_2 will be positive, since more books will be bought when income is higher, but that B_1 will be negative, since fewer books will be demanded when the price is higher. B_3 is called the constant term in the model. It is analogous to the y-intercept term in the simple linear regression model.

Again, e is a random variable called the *error term* that represents the effects of all possible factors other than price and income that might affect the demand for statistics books. Thus, e should be independent of x_1 and x_2. This can always be checked by plotting e against x_1 and e against x_2 to see if there is any pattern. The expectation of e is 0 and the variance of e is σ^2, which is unknown. We will assume that e has a normal distribution.

• *TWO DIFFERENCES BETWEEN SIMPLE REGRESSION AND MULTIPLE REGRESSION*

In principle, we will proceed exactly as we did with simple linear regression, when there was only one independent variable. We will calculate the values of B_1, B_2, and B_3 that minimize the sum of the squares of the errors between the values predicted by the equation and the actual values. There are two main differences between simple regression and multiple regression:

- We can't draw a picture of the relationship. Actually, if there are only two independent variables, we can attempt to draw a three-dimensional perspective drawing with x_1 and x_2 on the horizontal axes, y on the vertical axis, and one dot corresponding to each observation. Then the goal is to identify the plane that minimizes the sum of the squares of the vertical deviations between each observation and the plane. Drawing this type of diagram is very difficult, though. If there are more than two independent variables, it is impossible to draw a diagram. (Mathematicians still think of each observation as defining a point in multidimensional space, and the goal of multiple regression in that case is to find something called a *hyperplane* that best fits all the observations.)

- The calculation process is harder for multiple regression than it is for simple regression. But that won't worry us—we'll have the calculations done by a computer. In this book we will not attempt to explain how the multiple regression calculations are performed. Understanding those calculations requires a knowledge of matrix multiplication and matrix inversion. In a practical situation where you need to perform a multiple regression calculation, you will work with a computer statistical package on a microcomputer or mainframe computer.

In the rest of this chapter we will discuss how to interpret the results of a regression analysis.

YOU SHOULD REMEMBER

1. In multiple regression it is assumed that the true relation between the dependent variable y and the $m - 1$ independent variables $x_1, x_2, \ldots, x_{m-1}$ is given by the equation

$$y = B_1x_1 + B_2x_2 + \cdots + B_{m-1}x_{m-1} + B_m + e$$

where e is a normal random variable with mean 0 and unknown variance σ^2.

> **2.** If you have a list of observations for each of these variables, then a computer regression program will calculate estimated values for each of the coefficients $B_1, B_2, \ldots, B_m$.

MULTIPLE REGRESSION OUTPUT

Once we have fed the numbers into the computer and ordered it to grind out the calculations, we will be presented with a result something like this:

$$y = -7.738x_1 + 12.286x_2 - 2.765$$

This equation lists the estimated values for the coefficients. These particular numbers are for the statistics book sales example. The first thing we can learn from this result confirms what we expected: the coefficient for x_1 is negative, meaning that a higher price leads to lower sales. The coefficient of x_2 is positive, meaning that higher incomes lead to higher sales. We can use the estimated values of the coefficients to make predictions about the value of y (subject to all of the warnings discussed in Chapter 15). For example, if we knew of a town where average income was 20 and the price of statistics books was 6, then we would predict that the quantity of statistics books demanded would be

$$(-7.738 \times 6) + (12.286 \times 20) - 2.765 = 196.5$$

In general, we will use $\hat{b}_1$ to represent the computer's estimate of the coefficient B_1, $\hat{b}_2$ to represent the computer's estimate of the coefficient B_2, and so on. We'll consider a general situation where there are $m - 1$ independent variables. In that case we will have m coefficients to estimate (counting the constant term). Then the computer's regression output will look like this:

$$y = \hat{b}_1x_1 + \hat{b}_2x_2 + \cdots + \hat{b}_{m-1}x_{m-1} + \hat{b}_m$$

where $\hat{b}_m$ is the computer's estimate of the constant term.

• *THE* R^2 *VALUE*

The computer will also give an R^2 value for the regression. R^2 is called the **coefficient of multiple determination**. In this case the R^2 value is .9957. The interpretation of the R^2 value is similar to the interpretation of the r^2 value for a simple linear regression: R^2 measures the percent of variation in the dependent variable that can be explained by the regression. The value of R^2 will always be between 0 and 1.

To calculate R^2 we will first calculate several different sum of squares statistics. (See Chapter 14, where we calculated sum of squares statistics as part of analysis of variance.) The **total sum of squares (TSS)** is the sum of squares of the y values about $\bar{y}$:

$$\text{TSS} = \Sigma(y_i - \bar{y})^2$$

Again we will use Σ to mean $\Sigma_{i=1}^{n}$. Note that TSS is the same as the quantity we called SE_{av} in the simple linear regression case. We will use $\hat{y}_i$ to represent the regression's predicted value of y for the ith observation:

$$\hat{y}_i = \hat{b}_1 x_{i1} + \hat{b}_2 x_{i2} + \cdots + \hat{b}_{m-1} x_{i,m-1} + \hat{b}_m$$

For each observation we can calculate the difference between the value of y predicted by the regression line and the average value of y. Next we can sum the squares of all of those deviations and call it the **regression sum of squares (RGRSS)**:

$$\text{RGRSS} = \Sigma(\hat{y}_i - \bar{y})^2$$

Then we can calculate the residual for each observation, which is the difference between the actual value of y and the fitted value from the regression line:

$$(\text{residual}_i) = y_i - \hat{y}_i$$

The sum of the squares of all the residuals will be called the **error sum of squares (ERSS)**:

$$\text{ERSS} = \Sigma(y_i - \hat{y}_i)^2$$

This quantity is analogous to the quantity we called SE_{line} in the simple linear regression situation.

By performing the same type of calculation that we performed in Chapter 14, we can show that

$$\text{TSS} = \text{RGRSS} + \text{ERSS}$$

We can think of TSS as representing the total variation in the values of y. RGRSS is the amount of this variation that can be explained by the regression, and ERSS is the amount of remaining variation that cannot be explained by the regression. If the regression fits the data very well, then the value of RGRSS will be much bigger than the value of ERSS. We can calculate R^2 from either of these two formulas:

$$R^2 = 1 - \frac{\text{ERSS}}{\text{TSS}}$$

$$= \frac{\text{RGRSS}}{\text{TSS}}$$

For the statistics book example we have the following:

Town	y_i	x_1	x_2	$\hat{y}_i$	$(y_i - \hat{y}_i)$	$(y_i - \hat{y}_i)^2$	$y_i - \bar{y}$	$(y_i - \bar{y})^2$
1	166	10	20	165.579	0.421	0.177	8.733	76.265
2	180	9	21	185.603	-5.603	31.397	22.733	516.789
3	73	10	12	67.291	5.709	32.597	-84.267	7,100.927
4	81	14	16	85.484	-4.484	20.106	-76.267	5,816.655
5	229	8	24	230.199	-1.199	1.439	71.733	5,145.623
6	182	15	24	176.035	5.965	35.582	24.733	611.721
7	233	6	23	233.389	-0.389	0.151	75.733	5,735.487
8	102	10	15	104.149	-2.149	4.618	-55.267	3,054.441
9	190	7	20	188.793	1.207	1.457	32.733	1,071.449
10	150	10	19	153.293	-3.293	10.846	-7.267	52.809
11	221	11	25	219.272	1.728	2.985	63.733	4,061.895
12	137	15	21	139.177	-2.177	4.738	-20.267	410.751
13	173	8	19	168.769	4.231	17.902	15.733	247.527
14	150	12	20	150.104	-0.104	0.011	-7.267	52.809
15	92	10	14	91.863	0.137	0.019	-65.267	4,259.781
	total					164.025		38,214.933

From this table we can see the ERSS = 164.025 and TSS = 38,214.933. Then we can calculate

$$RGRSS = TSS - ERSS = 38,050.909$$

(Note: There will be slight discrepancies in some of these regression calculations because some intermediate results have been rounded.) Then we can calculate:

$$R^2 = \frac{38,050.909}{38,214.933} = .996$$

The estimated regression equation fits these points very well. However, there is an important caution you must keep in mind when interpreting the R^2 value of a multiple regression. It is always possible to increase the value of R^2 by adding more independent variables to the regression, whether or not they have anything to do with the dependent variable. For the regression results to be reliable the number of observations must be significantly greater than the number of coefficients you are estimating. Therefore, it is sometimes recommended to calculate the *adjusted R^2*:

$$(\text{adjusted } R^2) = 1 - \frac{(1 - R^2)(n - 1)}{n - m}$$

The adjusted R^2 will not necessarily go up if you add another variable because that will increase the value of m.

• THE F STATISTIC

Suppose some doubters tell you that they don't believe there is any connection between the dependent variable and the independent variables in your regression. They are making the following null hypothesis:

$$H_0: B_1 = B_2 = \cdots B_{m-1} = 0$$

In words, the person thinks that the true value of the coefficients for all $m - 1$ of the independent variables is zero. To test that hypothesis you may calculate this statistic:

$$F = \frac{\dfrac{RGRSS}{m - 1}}{\dfrac{ERSS}{n - m}}$$

This statistic is called the **F statistic** for the regression. If the null hypothesis is true, then it will have an F distribution with $m - 1$ degrees of freedom in the numerator and $n - m$ degrees of freedom in the denominator. (Remember that n is the number of observations and $m - 1$ is the number of independent variables.) If the null hypothesis is false, then we would expect RGRSS to be larger than ERSS, so the F statistic will be bigger than it would be if the null hypothesis was true. In our case the calculated value of the F statistic is 1392, which is much larger than 3.9 (the 95 percent critical value for an F distribution with 2 and 12 degrees of freedom). Therefore we can unquestionably reject the null hypothesis.

These results can be summarized in an analysis of variance (ANOVA) table (see Chapter 14):

Source of Variation	Sum of Squares	Degrees of Freedom	Mean Square	F Ratio
Regression	38,050.909	2	19,025.455	1,391.89
Error	164.025	12	13.669	
total	38,214.933	14		

We will call ERSS/$(n - m)$ the mean square error (MSE), and we will use it as an estimator of the unknown variance σ^2. In this case MSE = 13.669.

TESTING INDIVIDUAL COEFFICIENTS

Now we will proceed with the statistical analysis of the individual coefficients. Our procedure will be very similar to our simple regression procedure. In Chapter 15 we said that $\hat{m}$ (the least squares estimator of the slope of the regression line) had a normal distribution whose mean was equal to the true value of the slope and whose variance was equal to σ^2 divided by a complicated expression that depends on the x's. We can find exactly analogous results in the multiple regression case.

Let B_i represent the unknown true value of the ith coefficient, and let $\hat{b}_i$ represent the least squares estimator of that coefficient. Then $\hat{b}_i$ has a normal distribution whose mean is equal to the true value B_i and whose variance is equal to σ^2 divided by a complicated expression that depends on the x's. Once again we will use MSE to estimate σ^2, but we will not describe the nature of the complicated expression that depends on the x's in this book. Fortunately, a computer regression program will always include with its output the estimated value of the standard deviation for each coefficient. We'll call that standard error $s(\hat{b}_i)$:

$$s(\hat{b}_i) = \sqrt{\mathrm{Var}(\hat{b}_i)}$$

The values of the standard errors are often presented in parentheses below the coefficients:

$$y = -7.738x_1 + 12.286x_2 - 2.765$$
$$\quad\ (0.364) \qquad\ (0.257) \quad\ (6.39)$$

A smaller value of the standard error means that the coefficient estimate is more reliable.

As you might have expected, the quantity $(\hat{b}_i - B_i)/s(\hat{b}_i)$ has a t distribution with $n - m$ degrees of freedom. Therefore, we can calculate a confidence interval for B_i:

$$b_i \pm as(b_i)$$

where $\Pr(-a < t < a) = \mathrm{CL}$

CL is the confidence level

t is a random variable having a t distribution with $n - m$ degrees of freedom

We can also test the null hypothesis that $B_i = 0$. If $B_i = 0$, then x_i does not have an effect on y. If the hypothesis is true, then the quantity $\hat{b}_i/s(\hat{b}_i)$ will have a t distribution with $n - m$ degrees of freedom. This quantity is called the **t statistic** for the ith coefficient. The values of the t statistics are often calculated by a computer regression program. (If they are not, you can calculate them very easily once you know $\hat{b}_i$ and $s(\hat{b}_i)$.) For the statistics book example we have:

Variable	Coefficient	Standard Error	t Statistic
Price	− 7.738	0.364	−21.3
Income	12.286	0.257	47.8
Constant term	− 2.765	6.39	− 0.43

The 95 percent critical value for a t distribution with $15 - 3 = 12$ degrees of freedom is 2.179. The t statistics for both price and income are well outside this value, so we can safely reject the hypotheses that either the price or the income coefficient is zero. However, we cannot reject the hypothesis that the true value of the constant term is equal to zero.

YOU SHOULD REMEMBER

The computer will calculate, in addition to estimated values of B_1, B_2, . . ., B_m, the following:

- the R^2 value for the regression (a value close to 1 means that the values estimated by the regression equation are close to the actual values of y)

- the F statistic for the regression (used to test the hypothesis that the coefficients of all the independent variables are zero)

- the standard error for each coefficient

- the t statistic for each coefficient (used to test the hypothesis that the true value of that coefficient is 0)

FURTHER ANALYSIS OF REGRESSION MODELS

We will mention a few more topics that apply to the analysis of regression. Some of these topics are the same for multiple regression and simple regression, but there are some special issues that arise only with multiple regression.

- Residuals. Once you have calculated the regression residuals, you may perform visual analysis just as we did in the last chapter. You may make several scatter diagrams comparing the residuals to the dependent variable and to each of the independent variables. In each case there should be no apparent pattern. If you have time series data, it also helps to make a plot of the residuals with time. It may also help to plot the residuals against another independent variable that you have not included in the model. If the residuals seem to be related to that variable, then you should include it in the model.

- Transformations. Again, a model that is not linear to begin with can often be transformed into a linear model by using logarithms or some other type of transformation. For example, if the true relation is

$$y = b_0 x_1{}^{b_1} x_2{}^{b_2} x_3{}^{b_3}$$

then you may take the logarithm of both sides:

$$\log y = \log b_0 + b_1 \log x_1 + b_2 \log x_2 + b_3 \log x_3$$

and you may estimate the values of b_0, b_1, b_2, and b_3 by ordinary linear regression.

- Serial correlation. In the regression model we have assumed that all of the errors are independent. Suppose we have a set of time series observations where a positive value for the error for one period is more likely to be followed by a positive value for the next period. In that case the errors are not independent. This situation is called **serial correlation** or **autocorrelation**. The least squares estimators are less reliable in this situation. A computer regression program will normally calculate the value of a statistic called the **Durbin-Watson statistic**. A small value of this statistic indicates the presence of one specific type of serial correlation. (How small is small? If the computer does not tell you, you will need to look in a Durbin-Watson table.) In the case of serial correlation the regression results can be made more reliable by trying to find another independent variable to add to the model or by performing a transformation involving the differences between successive values of the variables.

- Multicollinearity. When two or more of the independent variables are closely correlated, then the problem of **multicollinearity** arises. If all of the independent variables are uncorrelated, then your regression model will still be able to accurately estimate the coefficients for the variables in the model even if some of the independent variables have been left out. However, the coefficient estimates become less reliable if some of the independent variables are highly correlated. In the extreme case where two of the independent variables are perfectly correlated, it is not even possible to calculate the least squares estimators. Also, the t statistics for the individual coefficients are not reliable when there is multicollinearity.

When two independent variables are highly correlated, it is not possible to accurately separate out their independent effects. For example, suppose you are trying to investigate the effects of income and education on demand for your product. You have collected information from a large sample of households. However, you are likely to find that the people with more income tend to be the people that have more education, so these two variables are highly correlated. A practical rule states that the problem of multicollinearity arises if the correlation coefficient between two variables is greater than .7. If you observe that people with higher income tend to buy your product more often, you do not know whether they do so because they have higher incomes or because they have more education.

Here is a similar type of example. Suppose you have observed that a particular chemical reaction occurs much more rapidly in warm, light environments than it does in cold, dark environments. However, you cannot tell whether it is the warmth or the light that speeds the reaction, since the two independent variables (temperature and amount of light) are correlated in your observations. The best way to solve the problem would be to obtain observations of the reaction in hot dark situations and cold light situations.

By analogy, the best solution in the income/education situation would be to obtain observations of people with high incomes/low educations and high educations/low incomes. However, it may be difficult to find these people. You may have no better alternative than removing either the education or the income variable from the regression and then remembering that the coefficient of the remaining variable represents the combined effect of the two variables.

- Dummy variables. Often some of the factors that affect the dependent variable are not quantitative factors that can be given by numbers. For example, suppose you are investigating consumption behavior with time series data for the period 1930 to 1950. You would expect that consumption behavior would have been significantly different during the years of World War II than it was before and after the war. To take this effect into account, you can create an artificial variable that will take the value 1 during each of the war years and the value 0 during each of the other years. This type of variable is called a **dummy variable** or an **indicator variable**. The coefficient of the World War II dummy variable indicates how much effect the war had on the constant term in the regression. Dummy variables can be used in many different situations with regressions.

• Simultaneous equations. Regression analysis is used to construct the economic models that try to predict economic activity. The branch of economics involved with this analysis is called **econometrics**. Econometric models require that regression analysis be applied to many different equations. The technique for applying regression in this situation is called *simultaneous equation analysis*.

YOU SHOULD REMEMBER

1. Analyzing the regression residuals is helpful in determining whether a particular multiple regression equation is appropriate.

2. Logarithms can often be used to transform a nonlinear model into a linear form.

3. The problem of *serial correlation* arises when the error terms are not independent for all observations.

4. The problem of *multicollinearity* arises when two or more of the independent variables are correlated with each other.

5. The use of *dummy variables*, which always have the value 0 or 1, makes it possible to include nonquantitative factors in regression equations.

6. The development of *econometric models* requires the application of regression methods to several equations that must be true simultaneously.

KNOW THE CONCEPTS

DO YOU KNOW THE BASICS?

Test your understanding of Chapter 16 by answering the following questions:

1. Is it correct to say that in a multiple regression the independent variable with the largest coefficient has the greatest effect on the dependent variable?

2. What hypothesis do you test with the F statistic?

3. When you perform hypothesis tests with the t statistics and the F statistic from a multiple regression, are they one-tailed tests or two-tailed tests?

4. What happens to regression calculations if an important independent variable is left out?

5. Suppose two of the independent variables in your regression analysis are perfectly correlated. Explain intuitively why this creates problems for the regression calculation.

6. Will adding another independent variable make the R^2 value for the regression go up or down?

7. Suppose someone reports a regression result for ten observations and nine independent variables with an R^2 value of 1. Do you think this regression calculation is reliable?

8. What does the graph of a regression equation with two independent variables look like?

TERMS FOR STUDY

coefficient of multiple determination
dummy variable
Durbin-Watson statistic
econometrics
error sum of squares (ERSS)
F statistic
multicollinearity

multiple regression
R^2
regression sum of squares (RGRSS)
serial correlation
standard error for coefficient
t statistic
total sum of squares (TSS)

PRACTICAL APPLICATION

COMPUTATIONAL PROBLEMS

For each exercise below you are given a list of three independent variables (X_1, X_2, and X_3) and the dependent variable (Y). You are also given the multiple regression coefficients and the standard error for each coefficient. Calculate the list of predicted values for the dependent variable ($\hat{y}_i$), the list of residuals, the R^2 value, the F statistic, and the t statistic for each coefficient.

1.

	X_1	X_2	X_3	Y
1:	30	16	58	2,426
2:	5	17	66	2,429
3:	9	29	49	1,609
4:	21	2	2	302
5:	50	27	17	857
6:	2	23	42	1,339

(continued)

	X_1	X_2	X_3	Y
7:	19	26	36	1,255
8:	49	24	25	1,213
9:	26	6	61	2,661
10:	0	6	74	2,869
11:	17	3	31	1,397
12:	40	24	92	3,779
13:	33	16	65	2,744
14:	39	8	65	2,944
15:	5	11	67	2,571
16:	12	3	63	2,623
17:	4	6	19	712
18:	14	16	34	1,280
19:	44	16	89	3,835
20:	40	23	11	555

Regression output:

Coefficient 1	X_1	12.031
	Standard error:	0.0403
Coefficient 2	X_2	-15.948
	Standard error:	0.0750
Coefficient 3	X_3	39.999
	Standard error:	0.0246
Coefficient 4	Constant	2.734

2.

	X_1	X_2	X_3	Y
1:	44	110	91	1,198
2:	40	115	77	1,143
3:	52	130	92	1,469
4:	53	106	78	1,681
5:	12	129	49	145
6:	57	92	76	1,875
7:	17	90	58	289
8:	6	96	54	-100
9:	29	109	59	836
10:	21	134	83	223
11:	59	149	81	1,913
12:	35	148	77	897
13:	49	104	45	1,733
14:	26	131	88	411
15:	48	94	64	1,533

	X_1	X_2	X_3	Y
16:	26	125	63	624
17:	36	113	90	887
18:	52	101	88	1,568
19:	40	140	68	1,188
20:	53	139	88	1,536

Regression output:

Coefficient 1	X_1	42.316
	Standard error:	0.4783

Coefficient 2	X_2	−0.689
	Standard error:	0.3704

Coefficient 3	X_3	−7.717
	Standard error:	0.5344

Coefficient 4	Constant	102.934

3.

	X_1	X_2	X_3	Y
1:	17	73	2	2,471
2:	29	89	35	2,008
3:	30	87	92	1,916
4:	34	46	19	2,537
5:	54	69	9	1,240
6:	58	39	59	2,840
7:	26	76	29	1,915
8:	5	91	97	328
9:	65	46	13	2,841
10:	59	6	58	1,838
11:	72	100	0	1,444
12:	85	71	50	3,238
13:	28	41	15	879
14:	34	56	61	1,319
15:	99	53	62	2,226
16:	35	11	63	702
17:	30	5	29	1,296
18:	1	68	9	1,734
19:	51	44	85	1,725
20:	27	44	35	2,290

Regression output:

Coefficient 1	X_1	13.617
	Standard error:	6.4905

(continued)

Coefficient 2	X_2	2.017
	Standard error:	5.9699
Coefficient 3	X_3	−5.282
	Standard error:	5.4609
Coefficient 4	Constant	1,372.812

4.

	X_1	X_2	X_3	Y
1:	31	33	39	20
2:	33	46	74	55
3:	12	3	114	50
4:	51	28	3	33
5:	8	30	32	34
6:	28	47	49	42
7:	39	29	7	57
8:	25	59	27	51
9:	59	24	111	18
10:	36	8	88	17
11:	9	27	37	55
12:	41	5	2	48
13:	59	43	54	37
14:	9	35	119	68
15:	9	2	49	71
16:	21	7	46	73
17:	8	6	11	81
18:	24	4	89	41
19:	13	26	23	58
20:	17	34	52	19

Regression output:

Coefficient 1	X_1	−0.573
	Standard error:	0.2388
Coefficient 2	X_2	−0.149
	Standard error:	0.2361
Coefficient 3	X_3	−0.116
	Standard error:	0.1065
Coefficient 4	Constant	71.257

5.

	X_1	X_2	X_3	Y
1:	62	136	163	2,120
2:	41	179	224	1,631
3:	61	123	191	2,135
4:	42	133	189	1,541
5:	3	122	219	328
6:	25	108	219	1,064
7:	62	172	207	2,222
8:	58	111	149	2,003
9:	80	96	196	2,652
10:	35	128	176	1,377
11:	43	96	209	1,570
12:	75	129	196	2,521
13:	3	158	156	443
14:	43	166	183	1,683
15:	79	150	216	2,751
16:	31	173	187	1,327
17:	55	122	170	1,890
18:	33	165	148	1,431
19:	18	106	163	772
20:	34	141	178	1,388

Regression output:

Coefficient 1	X_1	29.941
	Standard error:	0.4069
Coefficient 2	X_2	2.030
	Standard error:	0.3449
Coefficient 3	X_3	0.117
	Standard error:	0.3801
Coefficient 4	Constant	23.246

6.

	X_1	X_2	X_3	Y
1:	117	241	267	200
2:	142	234	258	158
3:	122	239	254	171
4:	116	297	251	207
5:	109	364	230	135
6:	63	247	238	232
7:	79	329	266	297
8:	149	223	224	168

(continued)

	X_1	X_2	X_3	Y
9:	115	299	223	226
10:	124	335	252	315
11:	82	249	255	252
12:	116	346	259	139
13:	126	287	255	273
14:	140	306	221	266
15:	75	332	257	150
16:	89	241	256	291
17:	140	271	237	305
18:	141	306	222	168
19:	115	287	263	329
20:	154	306	269	220

Regression output:

Coefficient 1	X_1	−0.200
	Standard error:	0.6114
Coefficient 2	X_2	−0.098
	Standard error:	0.3690
Coefficient 3	X_3	0.590
	Standard error:	0.9807
Coefficient 4	Constant	130.122

ANSWERS

KNOW THE CONCEPTS

1. In general, a larger coefficient value indicates a greater effect. However, remember that the numerical value of a coefficient depends on the unit that is used to measure the variable. For example, the coefficient value will increase if a variable is measured in feet instead of inches. You must examine the situation carefully before you make judgments based on the numerical values of the coefficients.

2. The hypothesis tested is that the true values of all coefficients are zero.

3. The t test is a two-tailed test; the F test is a one-tailed test.

4. The random error will be larger; the coefficient estimates may be biased.

5. If two independent variables are perfectly correlated, then you have no observations that allow you to distinguish the separate effects of each variable.

6. The R^2 value will go up.

7. This result is meaningless, since a regression equation with a perfect fit can always be found if there are as many coefficients as observations. Note that

in this case the number of degrees of freedom becomes 0, which makes no sense.

8. It is a plane in three-dimensional space.

PRACTICAL APPLICATION

Each t statistic is calculated by dividing the coefficient by the standard error. The 95 percent critical value for a t distribution with $20 - 4 = 16$ degrees of freedom is 2.12. An asterisk (*) is placed next to each t statistic that is outside this value. The R^2 value is $1 - \text{ERSS/TSS}$, and the F statistic is (RGRSS/3)/(ERSS/16). The 95 percent critical value for an F distribution with 3 and 16 degrees of freedom is 3.3. An asterisk is placed next to each F statistic when the regression as a whole is significant.

1. Coefficient 1 X_1 12.031
   t statistic = 298.6121*
   Coefficient 2 X_2 -15.948
   t statistic = -212.7053*
   Coefficient 3 X_3 39.999
   t statistic = 1627.6944*

	Y	Predicted	Residual	(Residual)^2
1:	2,426	2,428.461	−2.461	6.056
2:	2,429	2,431.721	−2.721	7.404
3:	1,609	1,608.481	0.519	0.270
4:	302	303.496	−1.496	2.237
5:	857	853.689	3.311	10.961
6:	1,339	1,339.955	−0.955	0.913
7:	1,255	1,256.649	−1.649	2.720
8:	1,213	1,209.497	3.503	12.274
9:	2,661	2,659.815	1.185	1.403
10:	2,869	2,866.989	2.011	4.045
11:	1,397	1,399.400	−2.400	5.759
12:	3,779	3,781.164	−2.164	4.681
13:	2,744	2,744.550	−0.550	0.302
14:	2,944	2,944.324	−0.324	0.105
15:	2,571	2,567.410	3.590	12.890
16:	2,623	2,619.219	3.781	14.298
17:	712	715.156	−3.156	9.958
18:	1,280	1,275.976	4.024	16.192
19:	3,835	3,836.877	−1.877	3.525
20:	555	557.173	−2.173	4.720

TSS = 20,813.334   ERSS = 120.712 R^2 = .99999
F statistic = 919,575*

2. Coefficient 1 X_1 42.316
           t statistic =   88.4666*
   Coefficient 2 X_2 −0.689
           t statistic =   −1.8598
   Coefficient 3 X_3 −7.717
           t statistic =   −14.4398*

	Y	Predicted	Residual	$(Residual)^2$
1:	1,198	1,186.833	11.167	124.707
2:	1,143	1,122.161	20.839	434.284
3:	1,469	1,503.867	−34.867	1,215.735
4:	1,681	1,670.752	10.248	105.031
5:	145	143.740	1.260	1.587
6:	1,875	1,865.093	9.907	98.148
7:	289	312.733	−23.733	563.263
8:	−100	−126.009	26.009	676.453
9:	836	799.721	36.279	1,316.187
10:	223	258.767	−35.767	1,279.276
11:	1,913	1,871.877	41.123	1,691.092
12:	897	887.849	9.151	83.745
13:	1,733	1,757.521	−24.521	601.300
14:	411	433.829	−22.829	521.178
15:	1,533	1,575.473	−42.473	1,803.994
16:	624	630.884	−6.884	47.387
17:	887	853.955	33.045	1,091.978
18:	1,568	1,554.711	13.289	176.595
19:	1,188	1,174.391	13.609	185.196
20:	1,536	1,570.851	−34.851	1,214.614

TSS = 7,343,037   ERSS = 13,231.75 R^2 = .9982
F statistic = 2,954*

3. Coefficient 1 X_1 13.617
           t statistic =   2.0979
   Coefficient 2 X_2 2.017
           t statistic =   0.3378
   Coefficient 3 X_3 −5.282
           t statistic =   −0.9673

	Y	Predicted	Residual	$(Residual)^2$
1:	2,471	1,740.937	730.063	532,992.518
2:	2,008	1,762.284	245.716	60,376.182

	Y	Predicted	Residual	(Residual)^2
3:	1,916	1,470.775	445.225	198,225.555
4:	2,537	1,828.176	708.824	502,431.852
5:	1,240	2,199.713	− 959.713	921,049.942
6:	2,840	1,929.568	910.432	828,885.940
7:	1,915	1,726.914	188.086	35,376.492
8:	328	1,112.011	− 784.011	614,672.679
9:	2,841	2,281.989	559.011	312,493.767
10:	1,838	1,881.923	− 43.923	1,929.200
11:	1,444	2,554.867	− 1,110.867	1,234,026.478
12:	3,238	2,409.289	828.711	686,761.304
13:	879	1,757.522	− 878.522	771,801.151
14:	1,319	1,626.483	− 307.483	94,545.498
15:	2,226	2,500.239	− 274.239	75,206.756
16:	702	1,538.792	− 836.792	700,220.561
17:	1,296	1,638.209	− 342.209	117,106.767
18:	1,734	1,476.010	257.990	66,558.864
19:	1,725	1,706.993	18.007	324.259
20:	2,290	1,644.308	645.692	416,917.993

TSS = 10,844,855   ERSS = 8,171,903.8 R^2 = .2465
F statistic = 1.744

4. Coefficient 1 X_1 − 0.573
           t statistic =   − 2.3973*
   Coefficient 2 X_2 − 0.149
           t statistic =   − 0.6302
   Coefficient 3 X_3 − 0.116
           t statistic =   − 1.0861

	Y	Predicted	Residual	(Residual)^2
1:	20	44.084	− 24.084	580.025
2:	55	36.954	18.046	325.662
3:	50	50.748	− 0.748	0.560
4:	33	37.542	− 4.542	20.632
5:	34	58.509	− 24.509	600.702
6:	42	42.561	− 0.561	0.315
7:	57	43.801	13.199	174.204
8:	51	45.038	5.962	35.540
9:	18	21.060	− 3.060	9.361
10:	17	39.271	− 22.271	496.007
11:	55	57.805	− 2.805	7.865

(continued)

	Y	Predicted	Residual	(Residual)^2
12:	48	46.806	1.194	1.425
13:	37	24.828	12.172	148.160
14:	68	47.125	20.875	435.754
15:	71	60.136	10.864	118.019
16:	73	52.869	20.131	405.273
17:	81	64.511	16.489	271.891
18:	41	46.622	−5.622	31.602
19:	58	57.283	0.717	0.514
20:	19	50.447	−31.447	988.884

TSS = 6996.8   ERSS = 4652.395 R^2 = .3351
F statistic = 2.688

5. Coefficient 1 X_1 29.941
   t statistic = 73.5922*
   Coefficient 2 X_2 2.030
   t statistic = 5.8862*
   Coefficient 3 X_3 0.117
   t statistic = 0.3067

	Y	Predicted	Residual	(Residual)^2
1:	2,120	2,174.714	−54.714	2,993.620
2:	1,631	1,640.368	−9.368	87.766
3:	2,135	2,121.642	13.358	178.424
4:	1,541	1,572.834	−31.834	1,013.398
5:	328	386.299	−58.299	3,398.730
6:	1,064	1.016.576	47.424	2,249.072
7:	2,222	2,252.935	−30.935	956.982
8:	2,003	2,002.560	0.440	0.194
9:	2,652	2,636.285	15.715	246.964
10:	1,377	1,351.580	25.420	646.180
11:	1,570	1,529.984	40.016	1,601.306
12:	2,521	2,553.581	−32.581	1,061.531
13:	443	452.047	−9.047	81.851
14:	1,683	1,669.077	13.923	193.858
15:	2,751	2,718.314	32.686	1,068.403
16:	1,327	1,324.463	2.537	6.434
17:	1,890	1,937.518	−47.518	2,257.986
18:	1,431	1,363.557	67.443	4,548.605
19:	772	796.400	−24.400	595.375
20:	1,388	1,348.266	39.734	1,578.752

TSS = 8,573,231   ERSS = 24,765.4 R^2 = .9971
F statistic = 1841*

6. Coefficient 1 X_1 -0.200
         t statistic $= -0.3279$
   Coefficient 2 X_2 -0.098
         t statistic $= -0.2661$
   Coefficient 3 X_3 0.590
         t statistic $= 0.6021$

	Y	Predicted	Residual	$(\text{Residual})^2$
1:	200	240.658	-40.658	1,653.078
2:	158	231.019	-73.019	5,331.746
3:	171	232.176	-61.176	3,742.491
4:	207	225.913	-18.913	357.706
5:	135	208.339	-73.339	5,378.589
6:	232	233.772	-1.772	3.140
7:	297	239.047	57.953	3,358.592
8:	168	210.619	-42.619	1,816.420
9:	226	209.384	16.616	276.084
10:	315	221.169	93.831	8,804.278
11:	252	239.804	12.196	148.736
12:	139	225.826	-86.826	7,538.777
13:	273	227.252	45.748	2,092.893
14:	266	202.504	63.496	4,031.776
15:	150	234.240	-84.240	7,096.357
16:	291	239.777	51.223	2,623.824
17:	305	215.387	89.613	8,030.414
18:	168	202.894	-34.894	1,217.571
19:	329	234.181	94.819	8,990.644
20:	220	228.039	-8.039	64.629

TSS $= 75,617.8$ ERSS $= 72,557.7$ $R^2 = .0405$
F statistic $= .2249$

17
NONPARAMETRIC METHODS

KEY TERM

nonparametric method a statistical method that does not make assumptions about the specific forms of distributions and therefore does not focus on estimating unknown parameter values; some examples are the sign test, the Friedman F_r test, the Wilcoxon rank sum test, the Kruskal-Wallis H test, and the Wilcoxon signed rank test

You may have noticed a common thread running through the last few chapters. When testing hypotheses or generating confidence intervals, we've assumed that the population(s) we were sampling had a given distribution, usually normal. Although we may not have known what the parameters (that is, the mean and variance) were, we could estimate them from the samples and go on from there.

Often, we've assumed also that the data were continuous—that is, someone's height in centimeters could be 180.1, or 180.11, or 180.109, etc. What should we do if the data consist of people's ratings of a given product or service on a scale from 1 to 10? How do we allow for the fact that numerical ratings are subjective, that a product that is given an 8 is not necessarily "twice as good" as a product that is given a 4 by the same person?

In these instances we don't have the usual distributions and parameters to fall back on, and they call for different procedures, called **nonparametric methods**.

YOU SHOULD REMEMBER

1. Most of the statistical methods we discussed in preceding chapters involve estimating values for unknown parameters when a particular form of distribution is assumed to apply.

> **2.** Other statistical methods, called *nonparametric* methods, can be applied to situations where it is not necessary to make an assumption about the form of the distribution.

SIGN TEST

One of the simplest nonparametric methods is called the **sign test**. It is useful when evaluating a survey that tests which option the surveyee prefers, but not by how much. For example, suppose Jones Cola Company conducts a blind taste test matching its brand against that of its rival, Smith Cola Company. We will test the null hypothesis that among the population as a whole there is no difference in preferences between the two colas. If each person in the survey provided an accurate estimate of his or her liking for each cola on a scale of 1 to 10, we could use the statistical methods employed before with the *t* statistic. However, we might not have these rankings available; moreover, we might doubt the accuracy of interpersonal rating comparisons even if they were available. Therefore, we will use the sign test. We will simply ask 20 people in the survey which cola they prefer, using a plus sign to indicate people who prefer Jones cola and a minus sign to indicate the people who prefer Smith cola. Suppose the results are as follows:

$$+, -, +, +, +, +, -, +, +, -, +, +, +, -, +, +, -, +, +, +$$

Of the 20 people in the survey, 15 favored Jones Cola and 5 favored Smith Cola. If the null hypothesis is true, then each individual had a 50 percent chance of selecting Jones Cola and a 50 percent chance of selecting Smith Cola. Therefore, the number of plus signs would come from a binomial distribution with $n = 20$ and $p = .5$. We will use n_+ to stand for the number of plus signs. Then

$$E(n_+) = np = 10 \quad \text{and} \quad \text{Var}(n_+) = np(1 - p) = \frac{20}{4} = 5$$

We can use the normal distribution as an approximation for the binomial distribution. (For the sign test this approximation will generally be valid when $n > 10$.) Therefore, we'll use a test statistic Z:

$$Z = \frac{n_+ - np}{\sqrt{np(1 - p)}} = \frac{n_+ - n/2}{\sqrt{n}/2} = \frac{2n_+ - n}{\sqrt{n}}$$

which will have a standard normal distribution if the null hypothesis is true. In our case $n_+ = 15$ and $n = 20$, so we can calculate $Z = 2.236$. We can reject the null hypothesis at the 5 percent level, since 2.236 is outside 1.96.

One important advantage of the sign test is that you need not make restrictive assumptions about the nature of the population. For example, you do not need to assume that preferences are distributed according to the normal distribution. The disadvantage of the sign test is that it ignores some information that you may have in certain cases. If, as in the cola examples, the only information available tells you which brand was preferred, then you will have to use the sign test. If you also have available information that indicates the strength of each person's preferences, then the sign test is not as effective because it ignores that information.

FRIEDMAN F_r TEST

Above, where we were testing only two options, we cared only about which one people liked better. Given three or more options, we might also want to know which option people like second best, third best, and so on. Thus, given n options, we would ask people to rank their preferences. This would rule out a null hypothesis assuming a multinomial distribution (a straightforward generalization of the two-option case), which would take into account only which option a person liked best.

We therefore use the **Friedman F_r test** and construct a new test statistic:

$$F_r = \left[\frac{12}{bk(k+1)} \right] \left(\sum_{j=1}^{k} R_j^2 \right) - 3b(k+1)$$

where b is the sample size
k is the number of options
R_j is the rank sum of the jth option (that is, for the jth option, the sum of the ratings given by those surveyed)
Given the null hypothesis that there is no preference among the k options, and either $k > 5$ or $b > 5$, then F_r has approximately a chi-square distribution with $k - 1$ degrees of freedom, which can be used to test the null hypothesis.

Example: Application of the Friedman F_r Test

PROBLEM: Suppose 10 salesmen for a company compare their sales (working independently) in four cities. Test to see whether they do better or worse than average in any of the cities. Here is the data table:

Salesman	City No. 1	Rank No. 1	City No. 2	Rank No. 2	City No. 3	Rank No. 3	City No. 4	Rank No. 4
1	15	3	18	4	9	1	11	2
2	19	3	20	4	13	2	6	1
3	27	4	7	1	15	3	8	2
4	43	4	9	1	19	2	29	3
5	18	4	17	3	16	2	11	1
6	12	4	10	2	11	3	9	1
7	20	4	13	1	16	2	18	3
8	14	1	22	2	23	3	31	4
9	40	4	14	2	25	3	13	1
10	9	1	14	2	26	3	19	4
	$R_1 = 32$		$R_2 = 22$		$R_3 = 24$		$R_4 = 22$	

SOLUTION   Then $b = 10$, $k = 4$, $F_r = 4.08$; and, by checking Table A3-3, we see that we can accept the null hypothesis at the 5 percent level.

Our general procedure, given survey results ranking preferences of choices (either because numerical data are not available, or because they don't fit a known distribution), is to test the null hypothesis that there is no preference among the choices, when either the size of the sample (b) or the number of options (k) is greater than 5.

Procedure for the Friedman F_r Test

1. Calculate the rank sums R_i.

2. Calculate

$$F_r = \left[\frac{12}{bk(k + 1)} \right] \left(\sum_{i=1}^{k} R_i^2 \right) - 3b(k + 1)$$

3. F_r has approximately a chi-square distribution with $k - 1$ degrees of freedom. Check Table A3-3 for $k - 1$ degrees of freedom. Accept or reject the null hypothesis at a chosen significance level.

Note: Ties may occur when ranking numerical data. If there are only a few ties, average the rankings among those tied; for example, 10, 20, 10, 15 would be ranked 1.5, 4, 1.5, 3 since there is a tie for first and second rankings. If there are many ties, your data may be unreliable.

YOU SHOULD REMEMBER

1. The sign test is used to test the hypothesis that there is no difference between two quantities when you have rankings instead of numerical values.

2. The Friedman F_r test is used to test the hypothesis that there is no difference between preferences when there are more than two possibilities.

WILCOXON RANK SUM TEST

Suppose now that you have samples (not necessarily the same size) from two populations that you think have the same (unknown) distribution. One way of testing whether they really do come from the same distribution is to pool the samples and see how they mix. If all the values of one sample are smaller than all the values of the other, then you should probably reject the null hypothesis (that the two populations have the same distribution). On the other hand, if they are well mixed, then you have no cause to reject the hypothesis.

A way of measuring how well the pooled samples are mixed is the **Wilcoxon rank sum test**. It consists of pooling the samples A and B, ranking them, and adding the ranks given to A to form the rank sum T. If the size of each sample is at least 8, then T has an approximately normal distribution with parameters

$$\mu_T = \frac{n_A(n_A + n_B + 1)}{2} \quad \text{and}$$

$$\sigma_T^2 = \frac{n_A n_B(n_A + n_B + 1)}{12}$$

where n_A = the size of sample A and n_B = the size of sample B. Then, by comparing $(T - \mu_T)/\sigma_T$ with Table A3-2, we can determine whether to accept the hypothesis.

Example: Application of the Wilcoxon Rank Sum Test

PROBLEM   Suppose nine Harvard and eight Yale economics professors make the following predictions for inflation for the coming year:

Harvard	Inflation Prediction (in percent)	Rank	Yale	Inflation Prediction (in percent)	Rank
1	8.6	1	1	15.8	15
2	9.7	3	2	13.4	11
3	11.8	6	3	12.2	7
4	17.1	16	4	17.3	17
5	12.9	10	5	12.6	9
6	11.7	5	6	14.3	14
7	12.4	8	7	14.2	13
8	10.3	4	8	13.5	12
9	8.9	2			

Test the null hypothesis that the professors' predictions have the same distributions.

SOLUTION   Here

$$T = 55, \qquad \mu_T = \frac{(18)}{2} = 81,$$

$$\sigma_T{}^2 = \frac{(9)(8)(18)}{12} = 108, \qquad \frac{T - \mu_T}{\sigma_T} = -2.5$$

and, by comparing with Table A3-2, we see that we can reject the hypothesis at the 95 percent level.

Thus we have the procedure shown below to test the null hypothesis that samples come from two populations with the same distribution.

Procedure for the Wilcoxon Rank Sum Test

1. Pool the samples.

2. Rank them (again, if there are relatively few ties, average their ranks).

3. Compute the rank sum T for sample A,

$$\mu_T = \frac{n_A(n_A + n_B + 1)}{2}, \qquad \text{and} \qquad \sigma_T{}^2 = \frac{n_A n_B(n_A + n_B + 1)}{12}$$

4. Compare the standard normal random variable $(T - \mu_T)/\sigma_T$ with Table A3-2 at a given level to determine whether to accept the hypothesis.

KRUSKAL-WALLIS H TEST

The **Kruskal-Wallis H test** is a generalization of the Wilcoxon rank sum test to more than two populations.

Example: Application of the Kruskal-Wallis **H** *Test*

PROBLEM Suppose that we have samples of k populations, and we want to test the hypothesis that the populations have the same distributions.

SOLUTION Again we pool the samples and rank them. Our test statistic is a bit more complicated, however:

$$H = \left[\frac{12}{n(n + 1)} \right]\left[\sum_{j=1}^{k} \left(\frac{R_j^2}{n_j} \right) \right] - 3(n + 1)$$

where

$$n = \sum_{j=1}^{k} n_j$$

n_j = the size of the sample of the jth population, and R_j = the jth rank sum. H has approximately a chi-square distribution with $k - 1$ degrees of freedom, if each of the n_j is larger than 5.

Thus, given a significance level and samples of k different populations (with every sample size larger than 5), calculate H, and check Table A3-3 under $k - 1$ degrees of freedom to decide whether to accept the null hypothesis (that the populations have the same distributions).

WILCOXON SIGNED RANK TEST

As you might quess, the **Wilcoxon signed rank test** is similar in some ways to the Wilcoxon rank sum test. It is applicable to situations similar to those discussed for the sign test. The null hypothesis states that there is no difference between the two populations.

Suppose we have a survey asking people to evaluate two options A and B. This time, rather than just stating which they prefer, they rate each on a scale (for example, from 1 to 10 or 100). We then rank the absolute values of the differences between the ratings for A and B, and sum the ranks for the positive differences and negative differences, giving T_+ and T_-. If we let T be the smaller of T_+ and T_-, and the sample size N is large, then T is a normal random variable with parameters

$$\mu_T = \frac{N(N+1)}{4} \quad \text{and} \quad \sigma_T = \frac{N(N+1)(2N+1)}{24}$$

if the null hypothesis is correct. We then have a standard normal random variable $(T - \mu_T)/\sigma_T$ that we can compare with Table A3-2 to decide whether to accept our hypothesis.

Example: Application of the Wilcoxon Signed Rank Test

PROBLEM Suppose we take a random sample of 10 people who have rented cars from both Able Rental Cars and Baker Rental Cars, and ask them to rate both rental services on a scale of 1 to 10. We then have the following table:

Person	A	B	A − B	|A − B|	Rank	+	−
1	4	7	−3	3	9		9
2	5	6	−1	1	3		3
3	9	6	3	3	9	9	
4	5	7	−2	2	6.5		6.5
5	4	5	−1	1	3		3
6	9	8	1	1	3	3	
7	5	4	1	1	3	3	
8	7	6	1	1	3	3	
9	3	5	−2	2	6.5		6.5
10	3	6	−3	3	9		9
				totals		18	37

Test the hypothesis that there is no difference between the two companies.

SOLUTION Then $T_+ = 18$, $T_- = 37$. Also,

$$T = T_+ = 18, \quad \mu_T = \frac{10(11)}{4} = 27.5,$$

$$\sigma_T{}^2 = \frac{10(11)(21)}{24} = 96.25.$$

Thus $(T - \mu_T)/\sigma_T = -0.97$, and we can accept the hypothesis at the 95 percent level.

Again in case of ties, average the ranks involved. If persons give the same rating to A and B (that is, $A - B = 0$), perform the test twice, first counting 0 as a positive difference, the second time counting 0 as a negative difference. If the results agree, fine. If not, your data may be unreliable.

We then have the procedure shown below, given ratings of two options by N people (companies, departments, etc.).

Procedure for the Wilcoxon Signed Rank Test

1. Rank the absolute values of the differences of the ratings.

2. Calculate the rank sums T_+ and T_-.

3. Let T be the smaller of T_+ and T_-,

$$\mu_T = \frac{N(N + 1)}{4}, \quad \text{and} \quad \sigma_T^2 = \frac{N(N + 1)(2N + 1)}{24}$$

4. Compare $(T - \mu_T)/\sigma_T$ with Table A3-2 to determine whether to accept the null hypothesis at a given significance level.

YOU SHOULD REMEMBER

1. The Wilcoxon rank sum test is used to test the hypothesis that there is no difference between the means of populations.

2. The Kruskal-Wallis H test is a generalization of the Wilcoxon rank sum test, used when there are more than two populations.

3. The Wilcoxon signed rank test is used to test the hypothesis that there is no difference between the means of populations.

KNOW THE CONCEPTS

DO YOU KNOW THE BASICS?

Test your understanding of Chapter 17 by answering the following questions:

1. In general, what is one of the main advantages of nonparametric statistical methods? What is one of the main disadvantages?

2. Can you explain intuitively why the sign test works?

3. Is the sign test a two-tailed test or a one-tailed test?

4. Which test makes more use of the available information: the sign test or the Wilcoxon signed rank test?

5. Why would the presence of too many ties in the rankings make the Friedman F_r test unreliable?

TERMS FOR STUDY

Friedman F_r test

Kruskal-Wallis H test

nonparametric method

sign test

Wilcoxon rank sum test

Wilcoxon signed rank test

PRACTICAL APPLICATION

COMPUTATIONAL PROBLEMS

1. Of 25 rental car users surveyed, 9 preferred Arthur's Rental Cars to Beaumont's Rental Cars, the other 16 preferring Beaumont's. Test the null hypothesis that rental car users have no preference between Arthur's Rental Cars and Beaumont's Rental Cars.

2. On a certain uneven piece of ground, Route 13 splits into two roads that eventually merge, one going over some mountains and the other down through a valley. If a random sampling of 30 drivers showed that 8 drivers took the high road and 22 drivers took the low road, test the null hypothesis that Route 13 drivers have no preference between the two roads.

3. In a taste test, 13 out of 18 people preferred spaghetti sauce made with ground oregano to sauce made with crushed oregano leaves, the other 5 people preferring the crushed leaves. Test the null hypothesis that people are indifferent to which form of oregano is used in their spaghetti sauce.

4. Fifteen frequent fliers are asked to rate in order of preference three airlines: Aircat (A), Bluebird (B), and Condor (C) Airlines. The results are as follows:

A	B	C	A	B	C	A	B	C
1	3	2	3	1	2	1	3	2
1	3	2	3	2	1	3	2	1
1	3	2	2	3	1	2	3	1
1	2	3	1	3	2	3	1	2
2	1	3	1	2	3	1	3	2

Test the null hypothesis that there is no preference among fliers between the three airlines.

5. Four people (A, B, C, D) are asked to rate in order of preference the 12 months of the year. The results are as follows:

	J	F	M	A	M	J	J	A	S	O	N	D
A	12	11	5	3	4	6	7	8	1	2	9	10
B	12	1	2	9	4	3	5	6	7	8	10	11
C	1	2	4	5	6	9	3	7	8	10	11	12
D	1	4	3	12	10	5	9	11	8	7	6	2

Use these data to test the null hypothesis that people have no preference between the months of the year.

6. Six companies show the following profits (in thousands of dollars) from sales in four different cities (A, B, C, D):

A	B	C	D
22	11	16	14
20	19	18	14
19	24	16	13
15	18	17	19
18	17	13	15
17	16	19	12

Use the Friedman F_r statistic to test the null hypothesis that the four cities are equally profitable for the companies.

7. Thirteen voters surveyed showed the following preferences between three presidential candidates (A, B, C):

A	B	C	A	B	C
3	2	1	1	2	3
2	3	1	3	1	2
3	1	2	3	2	1
1	2	3	2	1	3
1	2	3	2	1	3
3	2	1	1	2	3
1	2	3			

Test the null hypothesis that voters have no preference between the candidates.

8. Here are the sales figures (in hundreds) for a company in 10 towns in Arizona and 15 towns in Michigan:

   AZ: 16, 8, 7, 14, 22, 27, 19, 23, 18, 30

   MI: 13, 28, 17, 26, 12, 11, 33, 35, 5, 40, 39, 10, 6, 32, 31

   Test the null hypothesis that the distribution of sales in Arizona is the same as that in Michigan.

9. Here are the defense budgets (in millions of dollars, adjusted for inflation) of two countries (*A, B*) over a decade:

   A: 10, 12, 15, 16, 18, 22, 26, 28, 30, 29

   B: 9, 8, 17, 14, 13, 19, 24, 25, 27, 31

   Test the null hypothesis that the budgets have the same distribution.

10. Twelve light bulbs of brand *X* and fifteen of brand *Y* were tested to see how long they would last, with the following results (in hundreds of hours):

    X: 15, 19, 23, 25, 26, 28, 30, 29, 32, 31, 20, 14

    Y: 20, 18, 16, 19, 22, 24, 17, 21, 10, 11, 32, 27, 12, 14, 9

    Test the null hypothesis that the lifetimes of the two brands of light bulbs have the same distribution.

11. Here are the incomes (in thousands of dollars) of 15 plumbers in California and 13 plumbers in Maine:

    CA: 22, 19, 17, 24, 25, 30, 29, 31, 28, 37, 15, 40, 38, 39, 17

    ME: 21, 20, 18, 17, 23, 26, 28, 29, 32, 35, 19, 27, 16

    Test the null hypothesis that plumbers' incomes are distributed equally in both states.

12. Here are the benefits (in thousands of dollars) paid to employees of three yo-yo manufacturers (*A, B, C*) over a decade:

A	B	C	A	B	C
10	25	16	23	19	28
26	12	24	30	14	18
29	20	13	31	38	35
21	11	22	39	32	37
17	27	15	33	36	34

Test the hypothesis that the benefits expenditures of the three companies have the same distribution.

13. Four experimental precision scales (A, B, C, D) are tested on a fixed weight, with the following results:

   A: 103, 121, 106, 120, 114, 128, 116

   B: 112, 105, 132, 136, 109, 138, 135, 126, 124, 117

   C: 131, 104, 130, 108, 123, 119, 113, 133, 127, 134, 125, 115

   D: 129, 111, 122, 137, 107, 110, 139, 118

Test the null hypothesis that the distributions of values given by the four scales are the same.

14. Fifteen people are asked to rate on a scale of 1–100 two different stereo systems (A, B), with the following results:

A	B	A	B	A	B
85	78	95	70	65	83
60	99	99	65	85	72
70	85	93	62	89	95
75	99	75	80	90	60
88	90	60	78	95	50

Test the null hypothesis that people have no preference between the two kinds of stereos.

15. Twenty people rate on a scale of 1–10 two political candidates (A, B), with the following results:

A	B	A	B	A	B	A	B
2	8	8	4	6	5	9	7
5	6	9	8	7	4	8	9
3	5	8	9	8	5	4	6
7	8	7	8	8	6	10	7
4	5	5	6	9	6	8	5

Test the null hypothesis that people have no preference between the candidates.

16. Eighteen people rate on a scale of 1–20 two different brands (A, B) of water pistols, with the following results:

A	B	A	B	A	B
18	15	17	13	11	17
14	12	12	15	13	12
17	19	13	17	14	16
16	12	18	15	12	13
18	15	19	16	11	10
11	9	12	16	16	12

Test the null hypothesis that people have no preference between the pistols.

17. Twenty-five people rate on a scale of 1–10 two car models (A, B), with the following results:

A	B	A	B	A	B
4	6	6	4	8	2
8	7	8	7	9	8
5	4	5	8	8	6
4	2	7	6	7	8
3	7	4	8	6	7
6	5	5	9	3	6
9	8	6	3	5	6
7	8	7	3	4	2
2	5				

Test the null hypothesis that people have no preference between the car models.

ANSWERS

KNOW THE CONCEPTS

1. The main advantage of nonparametric methods is that it is not necessary to make assumptions about the specific form of the distributions you are dealing with. The main disadvantage is that nonparametric tests do not use all of the data you may have available.

2. If the null hypothesis is true, then the number of pluses will have a binomial distribution with $p = .5$.

3. The sign test is a two-tailed test.

4. The Wilcoxon signed rank test makes more use of the information.

5. If there were too many ties in the rankings for the Friedman F_r test, the data points that were not part of the ties would have too great an effect on the result. The situation would be similar to using too small a sample.

PRACTICAL APPLICATION

1. $n = 25, n_+ = 16, Z = 1.4$; accept the null hypothesis.
2. $n = 30, n_+ = 22, Z = 2.56$; reject the null hypothesis.
3. $n = 18, n_+ = 13, Z = 1.89$; accept the null hypothesis.
4. $R_A = 26, R_B = 35, R_C = 29,$

$$F_r = \left(\frac{12}{(15)(3)(4)}\right)(26^2 + 35^2 + 29^2) - 3(15)(4) = 2.8$$

and you would accept the null hypothesis at the 90 percent level.

5. $R_1 = 26, R_2 = 18, R_3 = 14, R_4 = 29, R_5 = 24, R_6 = 23, R_7 = 24,$
   $R_8 = 32, R_9 = 24, R_{10} = 27, R_{11} = 36, R_{12} = 35,$

$$F_r = \left(\frac{12}{(4)(12)(13)}\right)(26^2 + 18^2 + 14^2 + 29^2 + 24^2 + 23^2 + 24^2$$
$$+ 32^2 + 24^2 + 27^2 + 36^2 + 35^2) - 3(4)(13) = 8.77$$

and you would accept the null hypothesis at the 90 percent level.

6. $R_A = 19, R_B = 16, R_C = 14, R_D = 11,$

$$F_r = \left(\frac{12}{(6)(4)(5)}\right)(19^2 + 16^2 + 14^2 + 11^2) - 3(6)(5) = 3.4$$

and you would accept the hypothesis at the 90 percent level.

7. $R_A = 26, R_B = 23, R_C = 29$

$$F_r = \left(\frac{12}{(13)(3)(4)}\right)(26^2 + 23^2 + 29^2) - 3(13)(4) = 1.38$$

and you would accept the null hypothesis at the 90 percent level.

8. $T = 116, \mu_T = 10(26)/2 = 130, \sigma_T^2 = (10)(15)(26)/12 = 325,$ $(T - \mu_T)/\sigma_T = -0.776$, and you would accept the null hypothesis at the 90 percent level.

9. $T = 113, \mu_T = 10(21)/2 = 105, \sigma_T^2 = (10)(10)(21)/12 = 175,$ $(T - \mu_T)/\sigma_T = 0.605$, and you would accept the null hypothesis at the 90 percent level.

10. $T = 214$, $\mu_T = 12(28)/2 = 168$, $\sigma_T^2 = (12)(15)(28)/12 = 420$, $(T - \mu_T)/\sigma_T = 2.25$, and you would reject the null hypothesis at the 95 percent level.

11. $T = 240.5$, $\mu_T = 15(29)/2 = 217.5$, $\sigma_T^2 = (15)(13)(29)/12 = 471.25$, $(T - \mu_T)/\sigma_T = 1.06$, and you would accept the null hypothesis at the 90 percent level.

12. $R_A = 169$, $R_B = 144$, $R_C = 152$,

$$H = \left(\frac{12}{30(31)}\right)\left(\frac{169^2}{10} + \frac{144^2}{10} + \frac{152^2}{10}\right) - 3(31) = 0.421$$

and you would accept the null hypothesis at the 90 percent level.

13. $R_A = 94$, $R_B = 214$, $R_C = 238$, $R_D = 157$,

$$H = \left(\frac{12}{37(38)}\right)\left(\frac{94^2}{7} + \frac{214^2}{10} + \frac{238^2}{12} + \frac{157^2}{8}\right) - 3(38) = 2.44$$

and you would accept the null hypothesis at the 90 percent level.

14. $T_+ = 70$, $T_- = 50$, $T = 50$, $\mu_T = 15(16)/4 = 60$, $\sigma_T^2 = (15)(16)(31)/24 = 310$, $(T - \mu_T)/\sigma_T = -0.568$, and you would accept the null hypothesis at the 90 percent level.

15. $T_+ = 132$, $T_- = 78$, $T = 78$, $\mu_T = 20(21)/4 = 105$, $\sigma_T^2 = (20)(21)(41)/24 = 717.5$, $(T - \mu_T)/\sigma_T = -1.007$, and you would accept the null hypothesis at the 90 percent level. Note: In view of the large number of ties in this sample, the data may not be reliable.

16. $T_+ = 100$, $T_- = 71$, $T = 71$, $\mu_T = 18(19)/4 = 85.5$, $\sigma_T^2 = (18)(19)(37)/24 = 527.25$, $(T - \mu_T)/\sigma_T = -0.630$, and you would accept the null hypothesis at the 90 percent level.

17. $T_+ = 164$, $T_- = 161$, $T = 161$, $\mu_T = 25(26)/4 = 162.5$, $\sigma_T^2 = (25)(26)(51)/24 = 1,381.25$, $(T - \mu_T)/\sigma_T = -0.040$, and you would accept the null hypothesis at the 90 percent level. Note: In view of the large number of ties in this sample, the data may not be reliable.

18
BUSINESS DATA

KEY TERMS

gross domestic product (GDP) a measure of the total value of all goods and services produced in the country.

price index a measure of the price level

price level the average level of all prices in the country; can be measured by the GDP deflator, consumer price index, or producer price index

seasonal adjustment a procedure for adjusting time series data to compensate for seasonal variation in order to make nonseasonal changes in the data more apparent

time series data data that consist of several observations of a quantity at different points in time; can often be broken down into four components: a trend component, a cyclical component, a seasonal component, and an irregular component

Many times it is helpful to know about the current state of the economy. In this chapter we will discuss some of the most important statistical measures that describe the economy, as well as some general techniques for analyzing different types of business data.

GROSS DOMESTIC PRODUCT

The most common statistic for measuring the overall level of production in the entire economy is called the **gross domestic product** (**GDP** for short). It measures the total value of all goods and services produced in the country in a given year. The job of collecting the statistics used to compile GDP is handled by the Department of Commerce, and the results are published in a periodical called *Survey of Current Business*. Estimates of GDP are calculated for each quarter of the year. Measuring GDP requires an immense data collection effort. The process of analyzing these data is called *national income accounting*.

In reality, of course, the Department of Commerce cannot observe every single good or service produced, so it needs to use statistical procedures to estimate GDP based on the data it collects from surveys of businesses. Suppose that we could compile a list of all goods and services produced in the economy. To simplify our discussion, let's assume that our economy produces only three goods: pizza, ice cream cones, and statistics books.

Good	Quantity Produced This Year
Pizzas	500
Ice cream cones	615
Books	90

(We are assuming that all pizzas are identical, all ice cream cones are identical, and so on. In reality we would need separate measures for each type of pizza, each type of ice cream cone, and each type of every other good.) Now we must add up the total production. We cannot simply add the quantities, because then we would (literally) have the problem of adding up apples and oranges. Instead, what we will do is add up the total *value* of each type of good produced. Therefore, we also need a list of the price of each good:

Good	Quantity	Price
Pizzas	500	$5.00
Ice cream cones	615	2.00
books	90	7.00

To get the total value of each good we must multiply the price times the quantity:

Good	Quantity	Price	Value
Pizzas	500	$5.00	$2,500
Ice cream cones	615	2.00	1,230
Books	90	7.00	630
	GDP:		$4,360

This quantity is called the *nominal GDP*; later we will see how the *real GDP* corrects for inflation.

• SOME CAUTIONS IN MEASURING GDP
The GDP is found by adding up the total value. There are some complications that must be kept in mind while measuring GDP.

1. Intermediate goods cannot be counted. An intermediate good is a good produced by one firm and then sold to another firm for use in producing some other good. If intermediate goods were directly included in GDP, then they would be double counted, because they are counted as part of the final goods in which they are used.

   Here is an example. Suppose that our economy consists of a peanut farmer who produces $50 worth of peanuts. She sells $30 worth of peanuts to consumers and $20 worth to a peanut butter factory, which uses the peanuts to produce $60 worth of peanut butter. Here is the wrong way to calculate GDP.

$$\text{Value of peanuts} = 50$$
$$\text{Value of peanut butter} = \underline{60}$$
$$\text{Total GDP} = 110 \; (\textit{wrong})$$

   The peanuts that were sold to the peanut butter firm were counted twice in this calculation.

   There are two correct ways to calculate GDP. One way is to count only the final goods (sold to consumers) produced by each supplier:

$$\text{Value of peanuts sold to consumers} = 30$$
$$\text{Value of peanut butter sold to consumers} = \underline{60}$$
$$\text{Total GDP} = 90 \; (\textit{right})$$

   The other way is to subtract the value of the intermediate goods used by each firm from its output. (The value of a firm's output minus the value of its intermediate goods is called the *value added*.)

$$\text{Value added by peanut farmer} = 50 - 0 = 50$$
$$\text{Value added by peanut butter firm} = 60 - 20 = \underline{40}$$
$$\text{Total GDP} = 90 \; (\textit{right})$$

   (We have assumed that the farmer does not use any intermediate goods. In reality the farmer probably buys some fertilizer and other goods to use in the peanut farming process.)

2. Purely financial transactions are not included in GDP. If you put some money in the bank or buy shares of stock, you are buying a financial asset, not a currently produced good or service. However, if you pay a fee to a broker to handle the transaction, that fee is included in GDP because you are paying for broker service.

3. Transactions in old objects are not included in GDP. Suppose you buy a 1978 model used car. The value of the car cannot be counted as part of this year's GDP because it was already counted as part of GDP in 1978, and the same car cannot be counted twice. However, the profit that your used car dealer makes on the sale counts as part of the GDP because he is providing used car dealing service. (If you don't want to pay for his service, you can try on your own to find someone willing to sell the kind of car you want to buy.)

   If a good is produced this year but not sold until next year, it counts as part of GDP this year. It is counted as part of a category called "increase in inventory." If you are bothered by the question "How do we calculate the value of the good if it has not been sold yet?" we suggest you study accounting.

4. Government services are usually valued at their cost of production. Since most government services are not sold, it is harder to attach values to them. For example, if a U.S. Air Force pilot is paid $50,000, then it is assumed that the pilot is adding $50,000 to GDP. The salary of a commercial airline pilot, on the other hand, is not directly added to GDP; instead the value of the tickets sold by his airline is included. This procedure can sometimes lead to strange results. When the government pays $5,000 for a wrench, that wrench is counted as adding $5,000 to GDP.

It is important to remember that GDP is not intended to measure national welfare. For one thing, GDP leaves out some important parts of national production: nonmarket activities. All the work performed by do-it-yourselfers and housekeepers really should be counted as part of national production, but their work cannot be counted in GDP (unless they are paid) because it would be too difficult to measure the value of their services. The GDP does not include the value of leisure time, nor does it take into account the effects of harmful phenomena such as pollution. Also, if you are interested in the average goods and services available to each person in the country, you should look at per capita GDP rather then total GDP.

In the 1990s, the Commerce Department changed the way it reports these statistics. Previously, the main focus was on a figure called gross national product (GNP). GNP measures the total value of all production by factors (inputs) owned by U.S. citizens, whether or not the production occurs in the United States. The GDP and GNP will usually have similar values. Here is the difference between them in 1994:

gross domestic product: $6.931 trillion
  add receipts of factor income from the
    rest of the world: $0.159 trillion

subtract payment of factor income to
    rest of world: $0.168 trillion
equals gross national product $6.922 trillion

Part of the gross national product is used to replace capital goods that have worn out, so another statistic, the net national product, is calculated which subtracts out the value of capital goods that have worn out. (This amount is called depreciation, or capital consumption.) For 1994:

gross national product: $6.922 trillion
    subtract consumption of fixed capital: $0.818 trillion
    equals net national product $6.104 trillion

Another closely related concept is the national income. Every time a good is purchased, it shows up as part of national production, but it also generates income (wages or profits) for someone. Therefore, if we added up the value of all of the income in the country, we would get a figure close to net national product. The main difference is the amount of what are called indirect taxes, which are taxes such as sales taxes or excise taxes that are placed on the sales of goods, rather than directly on people or firms.

Here is the exact relation between net national product and national income for 1994 (source: *Economic Report of the President*, 1996):

net national product $6.104 trillion
    subtract indirect taxes $0.573 trillion
    subtract business transfer payments $0.030 trillion
    subtract statistical discrepancy $0.031 trillion
    add subsidies less current
        surplus of government enterprises $0.025 trillion
equals national income $5.495 trillion

PRICE INDEXES

If production in the country remained the same but all prices went up, then the value of GDP would go up. However, since GDP is supposed to measure national production, we need to find a way to separate the effects of a price increase from the effects of a production increase. There are many other reasons why we need to measure the average size of prices (called the **price level**) or the rate of change in the price level (called the **inflation rate**). Determining the price level is very difficult, but we can estimate it with a quantity called a **price index**. We will discuss three different price indexes: the **GDP deflator**, the **consumer price index**, and the **producer price index**.

Let's return to our pizza/ice cream/books economy. Suppose these are the figures for production in 1997.

	1997		
	Quantity	Price	Value
Pizza	500	5.00	2,500
Ice cream cones	615	2.00	1,230
Books	90	7.00	630
	GDP:		4,360

The value of GDP calculated this way, using the current year's prices and quantities, is called the **nominal GDP**. Compare these with the figures for 1998:

	1998		
	Quantity	Price	Value
Pizza	800	5.50	4,400
Ice cream cones	650	3.00	1,950
Books	100	8.00	800
	GDP:		7,150

The nominal GDP is higher than it was in 1997 for two reasons: the production of each good has gone up, but so has the price of each good. The **real GDP** is designed so that it increases only when production goes up, not when prices go up. If the production of each good went up by 10 percent, then we would say that the real GDP increased by 10 percent. However, since the percent increase varies for different products, it is more difficult to determine the increase in real GDP. Let's ask this hypothetical question: how much would the GDP have gone up in 1998 if the prices had all stayed the same as they were in 1997?

	Quantity (in 1998)	Price (in 1997)	Value
Pizza	800	5.00	4,000
Ice cream cones	650	2.00	1,300
Books	100	7.00	700
	GDP:		6,000

Because we are using 1997 prices, we say that 1997 is the **base year**. We can determine the growth factor for real GDP:

real GDP growth factor from 1997 to 1998 with 1997 as base =

$$\frac{\Sigma Q_{1998} P_{1997}}{\Sigma Q_{1997} P_{1997}} = \frac{6,000}{4,360} = 1.37615$$

The expression $\Sigma Q_i P_j$ means to multiply the quantity of each good in year i by the price of the good in year j, and then add up the total value for all goods. This growth factor means that real GDP grew about 37.6 percent from 1997 to 1998.

Prior to 1996, the Department of Commerce estimated growth in real GDP by arbitrarily calling one year the base year, and then evaluating the real GDP for all other years by using the prices from that base year. In the early 1990s, 1987 was the base year. However, this method has a big disadvantage. A few years after the base year the economy will have changed significantly, but the real GDP will be distorted since it keeps using prices from the base year. Therefore, in 1996 the department began using the **chain-weight** approach. Unfortunately, this approach is more complicated than simply sticking with a fixed base year, but it will give an improved picture of the economy.

Here is how to calculate the chain-weight real GDP growth for our example. We have already seen that growth from 1997 to 1998 was 37.6 percent, using 1997 prices as the base year. However, the choice of 1997 was arbitrary; we could also use 1998 prices as the base. To do this, we have to go back and recalculate what the 1997 GDP would have been if the prices had taken on their 1998 values:

	Quantity (in 1997)	Price (in 1998)	Value
Pizza	500	5.50	2,750
Ice cream cones	615	3.00	1,845
Books	90	8.00	720
	GDP:		5,315

Therefore, using 1998 prices, we estimate real GDP grew from 5,315 in 1997 to 7,150 in 1998, giving a growth factor of:

real GDP growth factor from 1997 to 1998 with 1998 as base =

$$\frac{\Sigma Q_{1998} P_{1998}}{\Sigma Q_{1997} P_{1998}} = \frac{7,150}{5,315} = 1.34525$$

Using 1998 as the base, we estimate real GDP grew only by 34.5 percent. Which estimate is better? you might wonder. There is no way that one is necessarily better than the other, so the solution is to compromise. We will combine the estimates by multiplying them together and taking the square root:

$$\text{estimated growth factor} = \sqrt{1.37615 \times 1.34525} = 1.36061$$

Therefore, according to the chain-weight approach, we estimate real GDP grew by about 36 percent from 1997 to 1998. Note that the production of some goods increased by more than 36 percent, and the production of other goods increased by less.

	Quantity (in 1997)	Quantity (in 1998)	Percent Increase
Pizza	500	800	60.0
Ice cream cones	615	650	5.7
Books	90	100	11.1

In general, the percentage growth in real GDP from year t to year $t + 1$ is given by this formula:

$$100 \times \left(\sqrt{\frac{\Sigma Q_{t+1}P_t}{\Sigma Q_t P_t} \times \frac{\Sigma Q_{t+1}P_{t+1}}{\Sigma Q_t P_{t+1}}} - 1 \right)$$

Each year, the Department of Commerce will estimate the growth in real GDP from the previous year, using the chain-weight method. For convenience, they still express real GDP in terms of a base year. However, this no longer means that real GDP is calculated by assuming that all prices stayed the same as in the base year; instead, it tells the value that GDP would have had if, starting from the base year, there was no inflation and instead the GDP grew each year at the rate given by the chain-weight index.

Currently, the base year is 1992. In the base year, the real GDP is always the same as the nominal GDP. This value is 6,244.4 billion dollars for 1992. The chain-weight approach indicates that GDP grew 2.2324 percent from 1992 to 1993; 3.4525 percent from 1993 to 1994; and 2.0684 percent from 1994 to 1995. These result in a value of real GDP in 1995 of 6,740.8 billion, measured in chained 1992 dollars.

In 1995, the nominal GDP (that is, the GDP using 1995 prices and 1995 quantities) was 7,247.7 billion dollars. If we divide the nominal GDP by the real GDP, we find a quantity called the **GDP deflator**:

$$\text{GDP deflator} = \frac{\text{nominal GDP}}{\text{real GDP}} = \frac{7,247.7}{6,740.8} = 1.0752$$

In the base year the GDP deflator is always 1. The GDP deflator is a measure of the average level of prices. Since the GDP deflator in 1995 is 1.0752, it means that, on average, prices in 1995 were 7.52 percent higher than they were in 1992.

• THE CONSUMER PRICE INDEX

The GDP deflator doesn't necessarily measure how much inflation affects you personally. You would be much more interested in a measure that told you how much the average prices of the things you buy had gone up. The Bureau of Labor Statistics calculates a quantity called the consumer price index that measures how much increases in the price level affect the average consumer. The percent change in the consumer price index is the most widely publicized measure of inflation. It usually makes the news when its value is reported each month.

Let's suppose that you are an average consumer and you always buy exactly the same goods each year: 30 pizzas, 25 ice creams, and 2 books. Then it is easy to calculate how price changes will affect you. Your cost of living is simply the amount of money it costs to buy that particular bundle of goods you consume each year.

	Cost of Living—1997		
	Quantity	Price	Cost
Pizzas	30	$5.00	$150
Ice cream cones	25	2.00	50
Books	2	7.00	14
	total cost of living:		$214

	Cost of Living—1998		
	Quantity	Price	Cost
Pizzas	30	$5.50	$165.00
Ice cream cones	25	2.10	52.50
Books	2	8.00	16.00
	total cost of living:		$233.50

It cost you $233.50 to maintain your standard of living in 1998 or 1.091 times as much as it did in 1997. Therefore, according to the consumer price index, the rate of inflation was 9.1 percent from 1997 to 1998.

The consumer price index is currently calculated so that the average level of prices in 1982 to 1984 is set equal to 100. In 1995, the value of the index was 152.4. For comparison, in 1950 the index was 24.1, meaning that on average consumer prices in 1950 were only one-quarter of what they became by 1982 to 1984.

The consumer price index is based on the cost of a bundle of specific consumer goods that represent the expenditures of an average consumer. The contents of this bundle were determined by a survey called the Consumer Expenditure Survey that checked with a sample of over 40,000 families to find out what they bought. This table lists the relative importance of the items in the CPI:

Good	Percent of Expenditures
Food and beverages	18.8
Housing	43.9
Apparel and upkeep	5.8
Transportation	18.0
Medical care	5.0
Entertainment	4.1
Other goods and services	4.4

(Source: Bureau of Labor Statistics, *Handbook of Labor Statistics*, December 1980, page 331.)

If you happen to buy exactly the same mix of goods, then the consumer price index will exactly reflect how price increases affect you. If you happen to spend relatively more on goods whose prices increase faster than average, then you will suffer from price increases more than the CPI indicates. On the other hand, if you tend to buy more goods with slower price increases, then you will not feel the effects of inflation as much.

An even more difficult problem arises because the expenditure patterns for the average consumer change with time. The Bureau of Labor Statistics (BLS) periodically conducts new expenditure surveys and revises the bundle of goods used to calculate the CPI. However, the BLS cannot do this too frequently because then it becomes difficult to compare CPI values from different years.

• THE PRODUCER PRICE INDEX

The BLS calculates another price index called the producer price index that measures changes in the prices of a basket of commodities used by firms in the production process. Values for the price indexes are released during the third week of each month for the preceding month.

The BLS periodical *Monthly Labor Review* is a good source of current economic statistics.

YOU SHOULD REMEMBER

1. Other important measures of the national economy are price indexes, which measure the average price level.

2. Three commnly used price indexes are the following:
   GDP deflator: a price index based on all goods that are part of the GDP
   consumer price index: a price index based on the cost of a specific market basket of consumer goods.
   producer price index: a price index based on the cost of a specific market basket of goods used by firms in the production process.

3. The *real gross domestic product* is a measure of GDP that corrects for inflation.

TIME SERIES DATA

When you have measurements for the same quantity for several consecutive time periods, you have **time series data**. Many interesting economic variables are in the form of time series. The GDP figures are calculated every quarter. Many other figures, such as the CPI and the unemployment rate, are calculated every month. Some statistics, such as the money supply, are calculated every week, and other time series, such as the Dow Jones industrial average closing level, have daily values. When analyzing time series data you often have two goals: to describe the patterns of the series in the past and to predict future values of a particular quantity. We will describe briefly some of the analytic techniques that are used to investigate time series data.

Many times series show definite trends up or down if you follow their values over long periods of time. For example, the value of real GDP in the United States has been increasing fairly steadily for the last few decades (Fig. 18-1). Here are some figures:

Year	Real GDP	Year	Real GDP
1969	3,388.0	1973	3,902.3
1970	3,388.2	1974	3,888.2
1971	3,500.1	1975	3,865.1
1972	3,690.3	1976	4,081.0

(continued)

Year	Real GDP	Year	Real GDP
1977	4,279.3	1987	5,648.4
1978	4,493.7	1988	5,862.9
1979	4,624.0	1989	6,060.4
1980	4,611.9	1990	6,138.7
1981	4,724.9	1991	6,079.0
1982	4,623.6	1992	6,244.4
1983	4,810.0	1993	6,383.8
1984	5,138.2	1994	6,604.2
1985	5,329.5	1995	6,740.8
1986	5,489.9		

Source: Economic Report of the President, 1996.

If you had to predict the value of next year's real GDP, you probably would guess that it would continue to rise. In that case you are assuming that the value of real GDP will continue to follow the same upward trend that it has been following. For the past few decades real GDP has been growing at a rate averaging about 3 percent per year. However, your prediction for next year would not necessarily be correct. Some years real GDP grows faster than its

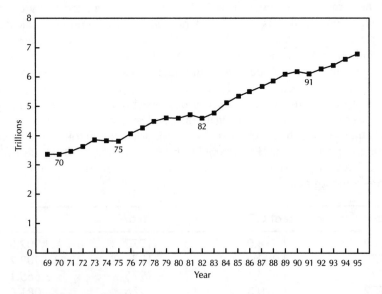

Figure 18-1. Real GDP

trend, and other years it does not grow as fast, or perhaps it even declines. In the period from 1929 to 1933 there was a major decline in the value of real GDP as the Great Depression hit. When the economy slows down, it is said that a **recession** has started. The United States suffered deep recessions in 1975 and 1982 and smaller recessions in 1970, 1980, and 1991.

The deepest time in a recession is called the *trough*. After the trough has been passed, the economy starts to grow more quickly. This phase is called the *expansion* or *recovery*. The economy continues to expand until it reaches the *peak*. These variations are called *cyclical variations* in the economy, because they seem to come in cycles of recession followed by expansion followed by recession. However, the cycles are not at all like the regular cycles in a clock or in the movements of the planets. You cannot use the lengths of past cycles to predict future cycles, but it is clear that these cyclical effects play a very important role in determining the level of business activity. At some times the economy is booming and there is a lot of spending, but at other times the economy is in a slump and unemployment is higher than usual.

One particularly interesting time series is the **index of leading indicators**. This index consists of a weighted average of several economic variables, such as construction permits issued, that tend to turn up before the whole economy turns up and tend to turn down before the whole economy turns down. The index can be used to forecast future upturns and downturns in the economy.

• *COMPONENTS OF TIME SERIES DATA*

We will view time series data as consisting of three components:

- the **trend component**
- the **cyclical component**
- the **irregular component**

The irregular component is included because there will always be some movements up or down in a time series that cannot be explained by a trend or by cyclical variations.

There is one other important part of many time series: the **seasonal component**. There is no seasonal component when you are looking at yearly data because in that case you have averaged over all seasons of the year, and the averaging process hides the seasonal variation. The seasonal component, however, is a very important part of quarterly, monthly, or weekly time series. We will discuss methods of analyzing seasonal variation in a later section of this chapter.

We will look at two methods for determining the trend of a set of data.

YOU SHOULD REMEMBER

1. Many important economic statistics are available in the form of *time series data*, consisting of observations of a variable for several consecutive time periods (such as every year or every week).

2. It is often possible to observe a *trend* component, a *cyclical* component, and a *seasonal* component in time series data.

3. The level of economic activity undergoes cyclical variations. A slowdown in the overall level of economic activity is called a *recession*.

4. The *index of leading indicators* is a time series consisting of a weighted average of several economic variables that tend to turn up or down before the whole economy changes direction. This index can sometimes be used to forecast future movement in the economy.

• DETERMINING THE TREND BY CALCULATING MOVING AVERAGES

The first method for calculating the trend component is the moving average method. Suppose we're trying to see whether the value of real GDP in 1982 was higher or lower than the trend value. Let's calculate the average real GDP for the 5 years closest to 1982 (1980, 1981, 1982, 1983, and 1984). This type of calculation is called a moving average (because the periods you average together "move" when you move your attention to a new period).

The table shows that the value of real GDP in 1982 was 4,623.6, but the moving average about this value was 4,781.7. Therefore, the actual real GDP in 1982 was below its trend value. This was a year when the economy was in a severe recession. We can calculate the ratio 4,623.6/4,781.7 to find that the 1982 real GDP was 96.7 percent of the trend value.

Year	Real GDP	5-Year Moving Average	Ratio
1969	3,388.0	—	—
1970	3,388.2	—	—
1971	3,500.1	3,573.8	0.979
1972	3,690.3	3,673.8	1.004
1973	3,902.3	3,769.2	1.035
1974	3,888.2	3,885.4	1.001
1975	3,865.1	4,003.2	0.966

(continued)

Year	Real GDP	5-Year Moving Average	Ratio
1976	4,081.0	4,121.5	0.990
1977	4,279.3	4,268.6	1.003
1978	4,493.7	4,418.0	1.017
1979	4,624.0	4,546.8	1.017
1980	4,611.9	4,615.6	0.999
1981	4,724.9	4,678.9	1.010
1982	4,623.6	4,781.7	0.967
1983	4,810.0	4,925.2	0.977
1984	5,138.2	5,078.2	1.012
1985	5,329.5	5,283.2	1.009
1986	5,489.9	5,493.8	0.999
1987	5,648.4	5,678.2	0.995
1988	5,862.9	5,840.1	1.004
1989	6,060.4	5,957.9	1.017
1990	6,138.7	6,077.1	1.010
1991	6,079.0	6,181.3	0.983
1992	6,244.4	6,290.0	0.993
1993	6,383.8	6,410.4	0.996
1994	6,604.2	—	—
1995	6,740.8	—	—

Note that we cannot calculate the moving average for the first 2 years or the last 2 years. You may be wondering how many observations you should put in each moving average. We chose five observations. If you choose too short a period, then the moving average itself will be affected by the cyclical and irregular variations and it will be harder to identify these components. If you choose too long a period, then you lose too many data at the beginning and end of the series.

We can draw a graph to compare the actual value of the GDP with its trend value. (See Figure 18-2).

To investigate the cyclical component we would like to observe the deviations of the actual values from the trend values. There are two ways to do this.

We might think we have an *additive* model, in which case the actual value of real GDP is determined by a formula like this:

GDP = trend value + cyclical component + irregular component

Then we calculate the deviation from the trend by subtracting the trend value from the actual value.

We would use the other method if we think we have a multiplicative model, in which case the actual value is determined by a formula like this:

GDP = trend value × cyclical component × irregular component

In this case the cyclical and irregular components are represented by numbers whose average value is 1, and we calculate the deviation by dividing the actual value by the trend value: (See the previous table).

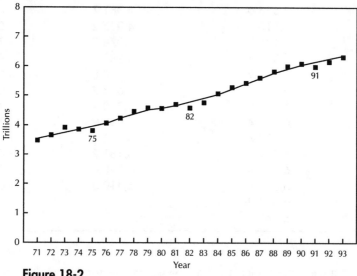

Figure 18-2

We can plot these ratios on a graph. (See Figure 18-3.) We can clearly see that the recessions in 1982 and 1975 caused the value of real GDP to be below its trend value. All of the movement in Figure 18-3 is attributed to either cyclical factors or irregular factors. There are more advanced techniques (known as *Fourier analysis*) that make it possible to analyze the cyclical components in a time series, but we will not discuss them in this book.

• *DETERMINING THE TREND BY REGRESSION*

It is also possible to use regression to determine the trend component in a time series. Consider this time series of yearly pizza sales in a small town:

50, 66, 81, 90, 98,
106, 115, 130, 146, 162,
177, 186, 194, 202, 211,
226, 242, 258, 273, 282,
290, 298, 307, 322, 338

Figure 18-4 shows a graph of these sales. There is a clear upward trend, and there also is a clear cyclical component. Apparently people in this town go on pizza binges that last about 4 years, but then they become tired of pizza and the growth of pizza consumption slows for the next 4 years. We can determine

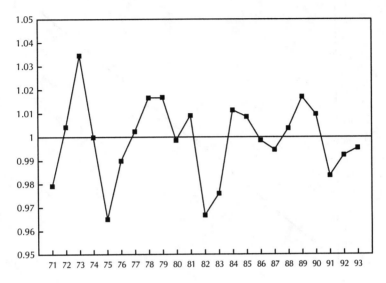

Figure 18-3

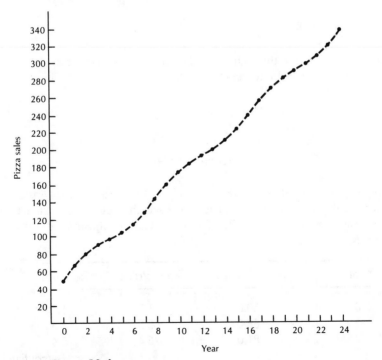

Figure 18-4

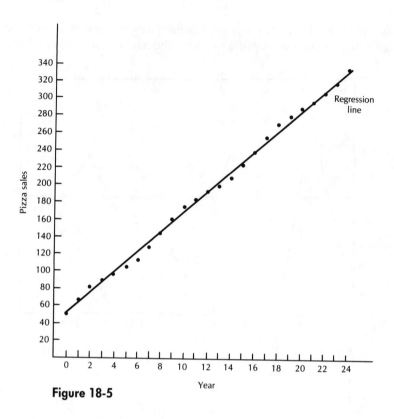

Figure 18-5

the trend by fitting a line through all of the data points (see Figure 18-5), and we can find the slope and intercept of the line by simple regression:

Trend line:

slope = 11.862

intercept = 51.662

Note that on Figures 18-4 and 18-5 we call the first year "year 0," instead of using the actual year number, in order to make the regression calculations easier. We can now make a graph of the residuals to illustrate the cyclical variation (Figure 18-6):

Year	Actual Sales	Trend Value	Residual
0	50.0	51.7	−1.7
1	66.0	63.5	2.5
2	81.0	75.4	5.6

(continued)

Year	Actual Sales	Trend Value	Residual
3	90.0	87.2	2.8
4	98.0	99.1	−1.1
5	106.0	111.0	−5.0
6	115.0	122.8	−7.8
7	130.0	134.7	−4.7
8	146.0	146.6	−0.6
9	162.0	158.4	3.6
10	177.0	170.3	6.7
11	186.0	182.1	3.9
12	194.0	194.0	0.0
13	202.0	205.9	−3.9
14	211.0	217.7	−6.7
15	226.0	229.6	−3.6
16	242.0	241.5	0.6
17	258.0	253.3	4.7
18	273.0	265.2	7.8
19	282.0	277.0	5.0
20	290.0	288.9	1.1
21	298.0	300.8	−2.8
22	307.0	312.6	−5.6
23	322.0	324.5	−2.5
24	338.0	336.3	1.7

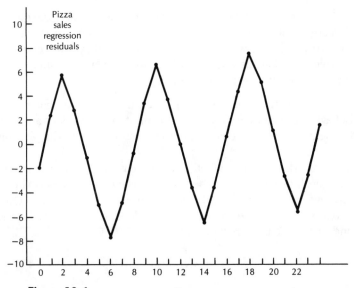

Figure 18-6   Year

EXPONENTIAL SMOOTHING

We will look at one more technique for forecasting time series values: **exponential smoothing**. For our example we will look at information on the saving rate (the percent of personal disposable income that is saved):

Year	Saving	Year	Saving
1981	9.1	1989	4.8
1982	8.8	1990	5.0
1983	6.6	1991	5.7
1984	8.4	1992	5.9
1985	6.9	1993	4.5
1986	6.2	1994	3.8
1987	5.0	1995	4.5
1988	5.2		

Source: Economic Report of the President, 1996.

Our goal is to forecast next year's saving rate by using the known past values of the saving rate. Suppose that we observe that this year's saving rate is higher than the rate we had forecasted for this year. There are two ways we could proceed:

1. We could assume that this year's saving rate was an abnormally high value. Perhaps something very unusual happened this year that caused people to save more, but we expect that next year they will return to normal behavior. In that case we should ignore the value for this year when we develop the forecast for next year.

2. We could assume that there has been a permanent change in people's behavior. In that case we will ignore all of the past values and our forecast for next year will be exactly the same as the actual value for this year.

In reality we do not know whether an increase in the saving rate in any one year is the result of a temporary aberration or is the start of a permanent change. Therefore, we will compromise between the two extreme plans. Our forecast of next year's value of the saving rate will depend both on this year's actual value and on this year's forecasted value:

$$F_{t+1} = aX_t + (1 - a)F_t$$

where X_t represents the actual value of the variable in
time period t (this year)
F_t represents our forecasted value in time period t
F_{t+1} represents our forecasted value for
period $t + 1$ (next year)
a is a number between 0 and 1

We can choose the value of a. Obviously we want to choose the value that will
lead to the best forecasts. Setting a equal to 1 means that our forecasted value
will always be equal to last period's actual value. This means that we are ignoring
all past values except the most recent value. Setting a equal to 0 means that
our forecasted value will always be equal to last year's forecasted value. In other
words, we will never change our forecast, no matter what happens in the real
world. We will regard each deviation from our forecasted value as being a
temporary random deviation.

Setting a value of a between 0 and 1 allows us to compromise between these
two extremes. For example, here are the results for the savings example, using
a value of $a = .5$:

Year	Actual	Forecast
1981	9.1	9.10
1982	8.8	9.10
1983	6.6	8.95
1984	8.4	7.78
1985	6.9	8.09
1986	6.2	7.49
1987	5.0	6.85
1988	5.2	5.92
1989	4.8	5.56
1990	5.0	5.18
1991	5.7	5.09
1992	5.9	5.40
1993	4.5	5.65
1994	3.8	5.07
1995	4.5	4.44

For the first year of the process (1981) we do not have any past forecasted
value, so we set the forecast for that year equal to the actual value. Note that
so far we have not really "forecasted" anything, since all we have done is
generate forecasted values for past years. However, we can now carry the process

forward 1 year at a time. Every year we can learn the new actual value of the saving rate, and then we can calculate the new forecasted value for the following year.

With a little mathematical manipulation we can show that the forecasted value can also be found from this equation:

$$F_{t+1} = aX_t + a(1 - a)X_{t-1} + a(1 - a)^2 X_{t-2} + a(1 - a)^3 X_{t-3} + \cdots$$

This year's forecasted value is equal to a weighted average of the past values. The weight attached to values in the recent past is greater than the weight attached to values from farther back.

One question remains: How do we choose the value of a? If our time series exhibits many random year-to-year fluctuations, then we should choose a relatively small value of a. In that case we don't want our forecasts to be too dependent on transitory swings in the variable. If long-term changes are more important, then we should choose a larger value of a so that the forecasted value will react more quickly to observed changes. In practice we can test several different values of a. Look at the past data and determine the error for each year. Then square all of the errors and add up all of the squares. Choose the value of a that leads to the smallest total squared error.

YOU SHOULD REMEMBER

1. The *trend component* in a time series is the component that increases or decreases smoothly with time. This component can be identified by using exponential smoothing, the moving average method, or regression analysis with time as the independent variable.

2. The *moving average* is the average value of the observations in a time series that are closest to one particular value. For example, a three-term moving average can be calculated that consists of the average value of one observation, its predecessor, and its successor.

SEASONAL ADJUSTMENT

• *THE NEED FOR SEASONAL ADJUSTMENT*

Suppose you are analyzing monthly time series data for retail toy sales. You will find that every year there is a big surge in toy sales in December. Or, if you are analyzing time series for construction, you will probably find that construction tends to be higher in the summer. Many other economic variables, such as employment, labor force participation, money demand, and heating oil demand, exhibit significant seasonal variation. It is not hard to understand why these variations occur, but they do make it more difficult to analyze time series data. If you observe that toy sales have increased in December, you need to investigate to determine whether there has been a real increase in demand for toys or whether sales have gone up just because they always go up in December. It is also possible that the toy business is in a slump. Sales may have gone up relative to what they were in November, but they may have risen less than you normally would have expected them to in December.

When time series data are used to analyze trends in the economy, it is necessary to separate out the seasonal variations. This process is known as **seasonal adjustment**. Many of the widely published economic statistics, such as the unemployment rate, are seasonally adjusted before they are published. For example, here is the 1995 monthly U.S. employment figures, in both seasonally adjusted and not seasonally adjusted form:

	Not Adjusted	Seasonally Adjusted
Jan	122.6	124.6
Feb	123.3	125.1
Mar	123.9	125.3
Apr	124.3	125.1
May	124.6	124.3
Jun	125.7	124.5
Jly	126.5	125.0
Aug	125.9	124.8
Sep	125.2	125.1
Oct	126.0	125.4
Nov	125.6	125.0
Dec	125.1	124.9

The actual (unadjusted) employment figures are highest in the summer. In each such case the adjusted figure is lower because it compensates for this seasonal effect. During the early months of the year, actual employment tends to be less than it is the rest of the year, so the adjusted figure is higher than the actual figure.

• *THE RATIO TO MOVING AVERAGE METHOD*

Calculating the seasonal adjustment factors is difficult, and the actual procedure used by the Bureau of Labor Statistics is very complicated. We will look at a relatively simple method of analyzing seasonal variation called the

ratio to moving average method. Here is a table for 5 years of monthly observations for employment:

	Employment (in millions)				
	1991	1992	1993	1994	1995
Jan	115.8	115.9	117.0	119.9	122.6
Feb	115.9	116.0	117.6	120.5	123.3
Mar	116.4	116.9	118.4	120.8	123.9
Apr	117.5	117.8	118.8	121.6	124.3
May	117.4	118.4	120.1	122.9	124.6
Jun	119.2	119.8	121.6	123.9	125.7
Jly	120.0	120.7	122.4	124.5	126.6
Aug	118.8	120.1	122.0	124.5	125.9
Sep	118.2	118.9	120.7	123.8	125.2
Oct	118.4	119.2	121.4	124.7	126.0
Nov	118.0	119.2	121.6	124.9	125.6
Dec	117.4	119.0	121.6	124.7	125.1

Source: Employment and Earnings, various years.

To estimate the effect of the seasonal variations we will first calculate a 12-period moving average for each figure. It is trickier to calculate a moving average with an even number of periods. For each entry we will calculate two moving averages:

- The average of the six preceding elements and the five following elements

- The average of the five preceding elements and the six following elements

For example, for July 1991 we first average all values from January 1991 to December 1991 (result: 117.7500); then we average all values from February 1991 to January 1992 (result: 117.7583). The value for the moving average for July is the average of these two values: 117.7542.

Here is a table of the moving averages for the employment data:

	1991	1992	1993	1994	1995
Jan	—	118.00	119.29	121.70	124.38
Feb	—	118.08	119.44	121.89	124.53
Mar	—	118.16	119.59	122.12	124.64

(continued)

	1991	1992	1993	1994	1995
Apr	—	118.23	119.76	122.39	124.75
May	—	118.31	119.95	122.66	124.84
Jun	—	118.43	120.16	122.93	124.88
Jly	117.75	118.54	120.39	123.17	—
Aug	117.76	118.65	120.63	123.40	—
Sep	117.79	118.78	120.85	123.65	—
Oct	117.82	118.88	121.07	123.89	—
Nov	117.88	119.00	121.30	124.07	—
Dec	117.94	119.14	121.51	124.22	—

Note that we cannot calculate a moving average for the first six values or the last six values.

Now we can calculate the ratio between each actual value and the corresponding moving average value. In this way we can see which months are lower or higher than the average for the 12-month period surrounding the month in question. We can call this ratio the **monthly trend ratio**.

	1991	1992	1993	1994	1995	Average	Standardized Average
Jan	—	0.9822	0.9808	0.9852	0.9857	0.9835	0.9835
Feb	—	0.9824	0.9846	0.9886	0.9902	0.9864	0.9865
Mar	—	0.9893	0.9900	0.9892	0.9940	0.9906	0.9907
Apr	—	0.9964	0.9920	0.9936	0.9964	0.9946	0.9946
May	—	1.0008	1.0013	1.0019	0.9981	1.0005	1.0005
Jun	—	1.0116	1.0120	1.0079	1.0065	1.0095	1.0095
Jly	1.0191	1.0182	1.0167	1.0108	—	1.0162	1.0162
Aug	1.0088	1.0122	1.0114	1.0089	—	1.0103	1.0103
Sep	1.0035	1.0010	0.9988	1.0012	—	1.0011	1.0011
Oct	1.0049	1.0027	1.0028	1.0066	—	1.0042	1.0042
Nov	1.0011	1.0017	1.0025	1.0067	—	1.0030	1.0030
Dec	0.9954	0.9988	1.0007	1.0039	—	0.9997	0.9997
					Sum:	11.99978	12

If there is a regular seasonal variation, then a month that is below its moving average value one year will tend to be lower than its moving average value every year. For example, the monthly trend ratios for January are all below 0.9857, and the monthly trend ratios for July are all above 1.0108.

Therefore, we will calculate the average monthly trend ratio for each month. These figures, shown in the column headed Average, will form the basis for our seasonal adjustment factor. If we add up the total of these figures, we obtain 11.99978. It would be convenient to have our seasonal factors add up to 12, so we will multiply each value by 12/11.99978. The final results, in the column headed Standardized Average, are the seasonal adjustment factors we will use. For example, the actual figure for July 1995 is 126.6. The seasonal adjustment factor for July is 1.0162, so the seasonally adjusted value for July 1995 is 126.6/1.0162 = 124.6. Here is a complete table of the seasonally adjusted values:

	1991	1992	1993	1994	1995
Jan	117.7	117.8	119.0	121.9	124.7
Feb	117.5	117.6	119.2	122.2	125.0
Mar	117.5	118.0	119.5	121.9	125.1
Apr	118.1	118.4	119.4	122.3	125.0
May	117.3	118.3	120.0	122.8	124.5
Jun	118.1	118.7	120.5	122.7	124.5
Jly	118.1	118.8	120.4	122.5	124.6
Aug	117.6	118.9	120.8	123.2	124.6
Sep	118.1	118.8	120.6	123.7	125.1
Oct	117.9	118.7	120.9	124.2	125.5
Nov	117.6	118.8	121.2	124.5	125.2
Dec	117.4	119.0	121.6	124.7	125.1

Figure 18-7 compares the seasonally adjusted data with the original data. The bottom part of the figure clearly shows how the actual employment peaks every summer and falls every January. The seasonally adjusted data averages out the peaks and valleys in the actual data.

However, the bottom figure greatly exaggerates the amount of the seasonal change. That is because the vertical scale of that figure starts near the lowest point in the data and ends near the highest point. You can always distort the magnitude of any shift in your data by drawing the diagram in this fashion. The top part of Figure 18-7 shows the same data with the vertical scale starting at 0. This way you have a non-distorted view of how big the seasonal variation really is. In this diagram you can see there is not much difference between the seasonally adjusted data and the non seasonally adjusted data. If you look closely you can still see the actual data peaking in the summer. The exercises give some examples where there is much larger seasonal variation.

YOU SHOULD REMEMBER

1. The *seasonal component* in a time series is the component that varies on a regular basis throughout the seasons of the year. The seasonal adjustment process is used to compensate for the regular seasonal variations in the observations to make nonseasonal changes more clearly apparent.

2. The *ratio to moving average method* is one possible seasonal adjustment method.

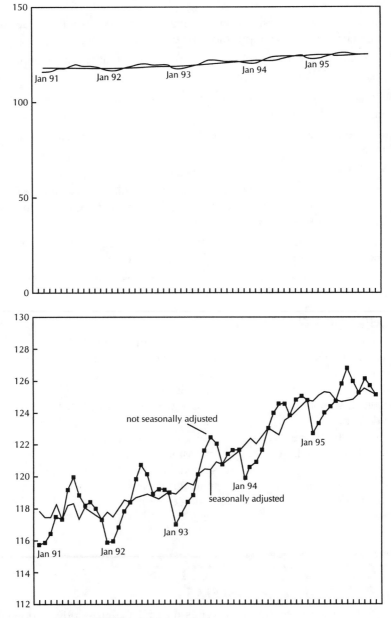

Figure 18-7. Employment (millions)

KNOW THE CONCEPTS

DO YOU KNOW THE BASICS?

Test your understanding of Chapter 18 by answering the following questions:

1. Why is the gross domestic product calculated by looking at the value added instead of the total value produced by each firm?
2. When a used car dealer sells a car, is the cost of that car included in GDP?
3. How is real GDP calculated?
4. Will the inflation rate be higher if it is measured by the GDP deflator or if it is measured by the consumer price index?
5. How are the values of government services measured when GDP is being calculated?
6. What is the purpose of seasonal adjustment?
7. If the actual value of a statistic for January is higher than the seasonally adjusted value, would we expect this statistic normally to be higher in January than it is the rest of the year?
8. When you analyze a trend with the moving average method, how many periods should you include in each moving average?
9. How do you choose the value of a, the constant used in exponential smoothing?

TERMS FOR STUDY

base year	net national product (NNP)
consumer price index (CPI)	nominal gross national product
cyclical component	price index
depreciation	price level
exponential smoothing	producer price index (PPI)
GDP deflator	ratio to moving average method
gross domestic product (GDP)	real gross national product
index of leading indicators	recession
inflation rate	seasonal adjustment
irregular component	time series data
moving average	trend component
national income	

PRACTICAL APPLICATION
COMPUTATIONAL PROBLEMS

1. The following lists present time series data for (a) gross private domestic investment, (b) personal consumption expenditures for services (in billions

of dollars), (c) total population (in thousands), (d) total employment (in thousands), (e) industrial production index, (f) the consumer price index (1982–84 = 100), and (g) the money supply narrowly defined (call M_1, measured in billions of dollars). For each list calculate a trend in two ways: with three-term moving averages and with simple linear regression. For each method calculate the ratio between each actual value and the trend value.

(The data for these exercises come from the Internet World Wide Web sites for the Federal Reserve Bank of St. Louis and the Bureau of Labor Statistics.)

Year	(a) Invest.	(b) Services	(c) Population	(d) Employment	(e) Production	(f) CPI	(g) M_1
1981	556.2	1,696.1	229,966	100,397	85.7	90.9	436.3
1982	501.1	1,728.2	232,188	99,526	81.9	96.5	474.3
1983	547.1	1,809.0	234,307	100,834	84.9	99.6	521.0
1984	715.6	1,883.0	236,348	105,005	92.8	103.9	552.1
1985	715.1	1,977.3	238,466	107,150	94.4	107.6	619.8
1986	722.5	2,041.4	240,651	109,597	95.3	109.6	724.4
1987	747.2	2,126.9	242,804	112,440	100.0	113.6	749.8
1988	773.9	2,212.4	245,021	114,968	104.4	118.3	786.9
1989	829.2	2,262.3	247,342	117,342	106.0	124.0	794.2
1990	799.7	2,321.3	249,913	117,914	106.0	130.7	825.8
1991	736.2	2,341.0	252,650	116,877	104.2	136.2	897.3
1992	790.4	2,409.4	255,419	117,598	107.7	140.3	1,024.4
1993	871.1	2,466.8	258,137	119,306	111.5	144.5	1,128.6
1994	1,014.4	2,519.4	260,660	123,060	118.1	148.2	1,148.0
1995	1,067.5	2,575.7	263,034	124,900	121.9	152.4	1,123.0

For each exercise below, you are given monthly time series data for an interesting economic quantity for 5 years. Use the ratio to moving average method to calculate seasonal adjustment factors for each series. Then use your spreadsheet or statistical program to graph the original data along with the seasonally adjusted data, and briefly comment on the nature of the seasonal variation for that quantity.

2.	Employment of 16- to 19-year-olds (thousands)				
	1991	1992	1993	1994	1995
Jan	5,488	5,173	5,063	5,507	5,758
Feb	5,539	5,106	5,160	5,511	5,734
Mar	5,577	5,035	5,155	5,499	5,959
Apr	5,634	5,180	5,230	5,645	5,956
May	5,663	5,359	5,564	5,902	6,147

(continued)

Employment of 16- to 19-year-olds (thousands)				
	1991	1992	1993	1994	1995
Jun	6,728	6,296	6,515	7,010	7,372
Jly	7,340	7,191	7,434	7,698	7,965
Aug	6,816	6,845	7,029	7,355	7,569
Sep	5,526	5,440	5,613	5,800	6,170
Oct	5,581	5,466	5,576	5,957	6,154
Nov	5,538	5,421	5,650	5,990	6,121
Dec	5,441	5,511	5,668	6,062	6,125

3.

Unemployment rate (percent)				
	1991	1992	1993	1994	1995
Jan	7.1	8.1	8.0	7.3	6.2
Feb	7.3	8.2	7.8	7.1	5.9
Mar	7.2	7.8	7.4	6.8	5.7
Apr	6.5	7.2	6.9	6.2	5.6
May	6.7	7.3	6.8	5.9	5.5
Jun	7.0	8.0	7.2	6.2	5.8
Jly	6.8	7.7	7.0	6.2	5.9
Aug	6.6	7.4	6.6	5.9	5.6
Sep	6.5	7.3	6.4	5.6	5.4
Oct	6.5	6.9	6.4	5.4	5.2
Nov	6.7	7.1	6.2	5.3	5.3
Dec	6.9	7.1	6.1	5.1	5.2

4.

Currency in circulation (billion dollars)				
	1991	1992	1993	1994	1995
Jan	249.7	267.7	293.5	324.0	355.8
Feb	252.6	269.4	295.3	327.2	357.0
Mar	255.5	271.0	297.9	330.6	361.3
Apr	256.0	273.3	301.3	334.3	365.5
May	257.3	275.7	304.4	337.2	367.9
Jun	259.1	277.2	307.4	340.5	368.2
Jly	260.8	280.8	311.0	344.7	369.1
Aug	262.0	282.9	312.8	345.6	369.1
Sep	261.7	284.6	314.7	347.0	369.3
Oct	263.1	287.0	317.3	349.5	370.0
Nov	266.3	290.0	319.8	353.2	371.7
Dec	269.9	295.0	324.8	357.5	376.1

5. Workers with a job but not at work due to bad weather (thousands)

	1991	1992	1993	1994	1995
Jan	453	368	651	511	386
Feb	194	388	245	616	259
Mar	207	136	327	187	168
Apr	118	85	162	59	99
May	90	11	56	77	137
Jun	37	127	30	55	26
Jly	41	43	93	47	36
Aug	36	53	28	59	42
Sep	30	42	33	19	35
Oct	19	16	40	118	58
Nov	79	64	68	94	76
Dec	121	198	108	140	149

ANSWERS

KNOW THE CONCEPTS

1. The value added is used to avoid double-counting intermediate goods.
2. The value of the car itself is not included, because that value was included in GDP in the year the car was produced. The profit of the car dealer is included, however, because that is a payment for a current service.
3. Real GDP is calculated with the chain-weight method; it indicates the amount of output produced after removing the effect of inflation.
4. In general, it is impossible to say.
5. Government services are valued at their costs of production.
6. The purpose of seasonal adjustment is to make changes arising from non-seasonal factors more apparent.
7. This statistic is usually higher in January than it is the rest of the year; its seasonal adjustment factor is greater than 1.
8. There is a trade-off. If you include too few periods, then the trend will not be clear. If you include too many periods, then the average will be less sensitive to each individual item and you will lose too many observations at the beginning and end.
9. You choose the value that seems to have provided the best forecasts in the past. A smaller value is appropriate if there are many random yearly fluctuations.

PRACTICAL APPLICATION

1. (a) Regression results for investment:
  slope = 31.706;   intercept = 537.202; $r^2 = 0.8301$

Year	Actual	Moving Average		Regression	
		Trend	Ratio	Trend	Ratio
1981	556.2	—	—	537.2	1.0354
1982	501.1	534.8	0.9370	568.9	0.8808
1983	547.1	587.9	0.9305	600.6	0.9109
1984	715.6	659.3	1.0854	632.3	1.1317
1985	715.1	717.7	0.9963	664.0	1.0769
1986	722.5	728.3	0.9921	695.7	1.0385
1987	747.2	747.9	0.9991	727.4	1.0272
1988	773.9	783.4	0.9878	759.1	1.0194
1989	829.2	800.9	1.0353	790.9	1.0485
1990	799.7	788.4	1.0144	822.6	0.9722
1991	736.2	775.4	0.9494	854.3	0.8618
1992	790.4	799.2	0.9889	886.0	0.8921
1993	871.1	892.0	0.9766	917.7	0.9492
1994	1014.4	984.3	1.0305	949.4	1.0685
1995	1067.5	—	—	981.1	1.0881

(b) Regression results for services:
  slope = 64.590;   intercept = 1705.881; $r^2 = 0.9896$

Year	Actual	Moving Average		Regression	
		Trend	Ratio	Trend	Ratio
1981	1,696.1	—	—	1705.9	0.9943
1982	1,728.2	1744.4	0.9907	1770.5	0.9761
1983	1,809.0	1806.7	1.0013	1835.1	0.9858
1984	1,883.0	1889.8	0.9964	1899.7	0.9912
1985	1,977.3	1967.2	1.0051	1964.2	1.0066
1986	2,041.4	2048.5	0.9965	2028.8	1.0062
1987	2,126.9	2126.9	1.0000	2093.4	1.0160
1988	2,212.4	2200.5	1.0054	2158.0	1.0252
1989	2,262.3	2265.3	0.9987	2222.6	1.0179
1990	2,321.3	2308.2	1.0057	2287.2	1.0149
1991	2,341.0	2357.2	0.9931	2351.8	0.9954
1992	2,409.4	2405.7	1.0015	2416.4	0.9971
1993	2,466.8	2465.2	1.0006	2481.0	0.9943
1994	2,519.4	2520.6	0.9995	2545.6	0.9897
1995	2,575.7	—	—	2610.1	0.9868

(c) Regression results for population:
  slope = 2,369.129; intercept = 229,209.8; $r^2 = 0.997$

Year	Actual	Moving Average		Regression	
		Trend	Ratio	Trend	Ratio
1981	229,966	—	—	229,210	1.0033
1982	232,188	232,154	1.0001	231,579	1.0026
1983	234,307	234,281	1.0001	233,948	1.0015
1984	236,348	236,374	0.9999	236,317	1.0001
1985	238,466	238,488	0.9999	238,686	0.9991
1986	240,651	240,640	1.0000	241,055	0.9983
1987	242,804	242,825	0.9999	243,425	0.9975
1988	245,021	245,056	0.9999	245,794	0.9969
1989	247,342	247,425	0.9997	248,163	0.9967
1990	249,913	249,968	0.9998	250,532	0.9975
1991	252,650	252,661	1.0000	252,901	0.9990
1992	255,419	255,402	1.0001	255,270	1.0006
1993	258,137	258,072	1.0003	257,639	1.0019
1994	260,660	260,610	1.0002	260,009	1.0025
1995	263,034	—	—	262,378	1.0025

(d) Regression results for employment:
  slope = 1,807.764; intercept = 99,806.58; $r^2 = 0.953469$

Year	Actual	Moving Average		Regression	
		Trend	Ratio	Trend	Ratio
1981	100,397	—	—	99,807	1.0059
1982	99,526	100,252	0.9928	101,614	0.9794
1983	100,834	101,788	0.9906	103,422	0.9750
1984	105,005	104,330	1.0065	105,230	0.9979
1985	107,150	107,251	0.9991	107,038	1.0010
1986	109,597	109,729	0.9988	108,845	1.0069
1987	112,440	112,335	1.0009	110,653	1.0161
1988	114,968	114,917	1.0004	112,461	1.0223
1989	117,342	116,741	1.0051	114,269	1.0269
1990	117,914	117,378	1.0046	116,076	1.0158
1991	116,877	117,463	0.9950	117,884	0.9915
1992	117,598	117,927	0.9972	119,692	0.9825
1993	119,306	119,988	0.9943	121,500	0.9819
1994	123,060	122,422	1.0052	123,308	0.9980
1995	124,900	—	—	125,115	0.9983

(e) Regression results for production:
   slope = 2.571;   intercept = 82.987; $r^2 = 0.9453$

		Moving Average		Regression	
Year	Actual	Trend	Ratio	Trend	Ratio
1981	85.7	—	—	83.0	1.0327
1982	81.9	84.2	0.9731	85.6	0.9572
1983	84.9	86.5	0.9811	88.1	0.9634
1984	92.8	90.7	1.0232	90.7	1.0231
1985	94.4	94.2	1.0025	93.3	1.0121
1986	95.3	96.6	0.9869	95.8	0.9943
1987	100.0	99.9	1.0010	98.4	1.0161
1988	104.4	103.5	1.0090	101.0	1.0338
1989	106.0	105.5	1.0051	103.6	1.0236
1990	106.0	105.4	1.0057	106.1	0.9988
1991	104.2	106.0	0.9833	108.7	0.9586
1992	107.7	107.8	0.9991	111.3	0.9679
1993	111.5	112.4	0.9917	113.8	0.9794
1994	118.1	117.2	1.0080	116.4	1.0145
1995	121.9	—	—	119.0	1.0245

(f) Regression results for CPI:
   slope = 4.461;   intercept = 89.857; $r^2 = 0.9930$

		Moving Average		Regression	
Year	Actual	Trend	Ratio	Trend	Ratio
1981	90.9	—	—	89.9	1.0116
1982	96.5	95.7	1.0087	94.3	1.0231
1983	99.6	100.0	0.9960	98.8	1.0083
1984	103.9	103.7	1.0019	103.2	1.0064
1985	107.6	107.0	1.0053	107.7	0.9990
1986	109.6	110.3	0.9940	112.2	0.9771
1987	113.6	113.8	0.9980	116.6	0.9741
1988	118.3	118.6	0.9972	121.1	0.9770
1989	124.0	124.3	0.9973	125.5	0.9877
1990	130.7	130.3	1.0031	130.0	1.0053
1991	136.2	135.7	1.0034	134.5	1.0129
1992	140.3	140.3	0.9998	138.9	1.0098
1993	144.5	144.3	1.0012	143.4	1.0077
1994	148.2	148.4	0.9989	147.9	1.0023
1995	152.4	—	—	152.3	1.0005

(g) Regression results for M1:
    slope = 53.057;   intercept = 415.660; r^2 = = 0.9732

Year	Actual	Moving Average		Regression	
		Trend	Ratio	Trend	Ratio
1981	436.3	—	—	415.7	1.0497
1982	474.3	477.2	0.9939	468.7	1.0119
1983	521.0	515.8	1.0101	521.8	0.9985
1984	552.1	564.3	0.9784	574.8	0.9605
1985	619.8	632.1	0.9805	627.9	0.9871
1986	724.4	698.0	1.0378	680.9	1.0638
1987	749.8	753.7	0.9948	734.0	1.0215
1988	786.9	777.0	1.0128	787.1	0.9998
1989	794.2	802.3	0.9899	840.1	0.9453
1990	825.8	839.1	0.9841	893.2	0.9246
1991	897.3	915.8	0.9798	946.2	0.9483
1992	1024.4	1016.8	1.0075	999.3	1.0251
1993	1128.6	1100.3	1.0257	1052.3	1.0725
1994	1148.0	1133.2	1.0131	1105.4	1.0385
1995	1123.0	—	—	1158.5	0.9694

2. Employment, 16- to 19-year-olds, monthly trend ratio:

	1991	1992	1993	1994	1995	Avg.	St. Avg.
Jan	–	0.9087	0.8846	0.9156	0.9101	0.9047	0.9056
Feb	–	0.8977	0.8987	0.9125	0.9034	0.9031	0.9039
Mar	–	0.8856	0.8956	0.9073	0.9353	0.9059	0.9068
Apr	–	0.9124	0.9067	0.9278	0.9314	0.9196	0.9204
May	–	0.9455	0.9623	0.9653	0.9592	0.9581	0.9589
Jun	–	1.1113	1.1236	1.1408	1.1489	1.1311	1.1322
Jly	1.2456	1.2696	1.2766	1.2473	–	1.2598	1.2609
Aug	1.1628	1.2090	1.2002	1.1879	–	1.1900	1.1911
Sep	0.9493	0.9596	0.9537	0.9325	–	0.9488	0.9496
Oct	0.9657	0.9630	0.9424	0.9528	–	0.9560	0.9568
Nov	0.9635	0.9533	0.9498	0.9545	–	0.9553	0.9562
Dec	0.9517	0.9661	0.9473	0.9621	–	0.9568	0.9577
					Sum:	11.989	12

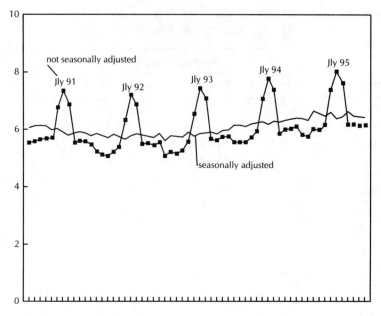

Figure 18-8. Employment, 16- to 19-year-olds (millions).

Seasonally Adjusted Data	1991	1992	1993	1994	1995
Jan	6,060	5,713	5,591	6,081	6,359
Feb	6,128	5,649	5,709	6,097	6,344
Mar	6,150	5,553	5,685	6,064	6,572
Apr	6,121	5,628	5,682	6,133	6,471
May	5,905	5,588	5,802	6,155	6,410
Jun	5,943	5,561	5,754	6,192	6,511
Jly	5,821	5,703	5,896	6,105	6,317
Aug	5,723	5,747	5,901	6,175	6,355
Sep	5,819	5,728	5,911	6,108	6,497
Oct	5,833	5,713	5,828	6,226	6,432
Nov	5,792	5,670	5,909	6,265	6,402
Dec	5,681	5,755	5,918	6,330	6,396

See Figure 18-8. The employment of teenagers is much higher in the summer than it is during the school year.

3. Unemployment rate, monthly trend ratio:

	1991	1992	1993	1994	1995	Avg.	St. Avg.
Jan	—	1.1166	1.1003	1.1260	1.0933	1.1090	1.1100
Feb	—	1.1195	1.0821	1.1058	1.0450	1.0881	1.0891
Mar	—	1.0552	1.0368	1.0695	1.0133	1.0437	1.0446
Apr	—	0.9675	0.9747	0.9867	0.9985	0.9819	0.9828
May	—	0.9766	0.9685	0.9510	0.9821	0.9696	0.9704
Jun	—	1.0667	1.0372	1.0122	1.0349	1.0378	1.0387
Jly	0.9915	1.0261	1.0188	1.0269	—	1.0158	1.0167
Aug	0.9514	0.9889	0.9688	0.9930	—	0.9755	0.9764
Sep	0.9286	0.9799	0.9470	0.9579	—	0.9533	0.9542
Oct	0.9214	0.9298	0.9546	0.9351	—	0.9352	0.9361
Nov	0.9426	0.9611	0.9341	0.9244	—	0.9405	0.9414
Dec	0.9617	0.9682	0.9301	0.8947	—	0.9387	0.9395
					Sum:	11.989	12

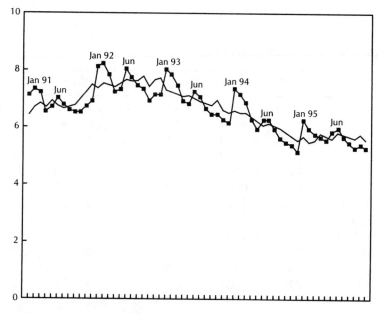

Figure 18-9. Unemployment rate (percent).

Seasonally Adjusted Data	1991	1992	1993	1994	1995
Jan	6.4	7.3	7.2	6.6	5.6
Feb	6.7	7.5	7.2	6.5	5.4
Mar	6.9	7.5	7.1	6.5	5.5
Apr	6.6	7.3	7.0	6.3	5.7
May	6.9	7.5	7.0	6.1	5.7
Jun	6.7	7.7	6.9	6.0	5.6
Jly	6.7	7.6	6.9	6.1	5.8
Aug	6.8	7.6	6.8	6.0	5.7
Sep	6.8	7.7	6.7	5.9	5.7
Oct	6.9	7.4	6.8	5.8	5.6
Nov	7.1	7.5	6.6	5.6	5.6
Dec	7.3	7.6	6.5	5.4	5.5

See Figure 18-9. The seasonal pattern in unemployment is more complicated; there is an increase in unemployment in January after the holiday season; there also is an increase in June when there is a large number of new workers.

4. Currency, monthly trend ratio:

	1991	1992	1993	1994	1995	Avg.	St. Avg.
Jan	—	0.9951	0.9963	0.9941	0.9963	0.9955	0.9955
Feb	—	0.9952	0.9939	0.9955	0.9941	0.9947	0.9947
Mar	—	0.9944	0.9943	0.9976	1.0008	0.9968	0.9968
Apr	—	0.9957	0.9973	1.0006	1.0074	1.0003	1.0003
May	—	0.9972	0.9993	1.0011	1.0095	1.0018	1.0018
Jun	—	0.9953	1.0009	1.0027	1.0061	1.0013	1.0013
Jly	1.0021	1.0006	1.0045	1.0071	—	1.0036	1.0036
Aug	1.0011	1.0004	1.0019	1.0022	—	1.0014	1.0014
Sep	0.9949	0.9987	0.9993	0.9990	—	0.9980	0.9980
Oct	0.9950	0.9991	0.9989	0.9988	—	0.9979	0.9980
Nov	1.0015	1.0013	0.9982	1.0020	—	1.0007	1.0007
Dec	1.0093	1.0100	1.0051	1.0072	—	1.0079	1.0079
					Sum:	12	12

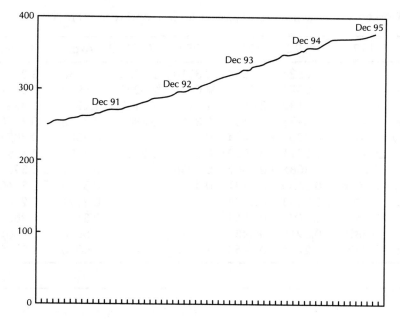

Figure 18-10. Currency (billions).

Seasonally Adjusted Data	1991	1992	1993	1994	1995
Jan	250.8	268.9	294.8	325.5	357.4
Feb	253.9	270.8	296.9	328.9	358.9
Mar	256.3	271.9	298.9	331.7	362.5
Apr	255.9	273.2	301.2	334.2	365.4
May	256.8	275.2	303.9	336.6	367.2
Jun	258.8	276.8	307.0	340.1	367.7
Jly	259.9	279.8	309.9	343.5	367.8
Aug	261.6	282.5	312.4	345.1	368.6
Sep	262.2	285.2	315.3	347.7	370.0
Oct	263.6	287.6	317.9	350.2	370.8
Nov	266.1	289.8	319.6	352.9	371.4
Dec	267.8	292.7	322.2	354.7	373.1

See Figure 18-10. There is very little seasonal variation, but there is a slight peak in December.

5. Workers not at work due to bad weather, monthly trend ratio:

	1991	1992	1993	1994	1995	Avg.	St. Avg.
Jan	—	3.0624	4.0858	3.3110	2.9952	3.3636	3.3597
Feb	—	3.2077	1.5277	4.0076	2.0281	2.6928	2.6897
Mar	—	1.1132	2.0571	1.2110	1.3159	1.4243	1.4227
Apr	—	0.6936	1.0151	0.3756	0.7868	0.7178	0.7170
May	—	0.0903	0.3484	0.4770	1.1176	0.5083	0.5077
Jun	—	1.0211	0.1909	0.3357	0.2128	0.4401	0.4396
Jly	0.3559	0.3085	0.6302	0.2938	—	0.3971	0.3966
Aug	0.3006	0.3650	0.1781	0.4218	—	0.3164	0.3160
Sep	0.2402	0.2853	0.1978	0.1530	—	0.2191	0.2188
Oct	0.1576	0.1010	0.2553	0.9434	—	0.3643	0.3639
Nov	0.6818	0.3915	0.4436	0.7273	—	0.5610	0.5604
Dec	1.0401	1.2276	0.6958	1.0725	—	1.0090	1.0078
					Sum:	12.014	12

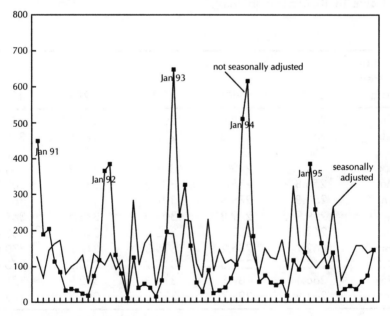

Figure 18-11. Workers not at work due to bad weather (thousands).

Seasonally Adjusted Data	1991	1992	1993	1994	1995
Jan	134.8	109.5	193.8	152.1	114.9
Feb	72.1	144.3	91.1	229.0	96.3
Mar	145.5	95.6	229.8	131.4	118.1
Apr	164.6	118.6	226.0	82.3	138.1
May	177.3	21.7	110.3	151.7	269.8
Jun	84.2	288.9	68.2	125.1	59.1
Jly	103.4	108.4	234.5	118.5	90.8
Aug	113.9	167.7	88.6	186.7	132.9
Sep	137.1	191.9	150.8	86.8	159.9
Oct	52.2	44.0	109.9	324.3	159.4
Nov	141.0	114.2	121.3	167.7	135.6
Dec	120.1	196.5	107.2	138.9	147.8

See Figure 18-11. There is a lot of random variation in this series, but the highest figures are in winter, as you would expect.

19
DECISION THEORY

KEY TERMS

decision theory the study of making decisions designed to achieve some objective, often under conditions of uncertainty

decision tree a diagram that illustrates all possible consequences of different decisions in different states of nature

expected payoff the expected value of the payoff resulting from a decision

Every day of our lives, we are faced with many decisions. Many of them involve outcomes where the difference between the possible results is inconsequential. If I use the "wrong" toothpaste, my teeth may not be as white. If I put my shoes on before my socks in the morning, I waste several minutes in correcting my error, and risk stretching my socks. In business, decisions usually have more important consequences. A company choosing one manufacturing process over another may go bankrupt if the choice was wrong. Thus it is important to come up with some rules, based on **decision theory**, that will help us to make the best decision possible, given the information that we have.

Statisticians distinguish between three different kinds of decisions:

1. Decisions under certainty. These are decisions where you have all the information you need and can calculate precisely the outcome of every choice you make. This doesn't necessarily mean that you know exactly what to do: there may be more data than a person can comprehend. For example, you may wish to maximize the production of a certain chemical while simultaneously satisfying 531 related Environmental Protection Agency regulations. In cases like this, it is usually best to use an appropriate algorithm, such as linear programming to evaluate the data and select the best outcome. You will often need a computer program to perform the calculations for you. We shall not deal with this category of decision here.

2. Decisions under uncertainty. These are decisions where you have to take chance into account, without worrying about what a competitor may be

doing, just you and Nature. Here, intuition about what random chance may be up to comes into play. A farmer may have to choose what crops to grow on the basis of what he thinks the weather will be like. He will probably base his decision on past weather and long-range forecasting.

3. Decisions under conflict. These are decisions where you have to take into account what a competitor may do. A company has to consider what its competitors may charge for an item when deciding its price. This category of decisions is dealt with by game theory and will not be discussed here.

DECISION TREE

Before making an intelligent decision, it is a good idea to have all of the possible choices and their outcomes charted for you. One way of doing this is to use a **decision tree**.

Suppose that you want to manufacture an item and have k different processes, $A_1, \ldots, A_k$, to choose from. Then you start out going from left to right with k branches (see Figure 19-1).

For each of the processes, there are different possible outcomes that we call **states of nature**. A state of nature is a chance result, usually unpredictable beforehand, such as the number rolled on a die or a particular day's weather.

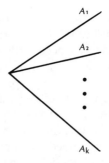

Figure 19-1

We'll call the states of nature $B_1, \ldots, B_l$. For example B_1 might be the outcome where the item came out perfect, B_2 the outcome where the item came out flawed, B_3 the outcome where the plant blew up, and so on. These are depicted as additional branches (see Figure 19-2).

Now you may have further decisions, $C_1, \ldots, C_m$, to make as a result of these outcomes, for example, to improve the process, rebuild the factory, or declare bankruptcy. These can be shown as more branches (see Figure 19-3).

The branching can be carried out as far as you wish, depending on how many levels you wish to analyze. The branches alternate between your choices and states of nature.

YOU SHOULD REMEMBER

1. Many business decisions must be made in environments of uncertainty. The field of decision theory provides guidance in such cases.

2. A decision tree is a diagram listing all possible outcomes for each possible decision and each possible state of nature.

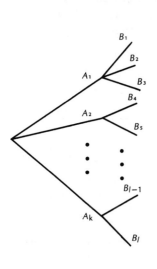

Figure 19-2

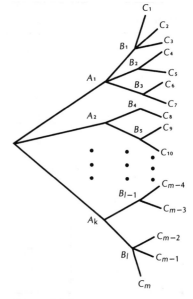

Figure 19-3

OBJECTIVE VARIABLES

We need to have some way of deciding which outcomes are better than others. A yo-yo manufacturer may think it best to make blue yo-yos because they sell better than yo-yos of other colors, or because blue is her favorite color.

Sales figures are examples of **objective variables**. They provide a way of comparing outcomes that mean the same to you, me, or anybody else. If item *A* sells more than item *B*, this is true no matter who reads the sales report. Other examples of objective variables include profits and production.

Color preference is not an objective variable.

PAYOFF TABLE

In the simplest of decision trees, we have one level of choices, one level of states of nature, and, for each choice, the same states of nature to consider. Then we can form a table (see Figure 19-4).

Of course, it looks a little empty now. If we have an objective variable X, then we can enter its values (see Figure 19-5). Here X_{ij} is the value of the objective variable corresponding to the ith choice and the jth state of nature. This array is then called the **payoff table**.

The payoff table may be huge, and it would be helpful if we could simplify it. One way is to reject *dominated* or *inadmissible actions*. These are choices that should not be made because there are other choices that are always better. More accurately, A_i is dominated by A_j if, for all k, X_{ik} is less than or equal to

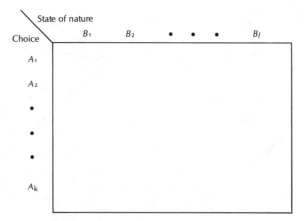

Figure 19-4

	B_1	B_2	• • •	B_l
A_1	x_{11}	x_{12}		x_{1l}
A_2	x_{21}	x_{22}		x_{2l}
•	•	•		•
•	•	•		•
•	•	•		•
A_k	x_{k1}	x_{k2}		x_{kl}

Figure 19-5

X_{jk}, and there is at least one value of k such that $X_{ik} < X_{jk}$. Thus we would never choose A_i because A_j would always be better for us. For example, see Figure 19-6, where A_4 is dominated by A_2 and A_3 and thus is inadmissible.

Once we've eliminated any inadmissible actions, we've done just about all we can do with complete certainty.

YOU SHOULD REMEMBER

1. An objective variable is a variable that you are trying to maximize or minimize in making a decision.

2. A payoff table lists the payoff (the value of the objective variable) that would result for each possible decision and each possible state of nature.

	B_1	B_2	B_3	B_4	B_5
A_1	1	2	5	8	3
A_2	2	2	5	2	7
A_3	3	9	8	7	6
A_4	0	1	3	2	4

Figure 19-6

EXPECTED PAYOFF

Even though we are uncertain about what state of nature will occur, we can often estimate the probability of one particular state occurring. A farmer can check past weather history, a company can survey consumer tastes, and so on. Then to each state of nature B_j we can associate a probability p_j. For the above example (Figure 19-6) we might have the probabilities given in Figure 19-7. Naturally we require that the p_j's add up to 1.

Even though the probabilities are at best estimates, we can still take the expected values of the objective variable for a given choice. Thus, for A_1,

$$E(X_{1j}) = 1(.3) + 2(.05) + 5(.5) + 8(.05) + 3(.1) = 3.6.$$

Similarly, $E(X_{2j}) = 4.0$, $E(X_{3j}) = 6.3$, and we don't bother to check choice A_4 since it is inadmissible.

The expected value of the objective variable for a given choice is called the **expected payoff**. If we compare expected payoffs, we see that A_3 has the highest, and we might want to choose it over A_1 and A_2 (again, A_4 is out of the running). Such a decision is said to be based on the *maximum expected payoff criterion*. This is not the only criterion for making a decision, but it is usually helpful.

Example: Using the Maximum Expected Payoff Criterion

PROBLEM   Suppose you have the following payoff table for a board game, where your choices are options available to you on your present turn, and the states of nature are the numbers your opponent may roll with two six-sided dice (see Figure 19-8). Which option should you choose, according to the maximum expected payoff criterion?

SOLUTION   It's reasonable to assume that the dice are fair, and we have the following probabilities for the states of nature (see Figure 19-9). Then

$$E(X_{1j}) = 10.14, \qquad E(X_{2j}) = 8.56,$$
$$E(X_{3j}) = 12.86, \qquad \text{and} \qquad E(X_{4j}) = 6.44$$

Thus, using the maximum expected payoff criterion, you would choose option A_3.

State	B_1	B_2	B_3	B_4	B_5
Probability	.30	.05	.50	.05	.10

Figure 19-7

State Choice	2	3	4	5	6	7	8	9	10	11	12
A₁	5	7	26	9	29	−6	8	15	4	−6	23
A₂	12	8	5	7	13	3	11	14	7	8	6
A₃	9	23	15	24	9	22	13	13	−5	−6	0
A₄	22	−3	−7	5	4	15	9	12	4	−2	6

Figure 19-8

State	2	3	4	5	6	7	8	9	10	11	12
Probability	1/36	2/36	3/36	4/36	5/36	6/36	5/36	4/36	3/36	2/36	1/36

Figure 19-9

YOU SHOULD REMEMBER

1. The expected payoff is the expected value of the payoff resulting from a decision. You can calculate it if you know the probability of occurrence for each state of nature.

2. The maximum expected payoff criterion is a helpful, but not the only, criterion for making a decision.

KNOW THE CONCEPTS

DO YOU KNOW THE BASICS?

Test your understanding of Chapter 19 by answering the following questions:

1. How do you determine the probabilities used when evaluating business decisions?

2. Why are inadmissible choices ruled out?

3. What is the value of a decision tree?

4. If two decisions provide the same expected return, how do you think you would choose between them?

5. Consider two stocks, one which is very safe and one which is very risky. Which do you think will provide the greater expected return?

TERMS FOR STUDY

decision theory   objective variable
decision tree   payoff table
expected payoff

PRACTICAL APPLICATION

COMPUTATIONAL PROBLEMS

1. In the following payoff table, which of the choices are inadmissible?

		State				
		I	II	III	IV	V
	A	4	3	6	9	7
	B	1	5	7	6	8
choice	C	0	2	-2	6	3
	D	1	5	4	6	8

2. If you estimate that the probabilities of the states in Exercise 1 are as follows:

I	II	III	IV	V
.2	.3	.1	.2	.2

what is the expected payoff for each choice?

3. Draw a decision tree relating the weather to the way you dress. How many outcomes can you list?

4. You are playing blackjack alone with the dealer, who has two 10's, and you have been dealt a queen and a 6. Draw a decision tree for your response.

5. Construct a payoff table for the first level of Exercise 4, with the state of nature being the next card to be dealt, using the following objective variables: -1 if you lose, 0 if your turn continues, and 1 if you win. (Remember that the dealer wins all ties and cannot draw another card on 20.)

6. Which is (are) the inadmissible choice(s)? in Exercise 4?

7. Assuming that the two 10's, the queen, and the 6 (Exercise 4) are the only cards drawn from a regular 52-card deck, calculate the probabilities for the states of nature (Exercise 5), and calculate the expected payoffs for the choices.

8. Suppose that you play the following game: You pick a number from 2 to 12 and roll two six-sided dice. If the number you pick is bigger than the number you roll, you lose in points the number you picked. If the number you pick is smaller than the number you roll, you gain in points the number you picked. If you pick the number you roll, you get 10 points. Construct a payoff table for this game.

9. Calculate the expected payoffs for the game described in Exercise 8.

10. List five objective variables.

ANSWERS

KNOW THE CONCEPTS

1. Often these probabilities are subjective estimates.

2. There is one choice that is better in all circumstances.

3. A decision tree illustrates the consequences of many different types of actions in different states of nature.

4. If you are like most people, you would probably choose the one with the least risk.

5. If they provided the same expected return, few people would want to buy the risky stock. Therefore, the risky stock must promise a higher expected return.

PRACTICAL APPLICATION

1. *C* and *D*

2. 5.5 for *A*, 5.2 for *B*, 2.2 for *C*, 4.9 for *D*

3.

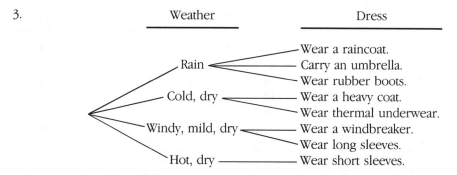

(Feel free to add more branches to this tree. There is no limit to the number of possibilities.)

4. Here is part of the tree:

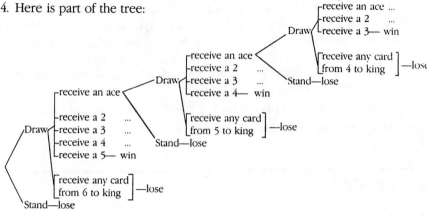

5.

	A	2	3	4	5	6	7	8	9	10	J	Q	K
Draw	0	0	0	0	1	−1	−1	−1	−1	−1	−1	−1	−1
Stand	−1	−1	−1	−1	−1	−1	−1	−1	−1	−1	−1	−1	−1

6. Stand

7. The probabilities of the states of nature are as follows:

A	4/48	6	3/48	J	4/48
2	4/48	7	4/48	Q	3/48
3	4/48	8	4/48	K	4/48
4	4/48	9	4/48		
5	4/48	10	2/48		

The expected value for draw is −1/2, and that for stand is −1.

8.

						Roll					
	2	3	4	5	6	7	8	9	10	11	12
2	10	2	2	2	2	2	2	2	2	2	2
3	−3	10	3	3	3	3	3	3	3	3	3
4	−4	−4	10	4	4	4	4	4	4	4	4
5	−5	−5	−5	10	5	5	5	5	5	5	5
6	−6	−6	−6	−6	10	6	6	6	6	6	6
7	−7	−7	−7	−7	−7	10	7	7	7	7	7
8	−8	−8	−8	−8	−8	−8	10	8	8	8	8
9	−9	−9	−9	−9	−9	−9	−9	10	9	9	9
10	−10	−10	−10	−10	−10	−10	−10	−10	10	10	10
11	−11	−11	−11	−11	−11	−11	−11	−11	−11	10	11
12	−12	−12	−12	−12	−12	−12	−12	−12	−12	−12	10

9. The expected payoff for each choice is as follows:

2	80/36	8	−38/36
3	116/36	9	−140/36
4	138/36	10	−240/36
5	140/36	11	−332/36
6	116/36	12	−410/36
7	60/36		

APPENDIX 1
GLOSSARY

alternative hypothesis the hypothesis that states, "The null hypothesis is false"

analysis of variance a method for testing the hypothesis that several different groups all have the same mean

ANOVA table a table that summarizes the results of an analysis of variance calculation

autocorrelation see **serial correlation**

average see **mean**

base year an arbitrary year relative to which a price index measures the average level of prices

Bayesian approach an approach to statistics where the researcher starts with a prior distribution and then uses results from observations to calculate an updated distribution

Bayes's rule a rule that tells how to calculate the conditional probability $\Pr(B|A)$ provided that $\Pr(A|B)$ and $\Pr(A|B^c)$ are known

Bernoulli trial an experiment that has only two possible results: one called "success" and one called "failure"

bimodal (a distribution) having two modes

binomial distribution the discrete probability distribution that applies when an experiment is conducted n times with each trial having a probability p of success and each trial being independent of every other trial

cell one of the locations in a contingency table

central limit theorem a theorem that states: the average of a large number of independent, identically distributed random variables will have a normal distribution

central tendency a measure that indicates the typical medium value of a distribution; the mean and median are examples of measures of central tendency

Chebyshev's theorem a theorem that states: for any group of numbers, the fraction that will be within k standard deviations of the mean will be at least $1 - 1/k^2$; also spelled **Tchebysheff's theorem**

chi-square distribution a continuous probability distribution related to the normal distribution; used in the chi-square test

452

chi-square test a statistical method to test the hypothesis that two factors are independent

class limits the two boundaries of a class used to group data

class mark the midpoint of a class

classical approach to probability an approach to probability based on defining a probability space consisting of all possible outcomes of an experiment where it is assumed that each outcome is equally likely

cluster sampling a sampling method where the population is divided into clusters; some clusters are selected at random, and then some members of the chosen clusters are selected at random to make up the sample

coefficient of determination a value between 0 and 1 that indicates how well variations in the independent variable in a regression explain variations in the dependent variable (symbolized by r^2)

coefficient of variation the standard deviation divided by the mean; this value indicates how big the dispersion is compared to the mean

combinations the number of different ways of selecting j objects from a group of n objects when the order in which the objects are chosen does not matter

common logarithm a logarithm to the base 10

complement for an event, all outcomes that are not part of that event

conditional density function the density function for one random variable when it is given that another random variable has a specified value; it can be calculated from the joint density function for the two random variables

conditional probability the probability that a particular event will occur when it is given that another event has occurred

confidence interval an interval based on observations of a sample and so constructed that there is a specified probability that the interval contains the unknown true value of a parameter (for example, it is common to calculate confidence intervals that have a 95 percent chance of containing the true value)

confidence level the degree of confidence associated with a confidence interval; the probability that the interval contains the true value of the parameter

consistent estimator an estimator that tends to converge toward the true value as the sample size becomes larger

consumer price index (CPI) a measure of the average price level at a particular time based on the cost of purchasing a fixed market basket of consumer goods

contingency table a table that shows how many observations fall within each cell, where the cells represent all of the possible combinations of two factors; a contingency table is used in the chi-square test to determine whether two factors are independent

continuous random variable a random variable that can take on any real-number value within a certain range; it is characterized by a density function curve such that the area under the curve between two numbers represents the probability that the random variable will be between those two numbers (for contrast, see **discrete random variable**)

convenience sampling a method of sampling where the items that are most conveniently available are selected as part of the sample; it is not appropriate to apply statistical analysis to samples selected in this manner

correction factor see **finite population correction factor**

correlation an indication of the degree of association between two quantities; its value is always between −1 and 1

covariance an indication of the degree of association between two quantities; it is related to the correlation but is not constrained to be between −1 and 1

critical region if the calculated value of the test statistic falls within the critical region, the null hypothesis is rejected (same as **rejection region**)

critical value the value (or values) at the boundary of the critical region

cumulative distribution function the function that gives the probability that a random variable will be less than or equal to a specific value

Current Population Survey a monthly survey of over 100,000 people conducted by the Census Bureau

cyclical component the component of a time series that moves up or down with the overall level of business activity in the economy

decision theory the study of making decisions designed to acheive some objective, often under conditions of uncertainty

decision tree a diagram that illustrates all possible consequences of different decisions in different states of nature

degrees of freedom the nature of a chi-square or a t-distribution is characterized by its degrees of freedom; the F distribution is characterized by the degrees of freedom in the numerator and the degrees of freedom in the denominator

density function see **probability density function**

dependent variable the variable in a regression that is assumed to be caused by the independent variable

depreciation the loss in value of capital goods as they wear out

descriptive statistics the study of ways of summarizing data

discrete random variable a random variable for which it is possible to make a list of all possible values (for contrast, see **continuous random variable**)

disjoint events two events that cannot both happen

dispersion the degree to which a distribution is spread out (same as *spread*)

dummy variable an independent variable used in regression analysis that has the value 1 when a specified condition is met and otherwise has the value 0 (same as *indicator variable*)

Durbin-Watson statistic a statistic used in regression analysis to test for the presence of serial correlation

econometrics the branch of economics that uses statistical analysis to analyze and forecast the economy

error a random term included in regression and analysis of variance that represents the effects of all factors other than those that have been specifically included in the analysis

error sum of squares a quantity that indicates the degree of random error in analysis of variance or regression

estimate the value of an estimator in a particular circumstance

estimator a quantity based on observations of a sample whose value is taken as an indicator of the value of an unknown population parameter (for example, the sample average $\bar{x}$ is often used as an estimator of the unknown population mean μ)

expectation the average value that would appear if a random variable was observed many times; also called the *expected value* or *mean* and symbolized by μ

expected payoff the expected value of the payoff resulting from a decision

expected value see **expectation**

exponential smoothing a method for analyzing the trend in a time series

extrapolation a forecast that attempts to predict what will happen in a situation that is outside the range of previously observed data

F **distribution** a continuous random variable distribution related to the chi-square distribution; it is used in the analysis of variance procedure and in regression analysis

F **statistic** in analysis of variance, a statistic to test the null hypothesis that all the groups come from populations with the same mean; in regression analysis, a statistic to test the null hypothesis that there is no connection between the independent variables and the dependent variable

factorial for a particular whole number, the product of all the whole numbers from 1 up to that number

finite population correction factor when sampling without replacement, the formulas for the standard deviation of the sample mean and sample proportion need to be multiplied by $\sqrt{(N - n)/(N - 1)}$, where N is the population size and n is the sample size; this is not necessary if the population is much larger than the sample

fitted value for a given set of values of the independent variables, the value of the dependent variable that is predicted by the regression equation

frequency diagram a bar diagram that illustrates how many observations fall within each category

frequency polygon a diagram that illustrates a frequency distribution with the use of line segments instead of bars

frequency table a table showing how many observations fall within each category

Friedman F_r test a procedure to test the hypothesis that there is no difference between preferences when there are more than two possibilities

GDP deflator a measure of the average price level

goodness-of-fit test a statistical procedure to test the hypothesis that a particular probability distribution fits an observed set of data

grand mean the mean of all elements in an analysis of variance test

gross domestic product (GDP) a measure of the total value of all goods and services produced in the country

gross national product measure of national output related to gross domestic product; it includes production by inputs owned by U.S. citizens but not production from domestic inputs that are foreign owned

grouped data a way of arranging data where, instead of listing each individual value, a table is made that shows how many values fall within certain categories

heteroscedasticity a situation in regression analysis where the error terms do not all have the same variance

histogram a bar diagram that illustrates a frequency distribution; each bar is drawn so that its area is proportional to the number of items in the interval it represents

hypergeometric distribution the discrete probability distribution that applies when a group of items is sampled without replacement

hypothesis testing a statistical procedure that involves collecting evidence and then making a decision as to whether a particular hypothesis should be accepted or rejected

independent events two or more events that do not affect each other

independent random variables two random variables that do not affect each other; knowing the value of one of the random variables does not provide any information about the other variable

independent variable a variable in a regression that affects the dependent variable, but is assumed not to be affected by the other variables in the regression

index of leading indicators an index consisting of 12 different economic time series that tend to turn up or down before the whole economy shows the same trend

indicator variable see **dummy variable**

inferior good a good that people buy less of as their income increases

inflation rate the rate of change in the average price level

interquartile range the value of the third quartile minus the value of the first quartile; half of the items in a list fall within the interquartile range

intersection for two sets, the set that consists of all elements that are in both of these sets

interval estimate an estimate for a parameter that gives a range of likely estimates for the parameter, instead of just a single value (see also **confidence interval**)

irregular component the component of a time series that moves up or down with no apparent pattern

joint probability density function if $f_{x,y}(x,y)$ is the joint probability density function for two random variables X and Y, then $f_{x,y}(x,y) = \Pr[(X = x)$ and $(Y = y)]$

Kruskal-Wallis H test a generalization of the Wilcoxon rank sum test to the situation where there are more than two populations

law of large numbers a law stating that, if a random variable is observed many times, the average of those observations will tend toward the mean of that random variable

least-squares estimators the estimates for the coefficients found from regression analysis; these estimates are calculated because they minimize the square of the error between the actual values of the dependent variable and the values predicted by the regression equation

level of significance for a hypothesis-testing procedure, the probability of committing a type 1 error

logarithm if $a^x = y$, then x is the logarithm to the base a of y

marginal probability function the probability function for one particular random variable that is calculated from a joint probability function

maximum likelihood estimator an estimator with the following property: if the true value of the unknown parameter has this value, then the probability of obtaining the sample that was actually observed is maximized

mean the value that is equal to the sum of a list of numbers divided by the number of numbers (same as *average*); the mean of a random variable is the average value that would appear if you observed the random variable many times (same as *expectation*); it is symbolized by the Greek letter μ (mu)

mean absolute deviation the average value of the absolute values of the distances from all numbers in a list to their mean

mean square a quantity, used in analysis of variance, that is equal to a sum of squares divided by its degrees of freedom

mean square error (MSE) a quantity used in regression to estimate the unknown value of the variance of the error term

mean square variance see **mean square**

median the value such that half of the numbers in a list are above it and half are below it

mode the value that occurs most frequently in a list of numbers

moving average the average value of the observations in a time series that are closest to one particular value

multicollinearity a problem that arises in regression analysis when two or more of the independent variables are closely correlated

multiple regression a statistical method for analyzing the relation between several independent variables and one dependent variable

multiplication principle a principle that states: if there are m possible outcomes of the first experiment and n possible outcomes of the second experiment, then there are mn possible combined outcomes of both experiments (if the two experiments are independent)

national income the sum of all income earned in a country; equal to the gross national product minus depreciation and indirect taxes

natural logarithm a logarithm to the base e, where $e = 2.71828 \ldots$

negatively correlated (two quantities) related so that one tends to be large when the other tends to be small

net national product (NNP) a measure of the national economy that is equal to gross national product minus depreciation

nominal gross domestic product the gross domestic product measured in current dollars, not corrected for inflation (for contrast, see **real gross domestic product)**

nonparametric method a statistical method that does not make assumptions about the specific forms of distributions and therefore does not focus on estimating unknown parameter values

normal distribution the most important continuous random variable distribution; its density function is bell-shaped; many real populations are distributed according to the normal distribution

null hypothesis the hypothesis that is being tested in a hypothesis-testing situation; often the null hypothesis is of the form, "There is no relation between two quantities"

objective variable a variable that the researcher is trying to maximize or minimize in decision theory

ogive a graph showing cumulative distribution

one-tailed test a hypothesis test where the critical region consists of only one tail of a distribution; the null hypothesis is rejected only if the test statistic has an extreme value in one direction

open-ended class a class for grouping data with no upper class limit or no lower class limit

outcome one of the possible results of a probability experiment

outlier an observation that is significantly different from the other observations

parameter a quantity (usually unknown) that characterizes a population (for example, the population mean and population standard deviation are parameters)

payoff table a table showing payoffs in different states of nature in decision theory

percentile the pth percentile of a list is the number such that p percent of the elements in the list are less than that number

perfectly correlated two quantities X and Y are perfectly correlated if there is a relation between them of the form $Y = aX + b$, where a and b are two constants; their correlation will be 1 if a is positive and -1 if a is negative

permutations the number of different ways of selecting *j* objects from a group of *n* objects when each distinct way of ordering the chosen objects counts separately

pie chart a graph where a circle represents the whole amount and wedge-shaped sectors indicate the fraction in each category

point estimate a single value used as an estimator for an unknown parameter (for contrast, see **interval estimate**)

Poisson distribution the discrete probability distribution that gives the frequency of occurrence of certain types of random events; it can be used as an approximation for the binomial distribution

population the set of all items of interest

population standard deviation the standard deviation calculated when the values of all items in the population are known (for contrast, see **sample standard deviation**)

population variance the variance calculated when the values of all items in the population are known (for contrast, see **sample variance**)

positively correlated two quantities are positively correlated if they are related so that, if one quantity is large, the other tends to be large; their correlation is greater than 0

predicted value a value of a dependent variable that is calculated from a regression equation and values of the independent variables

prediction interval an interval so constructed that there is a specified probability that a certain random variable will be within that interval

price index a measure of the price level

price level the average level of all prices in the country; it can be measured by the GNP deflator, consumer price index, or producer price index

probability the study of chance phenomena

probability density function for a discrete random variable, the probability density function at a specific value is the probability that the random variable will have that value; for a continuous random variable, the probability density function is represented by a curve such that the area under the curve between two numbers is the probability that the random variable will be between those two numbers

probability function for a discrete random variable, the probability function at a specific value is the probability that the random variable will have that value

probability of an event the number of outcomes that corresponds to that event divided by the total number of possible outcomes

probability space the set of all possible outcomes from a random experiment (same as *sample space*)

producer price index (PPI) a measure of the price level based on a fixed market basket of goods used by producers

quartile the first quartile of a list is the number such that one quarter of the numbers in the list are below it; the third quartile is the number such that three quarters of the numbers are below it; the second quartile is the same as the median

r^2 a measure of how well the independent variable in a simple linear regression can explain changes in the dependent variable; its value is between 0 (meaning poor fit) and 1 (meaning perfect fit)

R^2 a measure of how well a multiple regression equation is able to explain changes in the dependent variable

random sample a sample chosen by a method such that each possible sample had an equal chance of being selected; also, a collection of independent random variables all chosen from the same distribution

random variable a variable whose value depends on the outcome of a random experiment

range the largest value in a list minus the smallest value

ratio to moving average method a method for seasonally adjusting the values in a time series

raw data the initial data the researcher starts with before beginning analysis

real gross domestic product a measure of the GDP that corrects for inflation; it only increases when there is an increase in the goods and services produced in a nation

recession a time when there is a slowdown in business activity in the country

regression line a line calculated in regression analysis that is used to estimate the relation between two quantities (the independent variable and the dependent variable)

regression sum of squares a quantity that indicates the degree to which a regression equation can explain the variations in the dependent variable

rejection region the set of values of the test statistic for which the hypothesis will be rejected

relative frequency view of probability a view that regards the probability of an event as being the fraction of times the event would occur if an experiment was repeated many times

residual the difference between the actual value of the dependent variable and the value predicted by the regression equation

sample a group of items chosen from the population and used to estimate the properties of the population

462 *GLOSSARY*

sample space see **probability space**

sample standard deviation the standard deviation calculated from the values in a sample; it is used to estimate the value of the population standard deviation

sample variance the variance calculated from the values in a sample; it is used to estimate the value of the population variance

sampling distribution the probability distribution of a statistic, such as the sample average $\bar{x}$

sampling with replacement a method for choosing a sample where an item that has been selected is put back in the population and therefore has a chance of being selected again

sampling without replacement a method for choosing a sample where an item that has been selected is not put back in the population and therefore cannot be selected again

scatter diagram a diagram showing the relation between two quantities; one quantity is measured on the vertical axis, the other quantity is measured on the horizontal axis, and each observation is represented by a dot

seasonal adjustment a procedure for adjusting time series data to compensate for seasonal variation in order to make nonseasonal changes in the data more apparent

serial correlation a problem that arises in regression analysis involving time series data when successive values of the random error term are not independent (same as *autocorrelation*)

set a well-defined collection of objects

sign test a procedure to test the hypothesis that there is no difference between two quantities for which there are rankings instead of numerical values

simple linear regression a method for analyzing the relation between one independent variable and one dependent variable

skewed not symmetrical

slope a number that describes the orientation of a line; a horizontal line has zero slope and a vertical line has infinite slope; the slope can be found by calculating the vertical distance between any two points on the line and dividing by the horizontal distance between those two points

spread see **dispersion**

standard deviation the square root of the variance

standard error the estimated standard deviation for a sample statistic

standard error for coefficient in regression analysis, the estimated standard deviation of the estimated coefficient; a small value of the standard error means that the estimated coefficient is a more precise estimate of the true coefficient

standard normal distribution the normal distribution with mean 0 and variance 1

state of nature in decision theory, an unpredictable outcome

statistic a quantity calculated from the items in a sample, such as the sample average $\bar{x}$; a statistic is often used as an estimator of an unknown population parameter

statistical inference the process of using observations of a sample to estimate the properties of the population

statistics the study of ways to analyze data; it consists of *descriptive statistics* and *statistical inference*

stratified sampling a sampling method where the population is divided into strata that are as much alike as possible

subjective view of probability a view that regards the probability of an event as being an individual's estimate of the likelihood of occurrence of that event

subscript a small number or letter written next to and slightly below another character

subset set A is a subset of set B if all elements in A are also in B

sum of squares the sum of the squared deviations of the elements in a list about a specified quantity; several sum of squares statistics are used in analysis of variance and regression

summation notation a notation that uses the Greek capital letter sigma (Σ) to mean "add up the values"; for example, ΣX means "add up all values of X"

symmetrical a distribution is symmetrical about a particular line if the halves of the distribution on either side of that line are mirror images of each other; for example, a normal distribution is symmetrical about its mean

_t_ distribution a continuous random variable distribution related to the normal and chi-square distributions; it is used for confidence interval calculations and hypothesis testing for small samples

_t_ statistic a statistic used to test the null hypothesis that the true value of a coefficient in regression analysis is zero

tail the extreme upper or lower end of a distribution

test statistic a quantity calculated from observed quantities used to test a null hypothesis; the test statistic is constructed so that it will come from a known distribution if the null hypothesis is true; therefore the null hypothesis is rejected if it seems implausible that the observed value of the test statistic could have come from that distribution

time series data data that consist of several observations of a quantity at different points in time

total sum of squares the sum of the squares of the deviations of all numbers in a list from the mean; used in analysis of variance and regression

treatment sum of squares in analysis of variance, a quantity that indicates the degree to which differences in the observations can be explained by the fact that the observations come from different groups

trend component the component in a time series that increases or decreases smoothly with time

two-tailed test a hypothesis test where the critical region consists of both tails of the distribution, so that the null hypothesis is rejected if the test statistic value is either too large or too small

two-way analysis of variance a test procedure that can be applied to a table of numbers in order to test two hypotheses: (1) there is no significant difference between the rows; and (2) there is no significant difference between the columns

type 1 error an error that occurs in hypothesis testing when the null hypothesis is rejected when it is really true

type 2 error an error that occurs in hypothesis testing when the null hypothesis is accepted when it is really false

unbiased estimator an estimator whose expected value is equal to the true value of the parameter it is trying to estimate

union for set A and set B, the set consisting of all elements that are in A or B or both

variance a measure of dispersion for a random variable or a list of numbers; it is symbolized by σ^2, where σ is the Greek lower case letter sigma; the standard deviation is the square root of the variance

vertical intercept the point where a line crosses the vertical axis

Wilcoxon rank sum test a procedure to test the hypothesis that there is no difference between the means of populations

Wilcoxon signed rank test a procedure to test the hypothesis that there is no difference between the means of populations

zone of acceptance if the calculated value of the test statistic falls within the zone of acceptance, the null hypothesis is accepted (for contrast, see **critical region**)

APPENDIX 2
CALCULATIONS

STATISTICAL CALCULATIONS ON A CALCULATOR

Important statistical calculations usually involve large amounts of data, more than anyone would want to try to analyze by hand. Fortunately, calculators and computers are designed to handle repetitive arithmetic calculations. Most scientific calculators available today have a wide range of statistical features. (Software packages for analyzing statistics will be discussed in the next section.)

What are the advantages and disadvantages of using a calculator instead of a computer software package? Calculators are portable and affordable, but there are limits to the amount of data that they can handle.

Since there are many different types of calculators on the market, we will not list the keystrokes for every single one; rather we will try to summarize the common procedures necessary to perform statistical analyses on a calculator. Once you purchase one yourself, it will be easier then to consult the owner's manual for your calculator.

1. Entering data. With improvements in the way calculators display data, it has become much easier to enter and edit data. Typically you will need to choose a STATISTICS menu on your calculator and then choose either a DATA or EDIT submenu. By using the arrow keys and where appropriate the DELETE key, you can enter your data into one or more lists or columns for analysis. Again, using the arrow keys, you can proofread and if necessary correct your data. (Remember this basic principle of statistical analysis: garbage in, garbage out.)

2. Saving data. One of the basic principles in working with calculators and computers is that you should never have to type in the same set of numbers twice. If there is any chance that you will want to analyze your data set further at a later date, it would be a good idea to save your data. Some calculators will provide you with a limited choice of names for your data set, while others will allow you to choose any name. You can then recall the data set at a later date when you need it.

3. Transferring data. If you are working with someone else who has the same kind of calculator, you may wish to share your data with them. This is possible with many scientific calculators, either through a cable that is plugged into both calculators or via an infrared data link.

4. Statistical calculations. Going back to the STATISTICS menu on your calculator, you will usually find the following options under a CALC sub-menu:

- basic descriptive statistics (mean, variance, standard deviation, min-
imum, maximum)

- various types of regression models

- hypothesis tests on some later model calculators

- statistical displays. If there is a STATISTICS PLOT option on your calculator, you will be able to graph frequency histograms, scatterplots
(and simple regression models superimposed over the data
points), and box-plots.

STATISTICAL CALCULATIONS USING A COMPUTER SOFTWARE PACKAGE

For most practical applications of statistics, it is best to use a computer software package such as SAS, Minitab, SPSS, etc. Why?

- A computer software package can handle much more in the way of
arithmetic calculations than you can by hand or with a calculator.

- A computer software package can offer a much wider array of statistical calculations and types of data displays than a calculator can.

- It is possible to insert the results of your work directly into a word-
processing document.

- It is easier to transmit your data and/or analysis to someone else,
either by floppy disk or via electronic mail.

- Computer displays can show more information in their screens at a
given time than a calculator can, making data review and display
much easier.

Statistical software packages are sold for a variety of operating systems
(DOS, Macintosh, Windows, UNIX, VMS, etc.), but all follow the same sort of
basic operations:

1. Entering data. Data can be entered at the keyboard (which can involve a
substantial amount of work for large data sets) or loaded from an existing
data file. Data can be entered or stored outside of the statistical software
package in an ASCII file, but it is usually easier to enter the data from
within the package itself. Typically the data will be displayed in a spreadsheet format; you may then use the arrow keys to move around to review
or edit data.

2. Saving data. One of the basic principles in working with calculators and computers is that you should never have to type in the same set of numbers twice. If there is any chance that you will want to analyze your data set further at a later date, it would be a good idea to save your data. Software packages will always prompt you before you exit them to save your data first into a data file; even so, it is best to save your data as soon as you have entered it in case of a systems crash.

3. Transferring data and results to a word-processing document. Within the Windows and Macintosh operating systems, it is possible to save your data and/or results onto a clipboard and then insert them into your document. You can also dynamically link the document to your data file so that if you change your data file at a later date the document will automatically be updated.

4. Statistical calculations. Statistical software packages will perform for you every calculation described in this book. The functions will typically be grouped together into menus. For example, descriptive statistics such as the mean, variance, standard deviation, etc., will typically be contained in a menu item called "One-variable statistics," regression models in a menu item called "Regression," etc.

5. Graphs. Statistical software packages will create for you every display discussed in this book. The different kinds of graphs will typically be contained in a menu item called "Plot."

If you are working from a DOS, VMS, or UNIX operating system, you will start your program by typing a command at the command prompt. If you are working from a Windows or Macintosh environment, you will start your program by clicking with your mouse on the program's icon. Once the program is running, you should be able to get specific advice on any topic from the program's Help feature (usually listed as a menu item).

STATISTICAL FUNCTIONS IN THE MICROSOFT EXCEL SPREADSHEET

A third possible choice of tools for statistical calculations is a general purpose spreadsheet, such as Microsoft Excel. If you use a spreadsheet for other purposes anyway, then learning the spreadsheet commands would be a good way to do statistical calculations.

The following table lists some examples of the statistical functions included with newer versions of Microsoft Excel.

Description of Function	Example	Result
BINOMDIST(k, n, p, FALSE) calculates the probability that X = k, where X has a binomial distribution with parameters n and p.	= BINOMDIST(6,10,0.75,FALSE)	0.14600
BINOMDIST(k, n, p, TRUE) calculates the cumulative probability that X is less than or equal to k; in other words, it sums the probabilities from X = 0 up to X = k.	= BINOMDIST(6,10,0.75,TRUE)	0.22412
CHIDIST(a, df) gives the probability that a chi square random variable with df degrees of freedom will be greater than a.	= CHIDIST(11.07,5)	0.05001
CHIINV(p, df) gives the value of a such that $\Pr(\chi^2_{df} > a) = p$, where χ^2_{df} is a chi square random variable with df degrees of freedom. This function is the inverse of the previous one.	= CHIINV(0.05,5)	11.07048
COMBIN(n, j) gives the number of combinations when j objects are selected from n objects.	= COMBIN(52,5)	2,598,960
FACT(n) gives n! (n factorial).	= FACT(9)	362,880
FDIST(a, df_{num}, df_{den}) gives the probability that an F random variable with df_{num} and df_{den} degrees of freedom will be greater than a.	= FDIST(5.96,10,4)	0.05006
FINV(p, df_{num}, df_{den}) gives the value of a such that $\Pr(F > a) = p$, where F is an F random variable with df_{num} and df_{den} degrees of freedom. This function is the inverse of the previous one.	= FINV(0.05,10,4)	5.96435
HYPGEOMDIST(k, ns, M, N) gives $\Pr(X = k)$, where X has a hypergeometric distribution with population size N, sample size ns, and M objects in the population of type M, and k objects in the sample of type M.	= HYPGEOMDIST(3,10,13,52)	0.27806
NORMDIST(x, mu, sigma, TRUE) gives the probability that a normal random variable (with mean mu and standard deviation sigma) will be less than x.	= NORMDIST(12,10,2,TRUE)	0.84134
NORMINV(p, mu, sigma) gives the value of a such that $\Pr(X < a) = p$ where X has a normal distribution. This function is the inverse of the previous one.	= NORMINV(0.841345,10,2)	12.00000

Description of Function	Example	Result
NORMSDIST(a) gives Pr(Z < a), where Z has a standard normal distribution.	= NORMSDIST(1.96)	0.97500
NORMSINV(p) gives the value a such that Pr(Z < a) = p, where Z has a standard normal distribution. This function is the inverse of the previous one.	= NORMSINV(0.975)	1.95996
TDIST(a, df, 1) gives the one-tail probability for a t distribution: Pr(T > a), where T is a random variable with the T distribution with df degrees of freedom.	= TDIST(2.365,7,1)	0.02499
TDIST(a, df, 2) gives the two-tail probability for a t distribution: Pr(T > a) + Pr(T < −a), or 2Pr(T > a).	= TDIST(2.365,7,2)	0.04997
TINV(p, df) give the value a such that Pr(T > a) = p, where T is a random variable with the T distribution with df degrees of freedom.	= TINV(0.05,7)	2.36462

Excel also provides several functions that calculate descriptive statistics for a range of data:

AVERAGE(*range*)	calculate average
STDEVP(*range*)	standard deviation of a population
STDEV(*range*)	standard deviation of a sample
MEDIAN(*range*)	calculate median
PERCENTRANK(*range, value*)	give the percentile rank of *value* within the list *range*

These functions can be used for simple regression and correlation:

SLOPE(*yrange, xrange*)	slope of simple regression line
INTERCEPT(*yrange, xrange*)	y intercept of simple regression line
RSQ(*yrange, xrange*)	r squared value of simple regression
CORREL(*yrange, xrange*)	correlation coefficient between two ranges

See the HELP menu for more details on these and other statistical functions.

APPENDIX 3
STATISTICAL TABLES

STANDARD NORMAL (Z) TABLE

TABLE A3-2: TWO-TAILED STANDARD NORMAL (Z) TABLE: $p = \Pr(-a < Z < a)$

This table gives the probability p that a standard normal random variable Z (mean 0, standard deviation 1) will be between $-a$ and a.

a	p	a	p
0.100	0.0796	1.400	0.8384
0.200	0.1586	**1.439**	**0.8500**
0.300	0.2358	1.500	0.8664
0.400	0.3108	1.600	0.8904
0.500	0.3830	**1.645**	**0.9000**
0.600	0.4514	1.700	0.9108
0.700	0.5160	1.800	0.9282
0.800	0.5762	1.900	0.9426
0.900	0.6318	**1.960**	**0.9500**
1.000	0.6826	2.000	0.9544
1.100	0.7286	2.500	0.9876
1.200	0.7698	**2.576**	**0.9900**
1.282	**0.8000**	3.000	0.9974
1.300	0.8064	3.100	0.9980

The values in boldface are those commonly used for confidence intervals and hypothesis testing.

The table on the next two pages gives the probability p that a standard normal random variable Z will be less than the specified value a: $p = \Pr(Z < a)$. Also, it gives the area under the standard normal density function to the left of the specified value a.

Here is the connection between the two tables. If $p_2 = \Pr(-a < Z < a)$ (the value from the two-tailed table), and $p_1 = \Pr(Z < a)$ (the value from the one-tailed table), then $p_2 = 2p_1 - 1$.

TABLE A3-1: ONE-TAILED STANDARD NORMAL (Z) RANDOM VARIABLE TABLE: PR(Z < a) = p

a	p	a	p	a	p	a	p	a	p	a	p
-2.99	.0014	-2.49	.0064	-1.99	.0233	-1.49	.0681	-0.99	.1611	-0.49	.3121
-2.98	.0014	-2.48	.0066	-1.98	.0239	-1.48	.0694	-0.98	.1635	-0.48	.3156
-2.97	.0015	-2.47	.0068	-1.97	.0244	-1.47	.0708	-0.97	.1660	-0.47	.3192
-2.96	.0015	-2.46	.0069	-1.96	.0250	-1.46	.0721	-0.96	.1685	-0.46	.3228
-2.95	.0016	-2.45	.0071	-1.95	.0256	-1.45	.0735	-0.95	.1711	-0.45	.3264
-2.94	.0016	-2.44	.0073	-1.94	.0262	-1.44	.0749	-0.94	.1736	-0.44	.3300
-2.93	.0017	-2.43	.0075	-1.93	.0268	-1.43	.0764	-0.93	.1762	-0.43	.3336
-2.92	.0018	-2.42	.0078	-1.92	.0274	-1.42	.0778	-0.92	.1788	-0.42	.3372
-2.91	.0018	-2.41	.0080	-1.91	.0281	-1.41	.0793	-0.91	.1814	-0.41	.3409
-2.90	.0019	-2.40	.0082	-1.90	.0287	-1.40	.0808	-0.90	.1841	-0.40	.3446
-2.89	.0019	-2.39	.0084	-1.89	.0294	-1.39	.0823	-0.89	.1867	-0.39	.3483
-2.88	.0020	-2.38	.0087	-1.88	.0301	-1.38	.0838	-0.88	.1894	-0.38	.3520
-2.87	.0021	-2.37	.0089	-1.87	.0307	-1.37	.0853	-0.87	.1922	-0.37	.3557
-2.86	.0021	-2.36	.0091	-1.86	.0314	-1.36	.0869	-0.86	.1949	-0.36	.3594
-2.85	.0022	-2.35	.0094	-1.85	.0322	-1.35	.0885	-0.85	.1977	-0.35	.3632
-2.84	.0023	-2.34	.0096	-1.84	.0329	-1.34	.0901	-0.84	.2005	-0.34	.3669
-2.83	.0023	-2.33	.0099	-1.83	.0336	-1.33	.0918	-0.83	.2033	-0.33	.3707
-2.82	.0024	-2.32	.0102	-1.82	.0344	-1.32	.0934	-0.82	.2061	-0.32	.3745
-2.81	.0025	-2.31	.0104	-1.81	.0351	-1.31	.0951	-0.81	.2090	-0.31	.3783
-2.80	.0026	-2.30	.0107	-1.80	.0359	-1.30	.0968	-0.80	.2119	-0.30	.3821
-2.79	.0026	-2.29	.0110	-1.79	.0367	-1.29	.0985	-0.79	.2148	-0.29	.3859
-2.78	.0027	-2.28	.0113	-1.78	.0375	-1.28	.1003	-0.78	.2177	-0.28	.3897
-2.77	.0028	-2.27	.0116	-1.77	.0384	-1.27	.1020	-0.77	.2206	-0.27	.3936
-2.76	.0029	-2.26	.0119	-1.76	.0392	-1.26	.1038	-0.76	.2236	-0.26	.3974
-2.75	.0030	-2.25	.0122	-1.75	.0401	-1.25	.1056	-0.75	.2266	-0.25	.4013
-2.74	.0031	-2.24	.0125	-1.74	.0409	-1.24	.1075	-0.74	.2296	-0.24	.4052
-2.73	.0032	-2.23	.0129	-1.73	.0418	-1.23	.1093	-0.73	.2327	-0.23	.4090
-2.72	.0033	-2.22	.0132	-1.72	.0427	-1.22	.1112	-0.72	.2358	-0.22	.4129
-2.71	.0034	-2.21	.0136	-1.71	.0436	-1.21	.1131	-0.71	.2389	-0.21	.4168
-2.70	.0035	-2.20	.0139	-1.70	.0446	-1.20	.1151	-0.70	.2420	-0.20	.4207
-2.69	.0036	-2.19	.0143	-1.69	.0455	-1.19	.1170	-0.69	.2451	-0.19	.4247
-2.68	.0037	-2.18	.0146	-1.68	.0465	-1.18	.1190	-0.68	.2483	-0.18	.4286
-2.67	.0038	-2.17	.0150	-1.67	.0475	-1.17	.1210	-0.67	.2514	-0.17	.4325
-2.66	.0039	-2.16	.0154	-1.66	.0485	-1.16	.1230	-0.66	.2546	-0.16	.4364
-2.65	.0040	-2.15	.0158	-1.65	.0495	-1.15	.1251	-0.65	.2578	-0.15	.4404
-2.64	.0041	-2.14	.0162	-1.64	.0505	-1.14	.1271	-0.64	.2611	-0.14	.4443
-2.63	.0043	-2.13	.0166	-1.63	.0516	-1.13	.1292	-0.63	.2643	-0.13	.4483
-2.62	.0044	-2.12	.0170	-1.62	.0526	-1.12	.1314	-0.62	.2676	-0.12	.4522
-2.61	.0045	-2.11	.0174	-1.61	.0537	-1.11	.1335	-0.61	.2709	-0.11	.4562
-2.60	.0047	-2.10	.0179	-1.60	.0548	-1.10	.1357	-0.60	.2743	-0.10	.4602
-2.59	.0048	-2.09	.0183	-1.59	.0559	-1.09	.1379	-0.59	.2776	-0.09	.4641
-2.58	.0049	-2.08	.0188	-1.58	.0570	-1.08	.1401	-0.58	.2810	-0.08	.4681
-2.57	.0051	-2.07	.0192	-1.57	.0582	-1.07	.1423	-0.57	.2843	-0.07	.4721
-2.56	.0052	-2.06	.0197	-1.56	.0594	-1.06	.1446	-0.56	.2877	-0.06	.4761
-2.55	.0054	-2.05	.0202	-1.55	.0605	-1.05	.1469	-0.55	.2912	-0.05	.4801
-2.54	.0055	-2.04	.0207	-1.54	.0618	-1.04	.1492	-0.54	.2946	-0.04	.4840
-2.53	.0057	-2.03	.0212	-1.53	.0630	-1.03	.1515	-0.53	.2981	-0.03	.4880
-2.52	.0059	-2.02	.0217	-1.52	.0642	-1.02	.1539	-0.52	.3015	-0.02	.4920
-2.51	.0060	-2.01	.0222	-1.51	.0655	-1.01	.1562	-0.51	.3050	-0.01	.4960
-2.50	.0062	-2.00	.0228	-1.50	.0668	-1.00	.1587	-0.50	.3085	0.00	.5000

TABLE A3-1: ONE-TAILED STANDARD NORMAL (Z)
RANDOM VARIABLE TABLE: PR(Z < a) = p

a	p	a	p	a	p	a	p	a	p	a	p
0.01	.5040	0.51	.6950	1.01	.8438	1.51	.9345	2.01	.9778	2.51	.9940
0.02	.5080	0.52	.6985	1.02	.8461	1.52	.9357	2.02	.9783	2.52	.9941
0.03	.5120	0.53	.7019	1.03	.8485	1.53	.9370	2.03	.9788	2.53	.9943
0.04	.5160	0.54	.7054	1.04	.8508	1.54	.9382	2.04	.9793	2.54	.9945
0.05	.5199	0.55	.7088	1.05	.8531	1.55	.9394	2.05	.9798	2.55	.9946
0.06	.5239	0.56	.7123	1.06	.8554	1.56	.9406	2.06	.9803	2.56	.9948
0.07	.5279	0.57	.7157	1.07	.8577	1.57	.9418	2.07	.9808	2.57	.9949
0.08	.5319	0.58	.7190	1.08	.8599	1.58	.9429	2.08	.9812	2.58	.9951
0.09	.5359	0.59	.7224	1.09	.8621	1.59	.9441	2.09	.9817	2.59	.9952
0.10	.5398	0.60	.7257	1.10	.8643	1.60	.9452	2.10	.9821	2.60	.9953
0.11	.5438	0.61	.7291	1.11	.8665	1.61	.9463	2.11	.9826	2.61	.9955
0.12	.5478	0.62	.7324	1.12	.8686	1.62	.9474	2.12	.9830	2.62	.9956
0.13	.5517	0.63	.7357	1.13	.8708	1.63	.9484	2.13	.9834	2.63	.9957
0.14	.5557	0.64	.7389	1.14	.8729	1.64	.9495	2.14	.9838	2.64	.9959
0.15	.5596	0.65	.7422	1.15	.8749	1.65	.9505	2.15	.9842	2.65	.9960
0.16	.5636	0.66	.7454	1.16	.8770	1.66	.9515	2.16	.9846	2.66	.9961
0.17	.5675	0.67	.7486	1.17	.8790	1.67	.9525	2.17	.9850	2.67	.9962
0.18	.5714	0.68	.7517	1.18	.8810	1.68	.9535	2.18	.9854	2.68	.9963
0.19	.5753	0.69	.7549	1.19	.8830	1.69	.9545	2.19	.9857	2.69	.9964
0.20	.5793	0.70	.7580	1.20	.8849	1.70	.9554	2.20	.9861	2.70	.9965
0.21	.5832	0.71	.7611	1.21	.8869	1.71	.9564	2.21	.9864	2.71	.9966
0.22	.5871	0.72	.7642	1.22	.8888	1.72	.9573	2.22	.9868	2.72	.9967
0.23	.5910	0.73	.7673	1.23	.8907	1.73	.9582	2.23	.9871	2.73	.9968
0.24	.5948	0.74	.7704	1.24	.8925	1.74	.9591	2.24	.9875	2.74	.9969
0.25	.5987	0.75	.7734	1.25	.8944	1.75	.9599	2.25	.9878	2.75	.9970
0.26	.6026	0.76	.7764	1.26	.8962	1.76	.9608	2.26	.9881	2.76	.9971
0.27	.6064	0.77	.7794	1.27	.8980	1.77	.9616	2.27	.9884	2.77	.9972
0.28	.6103	0.78	.7823	1.28	.8997	1.78	.9625	2.28	.9887	2.78	.9973
0.29	.6141	0.79	.7852	1.29	.9015	1.79	.9633	2.29	.9890	2.79	.9974
0.30	.6179	0.80	.7881	1.30	.9032	1.80	.9641	2.30	.9893	2.80	.9974
0.31	.6217	0.81	.7910	1.31	.9049	1.81	.9649	2.31	.9896	2.81	.9975
0.32	.6255	0.82	.7939	1.32	.9066	1.82	.9656	2.32	.9898	2.82	.9976
0.33	.6293	0.83	.7967	1.33	.9082	1.83	.9664	2.33	.9901	2.83	.9977
0.34	.6331	0.84	.7995	1.34	.9099	1.84	.9671	2.34	.9904	2.84	.9977
0.35	.6368	0.85	.8023	1.35	.9115	1.85	.9678	2.35	.9906	2.85	.9978
0.36	.6406	0.86	.8051	1.36	.9131	1.86	.9686	2.36	.9909	2.86	.9979
0.37	.6443	0.87	.8078	1.37	.9147	1.87	.9693	2.37	.9911	2.87	.9979
0.38	.6480	0.88	.8106	1.38	.9162	1.88	.9699	2.38	.9913	2.88	.9980
0.39	.6517	0.89	.8133	1.39	.9177	1.89	.9706	2.39	.9916	2.89	.9981
0.40	.6554	0.90	.8159	1.40	.9192	1.90	.9713	2.40	.9918	2.90	.9981
0.41	.6591	0.91	.8186	1.41	.9207	1.91	.9719	2.41	.9920	2.91	.9982
0.42	.6628	0.92	.8212	1.42	.9222	1.92	.9726	2.42	.9922	2.92	.9982
0.43	.6664	0.93	.8238	1.43	.9236	1.93	.9732	2.43	.9925	2.93	.9983
0.44	.6700	0.94	.8264	1.44	.9251	1.94	.9738	2.44	.9927	2.94	.9984
0.45	.6736	0.95	.8289	1.45	.9265	1.95	.9744	2.45	.9929	2.95	.9984
0.46	.6772	0.96	.8315	1.46	.9279	1.96	.9750	2.46	.9931	2.96	.9985
0.47	.6808	0.97	.8340	1.47	.9292	1.97	.9756	2.47	.9932	2.97	.9985
0.48	.6844	0.98	.8365	1.48	.9306	1.98	.9761	2.48	.9934	2.98	.9986
0.49	.6879	0.99	.8389	1.49	.9319	1.99	.9767	2.49	.9936	2.99	.9986
0.50	.6915	1.00	.8413	1.50	.9332	2.00	.9772	2.50	.9938	3.00	.9987

TABLE A3-3: CHI-SQUARE TABLE

The table gives the value of a such that $\Pr(\chi^2_{DF} < a) = p$, where χ^2_{DF} is a chi-square random variable with DF degrees of freedom. For example, there is a probability of .95 that a chi-square random variable with 6 degrees of freedom will be less than 12.6.

DF	p = .005	.01	.025	.05	.25	.5	.75	.9	.95	.975	.99
1	.000	.000	.001	.004	.10	.45	1.32	2.71	3.84	5.02	6.64
2	.010	.020	.051	.10	.58	1.39	2.77	4.61	5.99	7.38	9.21
3	.072	.11	.22	.35	1.21	2.37	4.11	6.25	7.81	9.35	11.3
4	.21	.30	.48	.71	1.92	3.36	5.39	7.78	9.49	11.1	13.3
5	.41	.55	.83	1.15	2.67	4.35	6.63	9.24	11.1	12.8	15.1
6	.68	.87	1.24	1.64	3.45	5.35	7.84	10.6	12.6	14.4	16.8
7	.99	1.24	1.69	2.17	4.25	6.35	9.04	12.0	14.1	16.0	18.5
8	1.34	1.65	2.18	2.73	5.07	7.34	10.2	13.4	15.5	17.5	20.1
9	1.73	2.09	2.70	3.33	5.90	8.34	11.4	14.7	16.9	19.0	21.7
10	2.16	2.56	3.25	3.94	6.74	9.34	12.5	16.0	18.3	20.5	23.2
11	2.60	3.05	3.82	4.57	7.58	10.3	13.7	17.3	19.7	21.9	24.7
12	3.07	3.57	4.40	5.23	8.44	11.3	14.8	18.5	21.0	23.3	26.2
13	3.56	4.11	5.01	5.89	9.30	12.3	16.0	19.8	22.4	24.7	27.7
14	4.07	4.66	5.63	6.57	10.2	13.3	17.1	21.1	23.7	26.1	29.1
15	4.60	5.23	6.26	7.26	11.0	14.3	18.2	22.3	25.0	27.5	30.6
16	5.14	5.81	6.91	7.96	11.9	15.3	19.4	23.5	26.3	28.8	32.0
17	5.70	6.41	7.56	8.67	12.8	16.3	20.5	24.8	27.6	30.2	33.4
18	6.26	7.01	8.23	9.39	13.7	17.3	21.6	26.0	28.9	31.5	34.8
19	6.84	7.63	8.91	10.1	14.6	18.3	22.7	27.2	30.1	32.9	36.2
20	7.43	8.26	9.59	10.9	15.5	19.3	23.8	28.4	31.4	34.2	37.6
21	8.03	8.90	10.3	11.6	16.3	20.3	24.9	29.6	32.7	35.5	38.9
22	8.64	9.54	11.0	12.3	17.2	21.3	26.0	30.8	33.9	36.8	40.3
23	9.26	10.2	11.7	13.1	18.1	22.3	27.1	32.0	35.2	38.1	41.6
24	9.89	10.9	12.4	13.8	19.0	23.3	28.2	33.2	36.4	39.4	43.0
25	10.5	11.5	13.1	14.6	19.9	24.3	29.3	34.4	37.7	40.6	44.3
26	11.2	12.2	13.8	15.4	20.8	25.3	30.4	35.6	38.9	41.9	45.6
27	11.8	12.9	14.6	16.1	21.7	26.3	31.5	36.7	40.1	43.2	47.0
28	12.5	13.6	15.3	16.9	22.7	27.3	32.6	37.9	41.3	44.5	48.3
29	13.1	14.3	16.0	17.7	23.6	28.3	33.7	39.1	42.6	45.7	49.6
30	13.8	14.9	16.8	18.5	24.5	29.3	34.8	40.3	43.8	47.0	50.9
35	17.2	18.5	20.6	22.5	29.0	34.3	40.2	46.1	49.8	53.2	57.3
40	20.7	22.2	24.4	26.5	33.7	39.3	45.6	51.8	55.8	59.3	63.7
50	28.0	29.7	32.4	34.8	42.9	49.3	56.3	63.2	67.5	71.4	76.2
60	35.5	37.5	40.5	43.2	52.3	59.3	67.0	74.4	79.1	83.3	88.4
70	43.3	45.4	48.8	51.7	61.7	69.3	77.6	85.5	90.5	95.0	100.4
80	51.2	53.5	57.2	60.4	71.1	79.3	88.1	96.6	101.9	106.6	112.3
90	59.2	61.8	65.6	69.1	80.6	89.3	98.6	107.6	113.1	118.1	124.1
100	67.3	70.1	74.2	77.9	90.1	99.3	109.1	118.5	124.3	129.6	135.8

TABLE A3-4: ONE-TAILED *t* DISTRIBUTION TABLE

If *T* is a random variable with a *t* distribution with *DF* degrees of freedom, then the table gives the value of *a* such that $Pr(T < a) = p$. For example, there is a .975 probability that a *T* random variable with *T* degrees of freedom will be less than 2.365.

DF	p = .750	p = .900	p = .950	p =. 975	p = .990	p = .995
1	1.000	3.078	6.314	12.706	31.821	63.657
2	0.816	1.886	2.920	4.303	6.965	9.925
3	0.765	1.638	2.353	3.182	4.541	5.841
4	0.741	1.533	2.132	2.776	3.747	4.604
5	0.727	1.476	2.015	2.571	3.365	4.032
6	0.718	1.440	1.943	2.447	3.143	3.707
7	0.711	1.415	1.895	2.365	2.998	3.499
8	0.706	1.397	1.860	2.306	2.896	3.355
9	0.703	1.383	1.833	2.262	2.821	3.250
10	0.700	1.372	1.812	2.228	2.764	3.169
11	0.697	1.363	1.796	2.201	2.718	3.106
12	0.695	1.356	1.782	2.179	2.681	3.055
13	0.694	1.350	1.771	2.160	2.650	3.012
14	0.692	1.345	1.761	2.145	2.624	2.977
15	0.691	1.341	1.753	2.131	2.602	2.947
16	0.690	1.337	1.746	2.120	2.583	2.921
17	0.689	1.333	1.740	2.110	2.567	2.898
18	0.688	1.330	1.734	2.101	2.552	2.878
19	0.688	1.328	1.729	2.093	2.539	2.861
20	0.687	1.325	1.725	2.086	2.528	2.845
21	0.686	1.323	1.721	2.080	2.518	2.831
22	0.686	1.321	1.717	2.074	2.508	2.819
23	0.685	1.319	1.714	2.069	2.500	2.807
24	0.685	1.318	1.711	2.064	2.492	2.797
25	0.684	1.316	1.708	2.060	2.485	2.787
26	0.684	1.315	1.706	2.056	2.479	2.779
27	0.684	1.314	1.703	2.052	2.473	2.771
28	0.683	1.313	1.701	2.048	2.467	2.763
29	0.683	1.311	1.699	2.045	2.462	2.756
30	0.683	1.310	1.697	2.042	2.457	2.750
35	0.682	1.306	1.690	2.030	2.438	2.724
40	0.681	1.303	1.684	2.021	2.423	2.704
45	0.680	1.301	1.679	2.014	2.412	2.690
50	0.679	1.299	1.676	2.009	2.403	2.678
55	0.679	1.297	1.673	2.004	2.396	2.668
60	0.679	1.296	1.671	2.000	2.390	2.660
80	0.678	1.292	1.664	1.990	2.374	2.639
100	0.677	1.290	1.660	1.984	2.364	2.626
120	0.677	1.289	1.658	1.980	2.358	2.617

TABLE A3-5: TWO-TAILED t DISTRIBUTION TABLE

If T is a random variable with a t distribution with DF degrees of freedom, then the table gives the value of a such that $\Pr(-a < T < a) = p$. For example, there is a 95 percent probability that a T random variable with 7 degrees of freedom will be between -2.365 and 2.365.

DF	$p = .900$	$p = .950$	$p = .99$
1	6.314	12.706	63.657
2	2.920	4.303	9.925
3	2.353	3.182	5.841
4	2.132	2.776	4.604
5	2.015	2.571	4.032
6	1.943	2.447	3.707
7	1.895	2.365	3.499
8	1.860	2.306	3.355
9	1.833	2.262	3.250
10	1.812	2.228	3.169
11	1.796	2.201	3.106
12	1.782	2.179	3.055
13	1.771	2.160	3.012
14	1.761	2.145	2.977
15	1.753	2.131	2.947
16	1.746	2.120	2.921
17	1.740	2.110	2.898
18	1.734	2.101	2.878
19	1.729	2.093	2.861
20	1.725	2.086	2.845
21	1.721	2.080	2.831
22	1.717	2.074	2.819
23	1.714	2.069	2.807
24	1.711	2.064	2.797
25	1.708	2.060	2.787
26	1.706	2.056	2.779
27	1.703	2.052	2.771
28	1.701	2.048	2.763
29	1.699	2.045	2.756
30	1.697	2.042	2.750
35	1.690	2.030	2.724
40	1.684	2.021	2.704
45	1.679	2.014	2.690
50	1.676	2.009	2.678
55	1.673	2.004	2.668
60	1.671	2.000	2.660
80	1.664	1.990	2.639
100	1.660	1.984	2.626
120	1.658	1.980	2.617

TABLE A3-6: *F* DISTRIBUTION TABLE

In each of the three tables below, the numerator degrees of freedom are read along the top, and the denominator degrees of freedom are read along the left side. The table gives the value of a such that $Pr(F < a) = p$, where F is a random variable with an F distribution with DF_{num} numerator degrees of freedom and DF_{den} denominator degrees of freedom. Each table has a different value of p; first .99; then .95; then .90. These are values commonly used for hypothesis testing. Another way to describe the table is to give the value of $1-p$, which is the right tail area (that is, the area to the right of the given value of a). For example, there is a .99 probability that an F random variable with 5 numerator degrees of freedom and 9 denominator degrees of freedom will be less than 6.06.

$Pr(F < a) = .99$; right tail area $= .01$:

DF_{den}	$DF_{num}=2$	3	4	5	10	15	20	30	60	120
2	99.00	99.16	99.25	99.30	99.40	99.43	99.45	99.47	99.48	99.49
3	30.82	29.46	28.71	28.24	27.23	26.87	26.69	26.50	26.32	26.22
4	18.00	16.69	15.98	15.52	14.55	14.20	14.02	13.84	13.65	13.56
5	13.27	12.06	11.39	10.97	10.05	9.72	9.55	9.38	9.20	9.11
6	10.92	9.78	9.15	8.75	7.87	7.56	7.40	7.23	7.06	6.97
7	9.55	8.45	7.85	7.46	6.62	6.31	6.16	5.99	5.82	5.74
8	8.65	7.59	7.01	6.63	5.81	5.52	5.36	5.20	5.03	4.95
9	8.02	6.99	6.42	6.06	5.26	4.96	4.81	4.65	4.48	4.40
10	7.56	6.55	5.99	5.64	4.85	4.56	4.41	4.25	4.08	4.00
15	6.36	5.42	4.89	4.56	3.80	3.52	3.37	3.21	3.05	2.96
20	5.85	4.94	4.43	4.10	3.37	3.09	2.94	2.78	2.61	2.52
30	5.39	4.51	4.02	3.70	2.98	2.70	2.55	2.39	2.21	2.11
60	4.98	4.13	3.65	3.34	2.63	2.35	2.20	2.03	1.84	1.73
120	4.79	3.95	3.48	3.17	2.47	2.19	2.03	1.86	1.66	1.53

$Pr(F < a) = .95$; right tail area $= .05$:

DF_{den}	$DF_{num}=2$	3	4	5	10	15	20	30	60	120
2	19.00	19.16	19.25	19.30	19.40	19.43	19.45	19.46	19.48	19.49
3	9.55	9.28	9.12	9.01	8.79	8.70	8.66	8.62	8.57	8.55
4	6.94	6.59	6.39	6.26	5.96	5.86	5.80	5.75	5.69	5.66
5	5.79	5.41	5.19	5.05	4.74	4.62	4.56	4.50	4.43	4.40
6	5.14	4.76	4.53	4.39	4.06	3.94	3.87	3.81	3.74	3.70
7	4.74	4.35	4.12	3.97	3.64	3.51	3.44	3.38	3.30	3.27
8	4.46	4.07	3.84	3.69	3.35	3.22	3.15	3.08	3.01	2.97
9	4.26	3.86	3.63	3.48	3.14	3.01	2.94	2.86	2.79	2.75
10	4.10	3.71	3.48	3.33	2.98	2.85	2.77	2.70	2.62	2.58
15	3.68	3.29	3.06	2.90	2.54	2.40	2.33	2.25	2.16	2.11
20	3.49	3.10	2.87	2.71	2.35	2.20	2.12	2.04	1.95	1.90
30	3.32	2.92	2.69	2.53	2.16	2.01	1.93	1.84	1.74	1.68
60	3.15	2.76	2.53	2.37	1.99	1.84	1.75	1.65	1.53	1.47
120	3.07	2.68	2.45	2.29	1.91	1.75	1.66	1.55	1.43	1.35

Pr($F < a$) = .90; right tail area = .10:

DF_{den}	DF_{num} = 2	3	4	5	10	15	20	30	60	120
2	9.00	9.16	9.24	9.29	9.39	9.42	9.44	9.46	9.47	9.48
3	5.46	5.39	5.34	5.31	5.23	5.20	5.18	5.17	5.15	5.14
4	4.32	4.19	4.11	4.05	3.92	3.87	3.84	3.82	3.79	3.78
5	3.78	3.62	3.52	3.45	3.30	3.24	3.21	3.17	3.14	3.12
6	3.46	3.29	3.18	3.11	2.94	2.87	2.84	2.80	2.76	2.74
7	3.26	3.07	2.96	2.88	2.70	2.63	2.59	2.56	2.51	2.49
8	3.11	2.92	2.81	2.73	2.54	2.46	2.42	2.38	2.34	2.32
9	3.01	2.81	2.69	2.61	2.42	2.34	2.30	2.25	2.21	2.18
10	2.92	2.73	2.61	2.52	2.32	2.24	2.20	2.16	2.11	2.08
15	2.70	2.49	2.36	2.27	2.06	1.97	1.92	1.87	1.82	1.79
20	2.59	2.38	2.25	2.16	1.94	1.84	1.79	1.74	1.68	1.64
30	2.49	2.28	2.14	2.05	1.82	1.72	1.67	1.61	1.54	1.50
60	2.39	2.18	2.04	1.95	1.71	1.60	1.54	1.48	1.40	1.35
120	2.35	2.13	1.99	1.90	1.65	1.55	1.48	1.41	1.32	1.26

INDEX

absolute value, 14
adjusted r squared, 364
airline, 138
alternative hypothesis, 48, 245
analysis of variance, 284–302
ANOVA table, 291, 299, 365
aptitude tests, 283
area, under curve, 156–157
arithmetic mean, 35
auditor, 4
autocorrelation, 368
average, 9, 22, 25, 35, 114, 126
  (*see also* mean)
axioms, 58

backup system, 98
bar chart, 21, 154
base, of logarithm function, 343
base year, 405
batting orders, 73
Bernoulli trial, 119, 136
bimodal distribution, 12
binomial coefficient, 77
binomial distribution, 133–140, 167,
  230, 237, 384

calculus, 343, 157
capital consumption, 403
cards, 61–62, 65–67, 72, 75–83, 94
cause and effect relationship, 335
cell, 267
central limit theorem, 166
central tendency, 11, 35
chain-weight approach, 405
Chebyshev's theorem, 17
chi, 172
chi-square distribution, 172–176, 219,
  268, 274, 385, 389
chi-square test, 267–272
class, 25, 27
class limit, 25
class mark, 25
classical approach to probability, 60
cluster sampling, 239
coefficient of determination, 323
coefficient of multiple determination, 362

coefficient of variation, 17
coefficients, in multiple regression, 362
coin tossing, 41–51, 114, 118, 136, 149,
  167–169, 246–247, 257
column sum of squares, 297
combinations, 75–84, 134
common logarithms, 343
complement, 63
conditional probability, 94–101
conditional probability function, 187
confidence interval, 212
confidence interval for difference of two
  means, 221
confidence interval for mean, 218
confidence interval for proportion,
  233–234
confidence interval for regression
  coefficient, 366
confidence interval for slope of
  regression line, 331–332
confidence interval for variance, 219
consistent estimator, 204
consumer price index, 404, 407
consumption spending, 314
contingency table, 267
continuous random variable, 151–158
control group, 5
convenience sampling, 240
correlation, 191, 325
covariance, 190, 193
critical region, 48, 248
critical value, 248
cumulative distribution function, 113,
  152, 184
Current Population Survey, 3, 241
cyclical variations, 411, 413

decision tree, 442
delta, 154
density function, 109, 155–157
dependent variable in regression, 317,
  360
depreciation, 403
descriptive statistics, 2
dice, 60–64, 110, 124, 170–172
difference between two means, 220, 258
difference between two proportions, 264
discrete random variables, 108
disjoint events, 65
dispersion, 13
disposable income, 314
draft lottery, 229
drug testing, 4